C000109773

Renewables

A Practical Handbook

Consulting Editors **Matt Bonass and Michael Rudd**

Consulting editors
Matt Bonass, Michael Rudd

Publisher
Sian O'Neill

Editors
Carolyn Boyle, Veronique Musson

Marketing manager
Alan Mowat

Production
John Meikle, Russell Anderson

Publishing directors
Guy Davis, Tony Harriss, Mark Lamb

Renewables: A Practical Handbook
is published by
Globe Law and Business
Globe Business Publishing Ltd
New Hibernia House
Winchester Walk
London SE1 9AG
United Kingdom
Tel +44 20 7234 0606
Fax +44 20 7234 0808
Web www.globelawandbusiness.com

Printed by CPI Antony Rowe

ISBN 978-1-905783-39-7

Renewables: A Practical Handbook
© 2010 Globe Business Publishing Ltd

Table of contents

Preface

Governor Gray Davis
37th governor of California (1999-2003)

This book addresses one of the most discussed and important topics of recent times: climate change. The United Nations' Climate Change Summit in Copenhagen in December 2009 generated unprecedented media coverage and, most importantly, created a global platform for leaders to consider fundamental steps to achieve meaningful carbon reductions.

Given the extraordinary attention now paid to climate change, it is hard to believe that nearly a decade ago, actions taken in this regard were considered controversial. Despite that, as governor of California I was proud to sign the first law in the United States to reduce global warming and greenhouse gases from cars and trucks. The automobile industry aggressively fought our law in court for seven years and lost. I was also pleased to create the first Greenhouse Gas Monitoring Registry and establish the United States' most ambitious commitment to renewable energy with the creation of the state-wide Renewables Portfolio Standard, requiring utilities substantially to increase their renewable electricity production.

Today, more than 30 US states and the District of Columbia have their own binding or voluntary Renewables Portfolio Standard programmes. At the federal level, under the leadership of President Obama, the American Recovery and Reinvestment Act provided in excess of $80 billion for clean energy investments and $11 billion for an improved and expanded energy grid, among other notable and related investments.

This impressive book features contributions from an array of thought leaders in both the public and private sectors. It serves as a useful guide to outline key policy issues and assist the reader in grasping the challenges and trends that will underpin the tremendous growth of these vital and emerging sectors. This book presents a thoughtful analysis of the issues relevant to financing, as well as risk identification and allocation. It also includes important guidance on project management and structuring.

Specific chapters are dedicated to providing technical, commercial and regulatory guidance in key areas such as wind, solar, hydro, embedded generation, tidal, wave and renewable gas. Finally, it also considers clean coal technologies, carbon capture and storage. Although these are not regarded as renewable projects, they nonetheless play an important role in reducing global emissions and preserving a diversified fuel source mix.

Developed as a tool for the business community, this book enables a better understanding of how these sectors will inevitably affect business. From an array of vantage points – financial, developmental or project oriented – this book serves as an invaluable guide for the business community as it participates in this growth area.

Having spent the better part of my professional life in government, I know first-hand that climate change has forever modified the 'business as usual' model. Where once, for example, renewable resources were considered fringe and beyond viable commercial application, we now see both the private and public sectors showing robust support for these, and other, burgeoning sectors. Notwithstanding this support, there remain many political, technical, economic and legal considerations to weigh in attempting to meet renewable energy and carbon-reducing targets imposed at the global, national and regional levels. It is clear, however, that tremendous opportunities exist for those that truly understand this evolving industry and seize the moment.

Gray Davis was elected the 37th governor of California in 1998. As governor, he made education a top priority by increasing accountability in schools and expanding access to higher education through record scholarships and loans. These reforms improved student scores for six consecutive years.

Governor Davis made record investments in infrastructure and created four centres of science and innovation on University of California campuses.

As governor, he demonstrated bold environmental leadership by signing the first law in the nation to reduce global warming and greenhouse gases. He also created the first Greenhouse Gas Monitoring Registry and established the nation's first and most ambitious commitment to renewable energy with a statewide Renewables Portfolio Standard.

In 2002 Governor Davis was re-elected.

He graduated from Stanford University (cum laude) and Columbia Law School, and was awarded the Bronze Star for his service during the Vietnam War. He served as lieutenant governor (1995-98), state controller (1987-95) and state assemblyman (1982-86).

Today Governor Davis is of counsel at Loeb & Loeb LLP and serves as a distinguished policy fellow at the UCLA School of Public Affairs. He regularly speaks before various groups and in 2009 was the keynote speaker for the Columbia Law School Graduation Ceremony. He serves on several boards, including the Saban Free Clinic and as the 2010 co-chair for the Southern California Leadership Council.

Foreword

Matt Bonass
Michael Rudd
SNR Denton UK LLP

We echo the words and thoughts of former Governor Gray Davis in the preface and are delighted that he has felt it appropriate to lend his name to this book.

In bringing the book together, we were reminded of how fast moving the renewables industry is and how deeply affected it is by world events. Since work commenced, the Labour government in the United Kingdom has been replaced by a Conservative/Liberal Democrat coalition which boasts that it is the "greenest government ever". Subsidy regimes, such as the Spanish feed-in tariff system, have faced challenges in what remains a difficult economic climate. Notwithstanding the world's most significant offshore oil drilling incident in the Gulf of Mexico, climate change legislation in the United States remains in a state of flux. Pakistan has suffered some of the worst flooding in living memory. We are on the verge of the next major Conference of the Parties meeting in Cancun in November 2010, following the Copenhagen Conference in December 2009. The debate on climate change and renewable energy remains as vibrant and polarised as ever, and is constantly evolving in response to events such as those mentioned above.

In determining the structure and substance of this book, we felt it was important that you, the reader, have the perspectives of legal, technical and commercial thought leaders from Europe and the United States (which, in addition to China, represent two of the largest renewable industry regions in the world). If you are new to the renewables industry, this book is intended to provide you with a broad and varied overview of this dynamic and multifaceted industry. For those of you with more experience in the renewables industry, we hope that the diversity of authors and topics will offer new perspectives and insights to expand your existing knowledge base. We hope that this book will be an ongoing resource for you to use, particularly for specific renewables transactions in which you may be involved.

Most of the legal, technical and/or commercial matters covered by chapters that focus on a particular country and/or technology have broader application to other countries and technologies. Therefore, we strongly encourage you to read all chapters in this book.

Together with our colleagues who have also contributed chapters, we are delighted to have been given the opportunity to work with Globe Law and Business on this publication. It has been an honour to collaborate with the contributors, who include accountants, consultants, lawyers and people at the heart of the renewables industry. We have really enjoyed debating and sharing thoughts with them on the topics covered.

We believe that we speak on behalf of all contributors when we say that we feel very fortunate to be able to spend the majority of our days working in this exciting sector. We hope that this book conveys our passion and enthusiasm for renewable energy, as well as providing the reader with a practical guide.

Why renewable energy?

Matt Bonass
SNR Denton UK LLP

1.   Introduction

Between December 7 and 18 2009 representatives from 193 countries met in Copenhagen for the 15th UN Climate Change Conference. The main purpose of the meeting was to formalise an agreement to extend or replace the 12-year old Kyoto Protocol, the principal international agreement governing climate change.[1] Attending the conference were countries representing:

- the leading developed nations, such as the United States and members of the European Union;
- the fast-developing nations, such as China and India; and
- other developing nations.

The idea of a Copenhagen gathering seemed unlikely even as recently as December 2007 at the Bali conference. The attendees then, as now, had vastly different agendas and aims.

At the time of writing more than 100 countries, which together produce more than 80% of energy-related greenhouse gas emissions, had become party to the Copenhagen Accord signed on December 18 2009.[2] Under the accord, developed nations have pledged over $30 billion in fast-track funding in favour of developing countries for the period 2010 to 2012. The accord also includes a consensus that the rise in global temperatures should be kept within two degrees centigrade of pre-industrial temperatures.

Feelings about the conference are mixed. To some, it represented an opportunity for debate and a fundamental reassessment of the place of climate change on the international political agenda. To others, it was an opportunity missed because of the actions of certain states to delay consensus or the richer nations in exploiting the position of the poorer nations.

Some referred to the requirement for states to submit emissions reduction targets by January 31 2010 as "a deadline that is not a deadline for an accord that is not an accord".[3]

---

1   For the full text of the protocol, see unfccc.int/resource/docs/convkp/kpeng.pdf. For the status of ratification, see unfccc.int/files/kyoto_protocol/status_of_ratification/application/pdf/kp_ratification _20091203.pdf.
2   For the full text of the accord, see unfccc.int/resource/docs/2009/cop15/eng/l07.pdf.
3   BBC website report, February 1 2010.

Most people reading this book will be familiar with the Copenhagen Conference and its outcome. However, following the unprecedented media and public scrutiny of the conference, most people across the world will link the word 'Copenhagen' with the climate change debate. As the homepage of the UN Framework Convention on Climate Change proudly pointed out at the time, the name Copenhagen became the most widely used search term on Google in December 2009, replacing the previously most popular search term 'Tiger Woods'.

The challenges posed by climate change and the need to move to a low-carbon economy are taking centre stage in international and governmental policies. That this is the case would have been barely conceivable less than 40 years ago.

2.   Historical perspective

In the 1970s the environment for what was referred to as 'alternative energy' was very different. The word 'cleantech' (meaning 'clean technology') did not exist. Alternative energy was seen as high risk/high return and the preserve of scientists or do-gooders. With the possible exception of hydropower, there was little support for alternative energy, whether from politicians, the private sector or investors. While general legislation and regulation governed the energy industry, there was very little bespoke climate change or renewable energy legislation in place. The only real investment in alternative energy was through government-funded research and development plans.

Times have changed. Events such as the Rio Earth Summit in 1992, the adoption of the Kyoto Protocol in 1997 and its coming into effect in 2005, the launch of the EU Emissions Trading Scheme in 2005 and the launch of the Stern Review by Sir Nicholas Stern – the former chief economist of the World Bank – have led to what some have described as the most dramatic industrial shift in more than a century: the "sixth revolution".[4]

3.   Growth drivers

The events described above do not explain on their own why the cleantech or renewable energy industry has become what it is today. This could not have happened without the interaction of several key drivers that are examined below.

3.1   Global warming realities

The World is on the brink of dangerous climate change and immediate action is needed to avert it.[5]

There has been much discussion over the years about whether climate change is happening and whether man-made emissions of greenhouse gases are causing climate change. This was brought to the fore in early 2010 by a number of scandals, including:

- 'Climategate', when the veracity of certain research undertaken by the United Kingdom's University of East Anglia's climatic research unit was questioned; and

---

4   Merrill Lynch equity note, November 2008.
5   International Scientific Congress on Climate Change, March 2009.

- 'Glaciergate', following admissions by the chairman of the Intergovernmental Panel on Climate Change that there was no scientific basis for claims that Himalayan glaciers could disappear by 2035.

However, the overwhelming consensus is that climate change is a reality. There is less talk about whether greenhouse gas emissions need to be reduced and more talk about:
- how great those reductions should be;
- who should pay for them; and
- how best to mitigate and adapt to climate change.

Climate change poses a vast risk to mankind's everyday lives. An increase in global temperatures beyond two degrees centigrade would be dangerous for human populations and too warm for many of the planet's natural systems to thrive.[6] If greenhouse gas emissions remain at their current levels, the planet is likely to experience a rise in temperature of five degrees centigrade. Therefore, even keeping within the two degrees centigrade band will require substantial adjustments.

As the planet warms, rainfall patterns are shifting. Rising sea levels, droughts, floods, hurricanes and forest fires are becoming increasingly common. In developing countries, both the climate and a lack of water are resulting in reduced agricultural productivity and crop failures, leading to hunger, malnutrition and disease. These natural events are visible to governments, organisations and individuals.

3.2 Price fluctuations

High fossil-fuel prices have traditionally encouraged the development of alternative sources of energy. Conversely, a lowering of fuel prices usually triggers a drop in interest for green-energy opportunities.

The price of oil has fluctuated wildly since mid-2008, rising to a high of $145 in July 2008, dropping to a low of $38 at the beginning of 2009, before settling on between $70 and $80 since October 2009.[7] Commentators have suggested that while oil prices may be considered low compared to the levels of mid-2008, they are high in real terms in the context of the recession. This should continue to drive the development of renewable energy.

However, the price of fossil fuels is not decisive in determining the growth of the renewable energy sector. The development of renewable energy in the Middle East does not appear to be linked to the price of oil. Investors are pursuing cleantech opportunities there, including the initiatives undertaken by the Abu Dhabi Future Energy Company (MASDAR), notwithstanding any fluctuation in the price of oil or the existence of plentiful reserves.

The price of fossil fuels is also volatile. Using renewable energy may reduce the need for hedging against long-term rises in fossil-fuel costs. This is considered further below.

---

6   Stern Review, October 2006.
7   Based on price per barrel of oil tracking Brent Crude.

3.3   Energy supply security

Natural resources are becoming scarcer. Some countries have little or no natural resources and rely on other countries for their imports. There is a general desire among nations to reduce the geopolitical and terrorist threats posed by a dependence on resources from regions of the world that are considered volatile. Therefore, the more that energy can be generated domestically, the less dependent a nation will be on fuel imports. This is another driver behind the popularity of renewable energy sources such as wind, solar, biomass and, where available, marine energy.

In addition, many developed nations are facing the challenge of an ageing electricity infrastructure. For instance, the UK Office of the Gas and Electricity Markets (Ofgem) estimates that by 2015 the United Kingdom's generating capacity could be cut by one-third as a generation of coal and nuclear power stations is closed.[8] There is therefore an overwhelming need to replace and build new capacity and grid infrastructure. While this is an obvious threat to the United Kingdom's energy supply, it represents a significant opportunity not only to restructure the power sector, but to do so using renewable sources.

There is a recognised need to develop alternatives to fossil fuels and a diverse low-carbon energy mix (including renewable and nuclear energy, and carbon capture and storage). However, this will not be cheap. Ofgem reported in October 2009 that £200 billion will be needed over the next 10 to 15 years to maintain reliable energy supplies in the United Kingdom while cutting carbon emissions.[9] The scale of investment is daunting, especially in light of:

- the lack of project finance and credit due to the economic climate; and
- the difficulties created by the UK planning legislation.

Despite this, many believe that the price will be worth paying.

Lastly, the world population is growing. There will be increasing competition for limited water (seen by some as the new oil), food and energy supplies, as well as a potential for food price inflation. Therefore, there needs to be a focus on energy efficiency and demand reduction, both to conserve what we use and to make what we use cleaner.

3.4   Advances in technology

Back in the 1970s, many saw the debate on climate change as a lot of hot air over hot air. Others considered alternative energy projects as pipe dreams, incapable of being commercialised – if they worked at all.

Without the technology boom of the late 1990s, it is unlikely that the cleantech revolution would have taken place. Many of those involved in the technology industry gravitated towards the cleantech industry, as they are complementary. As a result, investments in research and development for the cleantech sector have increased significantly. This in turn has led to an increasing ability to bring new clean technologies to market. What was previously a fantasy is fast becoming reality.

---

8   Ofgem report, February 3 2010.
9   Ofgem press release, October 9 2009.

4. New attitudes

As discussed above, many factors are pushing the growth of the renewable energy industry. The following section looks at the effects that these drivers have on the behaviour of governments, organisations and individuals.

4.1 The increasing role of governments

For many years, governments have been challenged by the need to ensure the supply of energy, both fundamentally and at the right price, and of food and water. They are now facing the challenges of climate change and the pressure to transition towards a low-carbon economy.

The solution, in the eyes of many governments, is to increase the use of renewable energy. Renewable energy can play a significant role in guarantying a country's energy supply while hedging against fluctuations in the price of fossil fuels. However, for governments, adapting to renewable energy offers additional benefits.

References to the Stern Review were usually accompanied in the UK press by a picture of either a melting glacier or a flooded village. What many people do not appreciate is that the Stern Review was a report on the economics of climate change. The review recognised that climate change is irreversible. Accordingly, it recommended committing 1% of gross domestic product (GDP) each year to combat climate change (Stern subsequently revised this upwards to 2%).[10] The review further recommended that money be spent now to avoid the jaw-dropping costs of dealing with adaptation amid catastrophic climatic events in the future. The costs will increase in GDP terms the longer that investment in renewable energy is delayed. Governments, and in particular the treasury sections of governments, have taken this on board.

Pure economics is not the only argument. There is vast political capital invested in climate change. Governments are in a race to dominate in this sector and are looking to attract inward investment, create new manufacturing industries and build the jobs of the future.

Politicians have also been quick to spot the growing public support for action on climate change. All around the world, governments are announcing climate change initiatives. Some may well feel strongly about the issue. Others may see it as either an opportunity to gain favour among their constituents or simply as something they must be seen to be saying for public relations reasons.

4.2 Regulation – the carrots and the sticks

Climate change is now shaping government action and there is growing political appetite for renewable energy policy. 'Green politics' has joined mainstream politics. Consequently, governments are committing (not always bindingly) to regulate the actions of companies and individuals within their territories.

The European Union has arguably taken the lead in the mitigation of climate change through regulation. The union aims to generate 20% of the energy used

---

10   Speaking at a conference in Copenhagen, March 12 2009.

within its borders from renewable sources by 2020.[11] Some countries have introduced more stringent, legally binding targets through domestic legislation. For example, the UK government has agreed to a 34% reduction of emissions by 2020 and an 80% reduction by 2050 (both based on 1990 levels).[12] In addition, in the United Kingdom alone, three key pieces of climate change-related legislation were passed in 2008 (the Climate Change Act, the Energy Act and the Planning Act) and at least five key strategy documents were published (including the Renewable Energy Strategy, the Low Carbon Transport Strategy, the Low Carbon Industrial Strategy and the Low Carbon Transition Plan). Finally, the UK government is making an increasing amount of grant funding available for key renewable energy sectors (eg, offshore wind and marine energy).[13]

Under the Obama administration, the United States has woken up to the threats and, perhaps more importantly, the opportunities linked to action on climate change. Under the US Stimulus Bill, $787 billion has been pledged to support the US economy, of which around $80 billion is earmarked for renewable energy. The US Clean Energy and Security Act envisages:

- a 17% cut in emissions by 2020 and an 83% cut in emissions by 2050 (both based on 2005 levels);
- that 15% of energy will come from renewable sources by 2020; and
- the introduction of a cap and trade scheme.[14]

Both China and South Korea have pledged significant percentages of their global stimulus packages towards renewable energy.

The renewable energy industry is still nascent: the Cleantech Group describes the sector's position as "the end of the beginning".[15] As a result, the industry relies on subsidies to enable it to compete with existing sources of energy. Subsidy regimes, such as feed-in tariffs, cash-back schemes, tradable certificates and banding of allowances, are examined elsewhere in this book. There remains a healthy concern about investing in propositions that are viable only because governments have altered market conditions.

It is generally recognised that long-term thinking about renewable energy is critical. There needs to be a secure, certain and consistent regulatory framework incorporating subsidy mechanisms to support the growth of renewable energy, most notably through the grandfathering of existing subsidies. There also needs to be a restructuring of the carbon markets that are currently not functioning effectively due to an over-allocation of allowances and uncertainty over the price of carbon post-2012.

Conversely, carbon taxes are likely to be imposed in certain jurisdictions in order

---

11   Directive 2009/28/EC on the promotion of the use of energy from renewable sources.
12   Targets set by the Climate Change Act 2008 and budgets proposed by the Committee on Climate Change under powers granted by this act.
13   The £950 million Strategic Investment Fund for New Business and the Carbon Trust Marine Renewables Proving Fund were announced on February 3 2010.
14   Under consideration before the US Senate as of Spring 2010.
15   Cleantech Report, "Ten Predictions for 2010", Cleantech Group, November 2009.

to provide a stick to drive a reduction in greenhouse gas emissions. One country that has looked carefully at this is France.[16]

4.3 Globalisation

Of course, the action of one or more governments is not enough to tackle the challenge of climate change. By its nature, climate change is a global phenomenon that needs to be addressed by all governments across the world: no country is immune. No previous technological revolution has been so global in nature. The Copenhagen Conference brought this fact into focus.

One of the biggest challenges facing climate change action on a global scale is the differing characteristics of the nations involved. We have referred above to some of the actions taken by developed countries such as EU member states and the United States.

Elsewhere in the world, China and India are becoming increasingly important players in the climate change debate. The great fear of the so-called BRIC nations (Brazil, Russia, India and China) is that as they strive to advance economic growth and alleviate poverty, climate change action could stifle their economic development. This could be in terms of:

- limiting their infrastructure development;
- holding back their sense of entrepreneurialism; or
- leading them to compromise their moves to confront energy security, resource scarcity and the ability to service their fast-growing populations.

There is therefore an absolute need to engage with developing nations to ensure that they can continue with their economic development and growth within a low-carbon framework, which could include technologies such as carbon capture and storage. Such action has historically been met with resistance from some countries (eg, the United States) which see any support given to China in this regard as eroding their competitive advantage in areas such as export and manufacturing. In the words of President Obama on the subject of energy generally: "I do not accept a future where the jobs and industries of tomorrow take root beyond our borders."[17]

That said, China is moving more rapidly than any other country towards the deployment of alternative energy. For example, as of January 2010 China was the largest manufacturer of solar panels in the world.[18] It has declared that by 2020, its carbon emissions intensity per dollar of GDP would decrease by between 40% and 45% on 2005 levels. At the time of writing, China was set to adopt a revised renewable energy law that:

- requires grid operators to buy all the power offered to them by renewable power providers; and
- imposes fines on grid operators which refuse to buy renewable power offered.

---

16 However, in March 2010 President Sarkozy rejected this idea because of concerns about French competitiveness.
17 State of the Union address, January 2010.
18 "China leading global race to make clean energy", in *New York Times*, January 30 2010.

China's actions are once again driven not just by a desire to be green and service the needs of its population, but also by an increasing realisation of the opportunities offered by action on climate change. China recognises that it will struggle to sustain its growth unless it embraces clean technology. It has also taken on board the economic advantages offered. The same can be broadly applied to India, which launched a national solar mission in November 2009 that represents one of eight core missions under the country's National Allocation Plan.

The Middle East countries are also at the forefront of developments to combat climate change. (This is perhaps surprising, given these countries' extensive resources of oil and gas and the availability within their borders of cheap, subsidised fuel and electricity.) Projects include the MASDAR initiative in Abu Dhabi, which proposes:

- the building of a carbon-neutral city;
- the formation of a number of equity investment funds backed by Credit Suisse; and
- the establishment of an educational facility based on the US Massachusetts Institute of Technology.

Elsewhere in the Middle East, one of the largest per-capita emitters of greenhouse gases, Qatar,[19] has established a cleantech fund with the United Kingdom's Carbon Trust to support the development of a low-carbon innovation centre in Qatar. Renewable energy is the subject of interest in the Middle East partly because of:

- the region's geography and climate;
- its requirement to service the energy needs of its population;
- its desire to reduce rapidly rising pollution levels;
- its ability to increase the export value of its traditional energy assets; and
- its desire to drive economic diversification and create jobs.[20]

Finally, the effects of climate change are and will be felt most deeply in the developing world – from Africa and South America to the small island nations that were so vociferous at the Copenhagen Conference. These countries cannot act alone and therefore must rely on the developed nations to assist them in reducing their greenhouse gas emissions. This may be through:

- mechanisms such as the Clean Development Mechanism and Joint Implementation established under the Kyoto Protocol; or
- a pledge to contribute to the $30 billion 'fast start' fund under Article 8 of the Copenhagen Accord.

Like China and India, the developing nations must be allowed to continue their economic development and to supply energy to their populations.

---

19   "Navigating the Numbers: Greenhouse Gas Data and International Climate Policy", www.wri.org/publications/navigating-the-numbers.
20   *Study on the potential for renewable energy in the MENA region*, Booz and Co, March 2010.

4.4   Individual social responsibility

We live in an age of increasing social responsibility, whether it comes to matters of climate change, foreign policy or the way in which we interact with our neighbours.

It is unlikely that a film such as Al Gore's *An Inconvenient Truth* would have been watched and debated in the 1970s to the extent that it has been over the past few years. The publication on the eve of the Copenhagen Conference of a common editorial, in 56 newspapers and 20 different languages, urging governments worldwide to act against climate change would also have been inconceivable.[21]

The significant proportion of the population that is sceptical about climate change or denounces the less attractive aspects of renewable energy (eg, onshore wind farms) cannot be ignored. Such reactions are to be expected at this relatively early stage of development of the renewables industry. However, there is also a general, deep-rooted concern about the effect of climate change and a desire to move to a more resource-efficient, low-carbon economy. A change in attitude is already perceivable, as demonstrated by the level of support for climate action at the London G20 conference in March 2009 and in Copenhagen in December 2009. The generations to come are likely to be increasingly intolerant of our generation's inaction on climate change.

Increasingly, individuals will try to reduce their carbon footprint by adapting their behaviour – be it through their choice of electricity supply or their participation in micro-generation schemes that benefit from subsidies, such as feed-in tariffs and voluntary offsetting.

4.5   Corporate social responsibility and adaptation

It is a very short step from individual action to the actions taken by the boards of directors of companies. Company boards of directors and, more importantly, shareholders around the world have been watching closely the initiatives on climate change taken by governments and the reaction of the public, their ultimate consumers.

Companies are embracing the fight against climate change. Some 960 multinationals signed the Copenhagen Communiqué to register their support for the actions to be debated at the international conference in December 2009.[22] In addition, the UK Carbon Disclosure Project has since 2002 been seeking and publishing information considered relevant to potential investors in relation to companies' greenhouse gas emissions.[23] Some of the world's largest companies, including General Electric, Google, Microsoft and Walmart, support actions to mitigate climate change. They have taken note of the threats – as well as of the opportunities.

While some shareholders may be ambivalent as to the environmental effects of climate change, they are well aware of the economic repercussions, in particular the financial risks of ignoring the impact of climate change. Accordingly, shareholders

---

21   Text drafted by *The Guardian* during months of consultations with the editors of the other newspapers involved.

22   The Copenhagen Communiqué is an initiative of the Prince of Wales's Corporate Leaders' Group on Climate Change, which is run by the University of Cambridge Programme for Sustainable Leadership.

23   www.cdproject.net.

are increasingly considering the need to factor climate change into company valuations.

Further, directors – particularly of public companies – are required under legislation and accounting policy to make statements about their compliance with and the effect of climate change on their businesses. For example, in January 2010 the US Securities and Exchange Commission issued an interpretative release on disclosure requirements relating to climate change which will have the effect of bringing climate change disclosures under the main umbrella of the commission's rules.[24] In layman's terms, the release recognises climate change as a material investor risk. This will include climate change legislation to seek compensation for damage (litigation is already being brought in the United States to seek compensation for the damages caused by Hurricane Katrina).

Companies must therefore consider whether to adapt to climate change. The choice is stark: they must either embrace the change and lead or pass up the opportunity to adopt, innovate and change, and fall behind their competitors. Todd Stern, the US special envoy on climate change, has stated that high-carbon goods and services are likely to become untenable in the near future.[25]

Companies are increasingly encouraged (one could even say coerced) to produce cleaner and more efficient products and to market and sell them more aggressively. The increase in consumption power of the middle classes has led to people buying cleantech products such as hybrid cars – partly to keep up with the Greens, rather than the Joneses, next door.

As discussed elsewhere in this chapter and throughout this book, governments are encouraging these actions through regulation and subsidies. Companies are also examining their supply chains closely to ensure that they fit with the impressive sustainability criteria that they no doubt promote through their website and marketing. For example, Walmart has announced its intention to eliminate 20 million tonnes of greenhouse gas emissions from its supply chain by the end of 2015.[26] What is more, business will score highly on procurement criteria that place increasing weight on sustainable resource efficiency. Companies are therefore assessing in detail the risk of climate change, adapting their business plans, reviewing suppliers, raising awareness within their organisations and, in some cases, taking out insurance against the impact of climate change on operations.

Companies are also investing in climate change actions as they see opportunities not just to minimise risk and protect their market as described above, but also to capitalise on the opportunities to expand their markets. Smart companies recognise that future carbon regulation is inevitable and are acting now so that they may gain 'first mover' advantage. Companies must also be aware that the public are becoming increasing adept at spotting examples of 'greenwashing' (ie, disingenuously spinning products and policies as environmentally friendly), and therefore any actions taken to combat climate change must be seen to be genuine or risk ridicule.

---

24   The release was published on February 2 2010 and is available at www.sec.gov/rules/interp/2010/33_9106.pdf.
25   Comments made at the US Climate Action Symposium, March 3 2009.
26   Press release, February 25 2010.

5.   Environmental investment

Forty years ago, the words 'environmental' and 'investment' were rarely mentioned in the same sentence. Today, there is a huge influx of capital into renewable energy through many dedicated investment venture capital and private equity funds, most of the world's largest pension funds, multinational private companies and high-net-worth individuals. The sector has grown from investments by $100 million funds to $500 million funds between 2004 and 2010.[27] Global investment in renewable energy stood at $35 billion in 2004; in 2008 it had risen to $155 billion.[28] To put these numbers into perspective, the International Energy Agency stated in 2009 that $10 trillion (or $125 billion a year) need to be spent over the next 20 years to stabilise greenhouse gas emissions. Investors now include individuals such as Warren Buffett, Sir Richard Branson and George Soros.

Investment in renewable energy is no longer seen as a novelty or the preserve of certain philanthropic funds. Rather, it is seen as a dynamic and lucrative business opportunity. As a leading US investor in cleantech once commented: "Investing in cleantech is all about the 'green', and this has nothing to do with the environment."[29]

As with any nascent industry, there are concerns about the lack of investment in the sector, particularly by way of project finance. However, given the current economic environment, this should not come as a surprise. Investors will need to adapt their strategies to deal with capital-intensive demands. In recent years infrastructure funds have been increasingly investing in renewable energy as they see opportunities for investment in a stable asset class. There is also a move to encourage the use of public-private partnerships to develop solutions to climate change around the world, such as the UK Private Finance Initiative.

It is not only the private sector that is investing in renewable energy: governments are following suit. As discussed above, numerous countries offer stimulus packages that allocate a significant percentage of money to renewable energy development. This is backed by further moneys being made available from government-backed entities such as the European Bank for Reconstruction and Development, the European Investment Bank and the World Bank. Hopefully, one of the effects of government stimulus will be to assist in unlocking private capital for investment.

6.   Money saver

As noted above, many organisations have realised that the drive towards a low-carbon economy can offer opportunities to increase revenues and enter new markets. In recessionary times, embracing renewable energy can also reduce costs in the long term.

Saving money, particularly at the end of the 2000s, has been vital to the general health of businesses. There is a real focus now on reducing energy demand and

---

27   Cleantech Report, "Ten predictions for 2010", Cleantech Group, November 2009.
28   New Energy Finance Report, September 2009.
29   Ira Ehrenpreis, general partner, Technology Partners.

increasing energy efficiency, which has been described as the "silent renewable".[30] This is particularly the case in the built environment. By using smart metering and grid technology, organisations can monitor and thereby choose to reduce their energy consumption. This means that they can save the planet at the same time as they save themselves money. Resource efficiency goes straight to the cost base of business. Companies are identifying fossil fuel-dependent assets that represent liabilities on their balance sheets and eliminating them. As a result, many companies are exiting the recession with an increasing awareness of their carbon footprint.

It should be possible to reduce costs in the long term by embracing the low-carbon economy. We have seen above that fossil-fuel pricing is hard to predict. The general trend in oil prices is upwards, while the cost of renewable energy is coming down. In the longer term, therefore, it seems likely that renewable energy will become cheaper than oil and gas. For example, many will argue that the long-term cost of fuel for renewable energy, such as wind, solar or marine, is nil once the initial investment in the technology has been made. The level of investment in technology required is reducing due to economies of scale and the ability to produce technology more cheaply.

7.   Conclusion

Many believe that the transition to a low-carbon economy will represent the greatest economic and technological shift in modern history. We are entering the sixth revolution, coming after:
- the industrial revolution;
- the age of steam and railways;
- the age of steel, electricity and engineering;
- the age of oil, automobiles and mass production; and
- the age of information and telecommunications.[31]

Renewable power capacity has expanded to 280 gigawatts globally in 2008, a 75% increase on 2004 levels.[32]

The drivers behind this growth include:
- overwhelming scientific evidence of climate change;
- significant advances in technology that are enabling the commercialisation of clean technologies;
- recurring fears about energy security (and indeed water and food security);
- fluctuations in fossil-fuel prices; and
- ageing electricity infrastructure.

These elements are influencing governments, organisations and individuals and forcing them to adapt their behaviour. Governments are taking action, either because of an awareness of the opportunities presented or because they cannot be seen to rest

---

30   Jonathan Johns, founder of consultancy Climate Change Matters, and former head of the renewable energy team at Ernst & Young, EY Country Attractiveness Index Report.
31   Merill Lynch equity note, November 2008.
32   REN21, "Renewables global status report", 2009.

idle. They are committing significant amounts of money to renewable energy schemes and creating subsidy regimes to encourage the transition to a low-carbon economy. Consequently, organisations and individuals are also adapting their behaviour. They are taking advantage of these government-driven subsidies. They are improving energy efficiency in their organisations and through their individual lifestyle choices. They are ensuring that they are not frozen out of their supply chains due to a failure to take account of climate change issues. Perhaps most importantly, they are investing in renewable energy and making significant returns.

Failure to take action against climate change will affect our ability to support growing populations. It will lead to conflict over energy, water and materials. Taking action will build economies and create jobs, increase domestic energy security, create business and investment opportunities, and drive forward technology developments.

The participation of the majority of states to the Copenhagen Conference in December 2009 is to be welcomed. All eyes will turn to Mexico in December 2010 for the next conference of the parties. However, we remain "at the end of the beginning" in terms of the transition to a low-carbon economy.[33] Tackling climate change, particularly during an economic recession, will remain a challenge. It is a challenge that governments, organisations and individuals must continue to embrace.

---

33   Cleantech Report, "Ten predictions for 2010", Cleantech Group, November 2009.

Renewable energy: at a growing age

Gil Forer
Joseph A Muscat
Ernst & Young LLP

1. Introduction

Adolescence is a time of emotional extremes. Sometimes, the future looks bright. Other times, even parents have their doubts.

All emerging innovations tend to go through similar mood swings. Renewable energy, one sector of the clean technology (cleantech) market, is no exception, as glowing reports of big plans and breakthroughs alternate with cynical follow-ups when the results fall short of expectations.

Sometimes, however, the reasons for the mood swings have little to do with the technology. After a strong recovery in 2009, most renewable energy stocks were at half their 2007 peak at the beginning of 2010. However, they have since bounced back.

The contribution of renewable electricity to the overall power grid has continued to rise. In Germany, for example, the share of electricity generated from renewable sources grew from 7.5% to 15.6% between 2000 and 2008.[1] This number is expected to double by 2015, especially for wind and solar power, according to an Ernst & Young global cleantech study.[2] Furthermore, the research suggests that corporations around the globe may be adopting cleantech innovations at the rate of 3% to 5% of annual revenues. This implies that many corporations continue to see in cleantech a combination of growth and efficiency.

Ironically, investors' reluctance towards cleantech is based not on the cleantech companies' performance, but on the company that the investors keep. When the price of fossil fuels is high, renewables become relatively more affordable. When fossil fuel prices fall – recently, because the world's economic reversals weakened demand for such fuels, despite the discovery of new reserves of fuels such as natural gas – the price that the market will pay for renewable energy falls too. Fortunately, energy prices are increasing, to an extent, and investor enthusiasm is beginning to return as well – albeit only to an extent too.

Such volatility is expected to continue. As traders traditionally say when asked where the stock market will go next: "It will fluctuate." Additionally, inexpensive energy options are a notion of the past. However, as the industry matures, the fluctuations should gradually diminish, with renewable energy becoming more cost competitive and a source of mainstream energy.

---

1   US Energy Information Administration, "International Energy Statistics", June 2010.
2   Ernst & Young, "Cleantech Matters: Going big: the rising influence if corporations on cleantech growth", November 2009.

A number of conditions favour cleantech in general and renewable power in particular. Energy demand is expected to rise by as much as 40% worldwide over the next 20 years.[3] Also, various specialists believe that we will reach peak oil by 2020. This growing demand for energy encourages the development of technologies to achieve greater efficiency and the advancement of alternative resources.

Consumers are attracted to renewables because of widely acknowledged environmental concerns. Many governments and businesses also support renewable energy developments, mainly because they see the strategic value in resource efficiency, energy independence and supply security. They also see the potential for cost savings in light of future increase in oil prices. Climate change and other environmental concerns also drive support for renewable energy. Increasingly, governments reflect such interests in laws, regulations and grants, while businesses reflect them through the installation of cleantechs and efforts to remove carbon emissions from their supply chains.

All around the world, the drive for increased resource efficiency and lowered carbon emissions is reflected in initiatives such as the billion-dollar stimulus packages that many countries set aside in 2009 for the development of renewable energy. This drive is further illustrated by the multiplication of corporate investments in cleantechs and partnerships with, and acquisitions of, cleantech companies.

The drive is also reflected in:
- the renewables portfolio mandates, alternative energy standards and the renewable or alternative energy goals of 36 US states,[4] which require that utility companies use a certain amount of electricity generated by renewable energy sources, purchased at a government-set price; and
- the US government's sponsorship of smart power grids – two-way, decentralised power networks – which are expected to create new opportunities for power generation and conservation.

This chapter examines:
- the obstacles that renewable energy developers encounter; and
- the strategies that renewable power executives and governmental energy planners are likely to use to try to circumvent these obstacles.

2. Points of light in an uncertain world

In 2009, one of the biggest challenges throughout the cleantech market was the recession-related flattening of energy demand. This demand has started to rebound, notably in the United States in early 2010.

The flattening demand affected not only energy prices, but also the prospects for cleantech projects. However, this is now considered a fluke. Many industry specialists expect that demand for all types of energy is likely to grow again soon – a surge that

3   International Energy Association, *World Energy Outlook*, November 2009.
4   "Renewable & Alternative Energy Portfolio Standards", Pew Centre on Climate Change (www.pewclimate.org), June 2010.

will lift cleantechs, including renewable energy, as well as conventional energy sources. Signs of this trend are already apparent.

Within the energy industry, most observers agree that the demand downturn of 2009 was a temporary knock-on effect of the recession of 2008 and 2009. They do not believe that this trend will last much longer. The International Energy Administration forecasts that the global demand for energy is likely to increase by 1.5% annually between 2007 and 2030 – an overall increase of 40%.

Certainly, this is the future that cleantech investors priced into their investments in 2009. The global cleantech market fell a little that year: total investments in clean energy declined by 6% from the record investment level of the previous year. However, the market stayed strong compared to the dramatic drop in business suffered by many industries, according to Bloomberg New Energy Finance, an analyst group specialising in clean energy and low carbon technology and carbon management. Despite the credit crunch, significant funds still found their way to the most mature renewable energy sub-sectors. Wind projects took the greatest share of funding, followed by solar and biomass, which took a distant second and third place respectively. That said, most commercial banks in Europe and the Americas reduced the amount of new credit offered to renewable energy projects, leaving public sector institutions to assume greater support and influence.

"We're in the eye of the hurricane. When we get out of this downturn, demand is going to come roaring back – and we'll need every resource we can muster to meet it," an executive recently told the authors during an executive roundtable on cleantechs.

Governments worldwide have started to offer these resources through incentivising job growth. In Europe, the United Kingdom announced a $100 billion investment to build 4,000 onshore and 3,000 offshore wind turbines by 2020, which is expected to create approximately 160,000 jobs.[5] Germany's renewable sector already:

- generates approximately $240 billion in annual revenue;
- employs 250,000 people; and
- is expected to provide more jobs than the country's automotive industry by 2020.[6]

Mexico has also projected new job opportunities in the renewable energy sector. The country's National Solar Water Heater Programme has the potential to create about 150,000 jobs by 2020.[7]

In the United States, President Obama announced in early 2010 the award of $2.3 billion in tax credits under the Recovery Act for cleantech manufacturing projects across the country. Further, he restated his administration's commitment to the clean energy sector during his 2010 State of the Union speech, when he said that "the nation that leads the clean energy economy will be the nation that leads the global economy". Throughout 43 US states, 183 projects are projected to create tens

5   "It ain't easy being", *Personnel Today*, Reed Business Information Limited, March 3 2009, via Factiva.
6   "Germany says green jobs will shorten recession", Reuters News Limited, February 24 2009, via Factiva.
7   United Nations, "Global Green New Deal – Environmentally-Focused Investment Historic Opportunity for 21st Century Prosperity and Job Generation" (2009).

of thousands of clean energy jobs, including opportunities in the solar, wind and efficiency and energy management technology sub-sectors.[8]

3.   Regulation and public support

A second and more serious challenge faced by the industry in some places is the rollback of certain special tax privileges. In Italy, for example, the feed-in premium for photovoltaic plants connected to a grid is due to end in December 2010. However, the Italian Ministry of Economic Development and the Ministry for the Environment, Land and Sea are proposing to renew incentive rates for photovoltaic systems from 2011 to 2015.[9] Other nations have cut or are considering cuts in tariffs. Even China is proposing reduced prices for large-scale solar farms to roughly one-quarter of the 2006 levels.[10]

After those in the United States, the German feed-in-tariffs remain the best value for consumers; they have consistently stimulated the industry – despite the recent downward readjustment to the solar tariff, which has triggered some protests. Germany determined that solar tariffs in particular should be reduced to reflect the swift reduction in costs, which in this case prompted new demand in the first half of 2010.[11]

Many in the business argue that feed-in tariffs will remain essential for large-scale development for years to come. As one renewables financier commented in Ernst & Young's global cleantech report: "If you have a good, strong, clear, transparent, enabling policy that allows the commercial value of renewable energy to compete aggressively on the current grid, you are going to get scale ... The proven method is to give project financiers something they can put in their spreadsheet, with some certainty."

Others argue that there are alternative ways to build in that kind of stability for developers. One method popular in some US states is the Renewable Portfolio Standard. Under this mechanism, power companies are required to generate a certain amount of their power using renewable resources.

In the end, as one renewable energy financier told the authors, the mechanism may not matter as long as developers have a guaranteed rate for the power that they generate. "There are different ways of skinning a cat. You can give tax rebates or you can mandate the feed-in tariff – or just set a tough Renewable Portfolio Standard, if you are serious about it. You send the price signal and the stuff gets built," he said.

Uncertainty about US policies on climate change is a continual source of anxiety for developers and investors. Many industry stakeholders say that when the climate change regulations in the United States become more certain, the US renewables generation market will experience significant growth. That said, the lack of regulation should not deter the rapid market growth that the authors forecast based on multiple worldwide market drivers.

Despite continued uncertainty about carbon policies, other market conditions

---

8   The White House, Office of the Press Secretary, "President Obama Awards $2.3 Billion for New Clean-Tech Manufacturing Jobs", January 2010,
9   Ernst & Young, "Renewable energy country attractiveness indices", June 2010.
10   *Ibid.*
11   *Ibid.*

favour strong growth of renewables in the world's two largest economies – the United States and China. Since June 2010 both countries can claim the number one spot in the list of most attractive nations for renewable energy investment.[12] These investment indices reflect the inclusion, in the US legislation of January 2009 on financial stimulus, of over $30 billion in loan guarantees and generous cash grants for renewables production.

However, the great deal of autonomy that US states enjoy in setting their power priorities suggests that what happens at the state level needs to be analysed. With that in mind, the top-ranked states for renewables generation are not just any states, but the largest states in the union by wealth and population: Texas, New York and California, in that order.[13]

With respect to carbon legislation, renewables companies should focus on the big picture. The long-term trend towards resource efficiency and lower carbon emissions remains strong. Whatever mix of carrots and sticks policymakers eventually choose, the strategy of most industrialised powers is clearly and firmly behind renewable energy. "Broadly speaking, the message is: forget Copenhagen," one renewable energy financier told the authors. "The long-term policy, the direction in which the world is moving, is toward de-carbonisation."

However, governmental commitment also creates a new set of risks. The challenge for the renewable energy sector and the broader cleantech market is that the gap between government rhetoric and regulatory reality remains wide. Companies should be careful not to pin their hopes on changes in tax structure or regulations. Instead, they need to work towards developing business models that do not depend on a public backstop. While that is not yet feasible for many cleantech products, maintaining a product focus is more likely to result in a genuine competitive advantage than relying on governmental largesse.

4.   Financing

At the moment, the most serious area of uncertainty for renewable energy is financing. Despite the growing consensus on the importance of renewables in future energy production, large buckets of capital remain uncommitted. Concerns about changing regulations, hydrocarbon price volatility and the risk that innovation will render a major investment obsolete have kept most investors and bankers extremely cautious of financing in the sector.

Global investment in cleantechs totaled $205 billion in 2009, including:

- $92 billion in asset finance;
- $60 billion in mergers and acquisitions;
- $11 billion in government research and development;
- $9 billion in corporate research and development;
- $4 billion in private equity; and
- $2 billion in venture capital.[14]

---

12   *Ibid*
13   Ernst & Young, "United States renewable energy attractiveness indices", November 2009.
14   Bloomberg New Energy Finance, February 2010.

As stated above, these numbers are down 6% from 2008 but still represent an impressive performance considering how rough the first quarter of 2009 was for many companies. In fact, if one makes abstraction of the results of the first quarter, that year's performance is close to that of the previous three years. This is partly thanks to strong growth in Asia, which experienced a 29% investment increase on the previous year.[15]

Worldwide, the largest cleantech investment in 2009 was $91.9 billion of asset finance – debt investments in projects such as wind farms, solar parks and biofuel plants. This was down 5% on the previous year. Some of the biggest developments consisted of offshore wind projects in UK waters – most notably the 1 gigawatt London Array scheme. However, China, with its stimulus programme worth $500 billion, bucked the trend. The total Chinese new-build asset finance investment in 2009 for wind was $21.8 billion, up 27%, with solar at $1.9 billion, up 97%.[16]

Early-stage venture financing for cleantech is now recovering as well, although longer-term financing for large-scale projects is still hard to find. Funding is still not assured, even for established, large-scale players. The struggle faced by small to mid-sized developers to find structured finance is likely to continue.

To an extent, governments have tried to fill this gap. In the United States, for instance, the $30 billion in loan guarantees included in the February 2009 stimulus legislation has gone a long way towards keeping the US renewables market moving. The European Investment Bank has also underwritten some projects. But even this support cannot make up for the relative dearth of commercial lending. As one commercial lender stated in early 2010 during a roundtable on cleantechs: "For projects that require long-term finance – where banks continue to very selectively deploy capital – I think we are still going to see a scarcity of funds."

At the same event, another banker argued that project financing will not improve until the debt markets or institutional investors begin to take some of these liabilities off the banks' balance sheets. "Until that happens, we are going to struggle to see significant project finance capital return to the market," she said.

For larger companies investing in renewables, it may seem tempting to wait until the sector matures and let the market reward them for managing old products in old ways. However, this would be a shortsighted response. Failure to take some steps towards cleaner energy use now will serve only to make the earnings calls of five years from now that much more painful.

5.   The kids are all right

It is clear that only a select group of the renewable energy companies around today will be the leaders of tomorrow. Some will fail when their technology fails to meet expectations. Others will flop because they lack one or more of the other elements necessary for success in marketing a new product – that is, the right investors, customers, business model and timing.

"The reality is that only a very few will survive on their own; a few others will

---

15   Bloomberg New Energy Finance, May 2010.
16   *Ibid*

successfully partner, more will be acquired at bargain basement prices and many will fail," said one venture capitalist quoted in the Ernst & Young global cleantech report.

Backers of many promising renewable energy technologies, especially in the United States, are caught in that classic first-jobber bind: getting hired without experience. Banks will not lend without a track record, but new technologies by definition do not have a track record. To make matters worse, most ordinary lenders are still standing by the sidelines in the United States. "Until we get that portion of the finance community back in lending, I think renewables development remains moribund," one venture capitalist told the authors.

If a robust and healthy market for renewable energy emerges, these failures will ultimately prove to be positive developments. For those companies that overcome this present, difficult period, the future looks bright.

Eventually, the industry's current challenges will likely appear as mere growing pains that are part of a historic shift away from a nearly exclusively fossil fuel-based economy to one based increasingly on clean technologies, including renewable energy. That transition will take time. The renewables market in 2008 made up about 3% of the total power generation and this share is expected to more than quadruple by 2030.[17] However, the authors are confident that the technology will have matured beyond its current awkward, adolescent age long before that.

---

17   International Energy Agency, *World Energy Outlook 2009*, November 2009.

Renewable energy support mechanisms: an overview

Katy Hogg
Ronan O'Regan
PricewaterhouseCoopers LLP

1. Introduction

1.1 What is renewable energy?

The definition of 'renewable energy' varies according to the context in which it is discussed. However, for the purposes of this chapter, it is assumed that the term covers energy that occurs naturally in the environment and, as such, is essentially inexhaustible. Renewable energy therefore encompasses energy harnessed from solar, wind, hydro,[1] biomass,[2] marine[3] and geothermal sources.

1.2 Why support renewable energy?

Harnessing natural energy resources in order to produce electricity and heat is considered to be more environmentally friendly than burning fossil fuels, largely because there is no direct increase in the levels of carbon dioxide in the earth's atmosphere as a result. Consequently, it is widely believed that increasing the use of such energy resources and reducing the reliance on carboniferous fossil fuels will ultimately reduce the carbon levels in the atmosphere and, therefore, any detrimental effects on the climate. In addition, the use of a region's natural, renewable resources increases the security of energy supply in that area by ensuring a constant (or at least a known and friendly) supply of energy, rather than relying on fuel imports.

In recognition of these issues, governments across the globe have set targets to reduce carbon emissions and increase renewable energy generation. (However, the Copenhagen summit of December 2009 failed to deliver a global agreement on climate change to succeed and build upon the Kyoto Agreement.) In parallel, governments have also accepted and acted upon the need to incentivise the market to invest in such technologies. As a result, governments have set up a multitude of financial support schemes for renewable energy.

---

1   'Hydro' refers to energy derived from flowing water. This can be from rivers or man-made installations, where water flows from a high-level reservoir down through a tunnel and away from a dam.

2   'Biomass' covers a broad range of natural materials including wood, energy crops and waste. It is defined as organic material derived from recent plant or animal matter. The carbon savings from using biomass to generate power or heat can vary widely because the they need to be considered in the light of the fossil-fuel energy that is used for cultivation, harvesting, processing and transportation. In addition, extensive change in the use of land can totally negate any carbon saving. Consequently, any biomass used needs to be grown and sourced sustainably in order to be considered 'renewable'.

3   'Marine energy' covers both wave and tidal energy (which can stem from tidal stream or tidal barrage).

1.3 Why is financial support necessary?

Investment in renewable energy projects requires financial incentivisation because such projects are typically more expensive on a cost per unit of energy generated basis[4] than the traditional methods of energy generation. In addition, they are in some cases also considered to be riskier investments due to technology or resource uncertainties.

The degree to which cost and risk factors apply varies according to technology and geography. However, as a general rule, these factors mean that investors in renewable energy projects need to invest greater sums on a pound per megawatt-hour basis than if they invested in more traditional forms of generation. Investors also need to get a higher return on their investment to compensate them for taking on additional risk. Alternatively, the risk will need to be reduced through the provision of more revenue certainty.

(a)   *Cost*

Renewable energy has historically been more expensive for a number of reasons:

- Renewable resources are often located in remote areas that require the costly installation of power lines in order to access the market (eg, offshore wind);
- Renewable sources are not always available due to weather effects; and
- Generation from natural resources relies (for the most part) on relatively novel and less efficient technology.

This additional cost varies significantly across technologies and geography. Accordingly, the level of support required to incentivise investment varies also.

Further, the detrimental climate impact of the carbon dioxide gases emitted from fossil-fuel combustion was until recently an unrecognised 'negative externality' – a societal cost resulting from energy generators' commercial operations and end users' consumption that was not quantified or factored into energy market economics. This omission contributed, and still contributes, to the lower cost of fossil-fuel generation. It is also the reason why regulators are attempting to put a cost on carbon emission through schemes such as the EU Emissions Trading Scheme.

On the flipside, as renewable technologies become more widespread and improve, it is expected that technology costs will fall. In addition, a long-term contributor to lower renewable energy costs is the low or zero cost of renewable fuel – biomass being the most obvious exception. Accordingly, it is not unreasonable to expect that at some point the cost of most renewable energies may be on a level with fossil-fuel energy and no longer require financial support, thus reaching grid parity.

(b)   *Risk*

Although the concept of harnessing natural resources for power and heat has been around for centuries, modern applications suitable for the large capacities required are generally still commercially uncompetitive with traditional sources of power.

---

4   Typically referred to in the United Kingdom as the pound per kilowatt-hour or pound per megawatt-hour measure.

Where a technology has an unproven or limited track record of performance, investors view the investment as more risky; accordingly, they require higher returns. The performance of wind energy depends on wind speeds and the availability of the turbines; the performance of solar energy depends on how much energy the sun gives out and how much can be converted into energy; biomass performance comes from ensuring compatibility between boiler and feedstock.

To date, renewable support mechanisms around the world have favoured technologies such as landfill gas, solar photovoltaic or large-scale wind. As a result, the installed capacities of these technologies are significantly greater than those of other technologies and costs have fallen significantly. Other technologies (eg, wave and tidal), however, are trickier and at an earlier stage; the risks are higher and the level of support to date has been minimal.

1.4   Scope of this chapter

This chapter provides an overview of the most common financial support mechanisms that have been or are available for renewable energy technologies around the world. Examples (primarily from Europe) are used to illustrate how such schemes work in practice.

The principal groups directly affected by the form and application of renewable energy support mechanisms include:

- the investors in renewable energy projects;
- the taxpayers that pay for the government's support through taxes or the consumers that are charged additional costs on their energy bills; and
- the regulators that design and manage the schemes.

The benefits and drawbacks of each mechanism are therefore considered from the point of view of these three groups.

This chapter does not attempt to be comprehensive, but is intended to give readers an understanding of the most prevalent support mechanisms used and their impact on key stakeholders.

An important route to reducing carbon emissions associated with energy consumption entails:

- managing demand;
- reducing energy usage; and
- introducing energy-efficient technologies and practices.

This chapter does not cover the incentive mechanisms that drive growth in this area.

2.   Stakeholder interests

2.1   Investors

The term 'investor' refers to all individuals or organisations that contribute capital and/or resources to the development of renewable energy projects of any size and expect a financial benefit.

This definition covers private equity and venture capital organisations, banks, wealthy individuals, ordinary consumers and communities, corporations, utilities and public sector bodies. These investor groups typically have different risk and return appetites, and evaluation criteria.

However, some general investment characteristics will be preferred by most, to a greater or lesser degree depending on context. These characteristics include expecting:

- as low as possible a risk with regard to technology performance and revenues; and
- as high as possible a cost and/or carbon-savings, and/or overall investment returns.

This translates into a preference for renewable energy support mechanisms that deliver satisfactory returns over the long term; regulatory certainty and transparency are therefore important factors in an investor's decision-making process.

2.2 Consumers

The broader consumer body includes individuals and organisations that consume energy and/or pay taxes. It is from this stakeholder group that the additional cost of supporting renewable energy is recovered – either through additions to energy bills or through taxes and levies on earnings and energy consumption. It is also this group that would ultimately bear the cost of climate change, although that is not so easily quantified.

As consumers are the 'bankrollers' for renewable energy support schemes, the following issues are of key concern to them:

- the total additional cost to their bills;
- the efficiency and effectiveness of the expenditure (ie, the total cost per tonne of carbon abated or per unit of installed capacity); and
- the wider benefits of supporting renewable energy (eg, positive climate effects and increased security of supply).

2.3 Regulators

Renewable support schemes can require significant upfront and ongoing administrative support and oversight. This includes the initial design process, public consultation, scheme administration and any additional treasury burden. The term 'regulator' is used here to cover all governmental or independent bodies that have an administrative or regulatory role in relation to renewable energy support schemes.

The success or failure of high-profile and costly support schemes can also have political repercussions, similarly to any other public expenditure programme. In recent years, as the sector has grown, governments have set up separate teams to build expertise and drive policy initiatives in this area. For example, the UK government made a number of reshuffles to bring together what is now the Department for Energy and Climate Change and the associated Office for Renewable Energy Deployment.

Regulators are largely concerned with the following:

- the effectiveness of the scheme in relation to carbon and renewables targets;
- the cost of oversight and administration of the scheme;

- the overall efficiency of the scheme in terms of expenditure and impact on end-consumer tariffs; and
- public perception.

3.   Typical support mechanisms

3.1   Summary

Renewable energy support mechanisms range from grants for early-stage technologies to tax incentives for capital invested in renewable energy plant, through to payments for each unit of energy generated. The objectives, scope, application and level of support of these schemes are based on, among other things, geography, technology and project scale.

Theoretically, all support schemes can be funded through tax revenues. In practice, however, the burden of the costs is typically placed on market participants such as electricity suppliers and transmission operators, which then pass on the additional costs to electricity consumers.

A summary of some of the most prevalent support mechanisms is set out below.

3.2   Research, development and demonstration

(a)   Research and development

New technologies, such as those used to harness wave and tidal energy, typically require considerable early-stage investment to get them from research and development to prototype stage and, for some, commercial operation. As the technologies progress through each stage, the required levels of funding often increase. Only a limited number of investors have the appetite for such high levels of funding coupled with technology risk.

In cases where private sector investment does not flow, governments must provide some level of support if they wish to see a technology-driven industry develop.

Research and development support programmes typically come in the form of grants or loans. In most cases they will not be designed with the expectation of financial returns. However, some research and development schemes incorporate a mechanism that enables the regulator (and the consumer) to share in any financial returns from technology successes. The United Kingdom's Launch Aid Scheme for civil aerospace is a globally recognised method for supporting research and development efforts on such terms. The scheme enables the government to contribute to the development of technologies. In return, it gives the government the right to a levy on each unit of the technology that is sold once certain thresholds are reached.

The typical recipients of Launch Aid funding are technology developers that use the funds to further their research and development experimentation, thereby (ideally) producing patentable, valuable technologies that can be sold on for commercial gain.

It is therefore crucial that the programme receiving such funds can be said to be one of research and development, and one that is working to develop identifiable, patentable technologies. This concept could well be applied to renewable energy technologies such as wave and tidal. However, the potential value of each device developed has to be weighed against the potential cost to get it to the next stage of development.

(b)   *Demonstration-scale grants*

Innovative technologies that make it through the research and development phase typically struggle to make the leap from prototype to commercial status. This is due to the investors' reluctance to spend money on demonstration-scale projects that have not yet been proven and offer no likely commercial returns in the near future, if at all.

As a result, governments and non-profit-making organisations are frequently called upon to assist the market in developing the technology to a commercial stage by providing funding for demonstration projects in order to prove that technologies work on a commercial scale. The levels of funding required can increase significantly between the research and development stage and demonstration stage.

The primary method employed to support demonstration projects is the grant scheme. Funds are set aside from government funds (which could be tax receipts and/or government debt) to develop a certain type or group of technologies. Grant award criteria are set. Bids are received and evaluated, and grant funding is awarded on certain conditions. The granting body does not typically expect to see a return on its funding, although it does expect its money to be spent wisely and according to the grant scheme rules.

A recent example is the Renewable Energy Demonstration Programme, a competitive grant scheme set up by the Australian Ministry for Resources and Energy in 2009. The scheme's aims include:

• demonstrating the viability of renewable energy technologies on a large scale;
• enhancing Australia's international leadership in renewable energy; and
• attracting private sector investment in renewable energy power generation.

At the time of writing, the programme had awarded A$235 million to support the development of four commercial-scale renewable energy projects, comprising geothermal energy, wave energy and an integrated mini-grid project involving wind, solar, biodiesel and storage technologies.

Grant funding was awarded on the basis that the applicants contributed funding in excess of A$500 million. This is a typical requirement of modern grant schemes (ie, that the private sector matches the grant funding or provides a majority funding in order for the grant funding to be awarded).

A number of grant schemes have proved too stringent or unrealistic in their award criteria for the market to gain access to the funds. For example, some grant schemes have required the technology developers to agree to make their intellectual property available to the public as a precondition to receiving funding. This can cause problems at a later stage in accessing private sector funding, where retention of intellectual property is a key part of investor diligence.

Investors' perspective: By participating in partially grant-funded projects, investors benefit from a financial partner which shares in the financial risk of the project, but which will typically not share in any future value gained through the successful delivery of the projects and resulting growth of the participants' respective businesses. A downside of publicly funded grants is that they typically place a restriction or additional burden on the investors' commercial operations so that the grant project will typically be unlikely or unable to yield significant returns.

Regulators' perspective: Regulators are typically either grant funding bodies or associated with the funding bodies through the provision of an administrative role. Governments generally use grant schemes for renewables not only to rearch targets for renewable energy, but also to support the broader market for renewable technologies and expertise. In doing so, governments promote trade in their region and raise their international profile.

It is important for the regulators awarding grants to:
- set clear award criteria;
- consider value for money; and
- ensure that the risks of supporting developing technologies are understood and mitigated appropriately.

Consumers' perspective: Grant funding for early-stage technologies is not typically as high profile as that for demonstration projects given the sums involved and the largely invisible nature of research and development programmes.

Consumer perspectives and pressure can come to bear more significantly on schemes that clearly lay out substantial sums of money for risky and expensive projects. In such cases consumers' fear of overspend and desire for accountability for any mistaken investment can lead to high-profile press coverage. It can be difficult for governments to justify expenditure where other, more economically viable investment projects are available.

3.3   Tax incentives

Tax incentives can be preferable to subsidies in the eyes of many governmental authorities: it is politically easier to defend tax incentives because there is no direct link to consumers through higher energy tariffs. However, tax-related incentive schemes have a cost to society that follows the reduction in tax revenues flowing through to governments and, therefore, public services.

Tax incentives used to promote the use of or investment in renewable energy technologies and projects include the following:
- reduced corporate tax rates;
- tax holidays;[5]
- investment allowances and tax credits;[6] and
- accelerated depreciation.[7]

---

5   No tax payable for a period of time.
6   Tax reduction based on the amount invested; they are in addition to normal depreciation.
7   Depreciation is applied to write down asset values more rapidly.

Certain countries combine one or more of the above incentives. For example, China's new corporate income regime provides a three-year exemption from and three-year, 50% reduction in corporate income tax for business income derived from qualified environmental protection, energy and water conservation projects.

Reduced tax rates, tax holidays and accelerated depreciation are relatively straightforward concepts, but investment allowances and credits are slightly more complex.

Investment allowances and tax credits are forms of tax relief based on the tax value of expenditures on qualifying investments. Qualifying investments can cover small renewable energy systems or efficient combined heat and power plant.

In order to ensure that a tax-relief scheme can be accessed and the objectives for renewable energy growth met, clarity is necessary with regard to:

- the definition of the eligible expenditure;
- the choice of the rate of allowance or credit;
- the restrictions on use of the credit or allowance; and
- the treatment of any incentives that cannot be used in the year in which they are earned as a result of insufficient taxable income.

As with most things tax related, such schemes can be very complicated and may require considerable technical expertise to ensure that the criteria are met. It is important that when a tax credit is in place, the taxpayer can use the tax credit. A tax credit on, for example, research and development should be drafted so that innovation is encouraged rather than stifled.

The United Kingdom runs a scheme to encourage investment in efficient combined heat and power plant called the CHP Quality Assurance Scheme. The benefits of qualifying as 'good quality' under the scheme include becoming eligible for:

- enhanced capital allowances;
- exemption from the climate change levy; and
- exemption from a proportion of applicable business rates.

The enhanced capital allowance aspect of the scheme was its main attraction until recently. Investment in good-quality combined and heat power qualifies for the enhanced capital allowance, as it is classified as 'energy-saving plant and machinery' – a definition that goes beyond combined heat and power into energy-efficiency technologies. The enhanced capital allowance is essentially a 100% first-year writedown against taxable profits made during the period of investment. Besides, where a combined heat and power plant is run on renewable fuels, the additional benefits of being deemed 'good quality' can be significant, with additional subsidies in the form of renewables obligation certificates for the renewable power produced. Looking further ahead, additional financial support is expected for the renewable heat produced.

Investors' perspective: Investors can derive significant value from tax allowances where investors integrate the benefits into their ongoing business effectively. However, navigating the scheme's complexities can be a barrier to accessing and releasing the value. There are also usually significant restrictions on how and when the tax incentive can be claimed. This may not always work to the investors' advantage where the outcome of other business ventures does not go as expected.

Consumers' perspective: Tax allowances or credits are relatively benign from the consumers' perspective, as they are not so easily identified as a 'cost' to consumers. However, where consumers are also eligible to benefit from the tax allowance, but are unable to do so due to the complexities of the scheme, the result is frustration and a lack of desired investment.

Regulators' perspective: A key question for authorities to consider when drafting tax incentives is whether the incentives influence the behaviour of taxpayers – that is, would the investment be made without the proposed tax incentive? Tax incentives also mean additional costs for the government (ie, in lost tax revenues). Another aspect to consider is the complexity that accompanies any tax legislation, for both the tax authorities and the associated governing bodies. This is also an issue for private sector participants that need to be able to navigate the regulations in order to attract the incentives.

3.4   Subsidies

(a)   Feed-in tariffs

Feed-in tariff schemes have contributed to considerable growth in renewable capacities in a number of countries, including Germany and Spain. Nevertheless, feed-in tariff schemes require careful design and oversight to ensure that they limit the potential for booms and busts, and excessive costs to consumers.

Feed-in tariffs come in two main guises: as a fixed tariff or a premium tariff. A fixed tariff provides a generator with an overall remuneration package per unit of energy output, independently of the electricity market price. Most European feed-in tariff schemes apply a fixed tariff model. A premium tariff is paid to a generator in addition to the market price for electricity. Premium tariffs are applied in the Czech Republic, the Netherlands and Spain.

The design of a feed-in tariff and the support levels allocated to different technologies depend on the overall aims of the policy. A scheme designed to meet a particular renewable generation target at the lowest cost to the economy will result in a technology mix that varies significantly from a scheme that is focused on driving a particular section of the market (eg, consumers and community-scale installations) to invest.

The most successful feed-in tariff schemes present a number of common characteristics in their design and the broader regulatory environment in which they operate. These characteristics include:
- low administrative and regulatory barriers;
- long-term revenue visibility;

- high levels of policy certainty; and
- high cost efficiency.

The chart below illustrates the high growth rates of solar photovoltaic capacities in Germany, Spain and Italy, where substantial feed-in tariff schemes were introduced at various stages. The UK picture is vastly different, primarily due to the lack of governmental support historically.

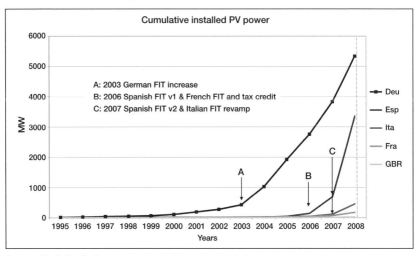

Source: PwC Analysis

The success of feed-in tariffs demonstrates that revenue certainty is hugely popular with investors, as it adds to the bankable nature of a project. However, in some cases the accelerated growth rates have also been an indicator that the regulators had set the tariffs attractively high, resulting in investors making above-expected returns.

On April 1 2010 the United Kingdom introduced feed-in tariffs for smaller-scale (under 5 megawatts) renewable electricity technologies such as solar photovoltaic and wind turbines. Electricity suppliers are obliged to make payments to their customers on the basis of the volume of renewable electricity produced, even if the electricity is consumed on site. This scheme has been introduced following an acknowledgement that the slow growth in renewables capacity in the United Kingdom, particularly for smaller-scale installations, could be significantly accelerated if the main support mechanism used (tradable renewable obligation certificates with variable, and therefore risky, value) was replaced by a feed-in tariff scheme. Complications around changing support mechanisms prior to achieving the 2020 targets has prevented a move to feed-in tariffs for larger-scale projects for the moment. However, some form of revenue stabilisation mechanism is likely in the long term.

To minimise subsidy costs, a feed-in tariff scheme can be banded by technology, thereby reducing overpayments to low-cost generators. Banding is particularly important among different technologies rather than different types of one technology, as the cost differences between technologies are greater. The UK feed-in tariff scheme incorporates banded tariffs that differ according to the type and scale of technology and have been calculated in order to give a 5% to 8% return on initial investment.

Further, generators that can supply electricity to the grid are paid an additional amount per unit of energy exported (all income for individual householders is exempt from income tax).

The feed-in tariff scheme may have a degradation rate built in to its future tariff levels, a capacity limit and/or a time limit for each phase of the scheme. All three features are designed to limit the exposure of the regulator and enable the feed-in tariff mechanism to flex in parallel with a maturing market where costs come down and risks fall. However, matching tariff levels to technology costs from the outset of the scheme results in higher policy costs than if a flat tariff were set and the scheme waited for costs to fall to a level that made the tariff economic. The downside of waiting for costs to fall is that uptake will be slower and targets may not be met on time. The UK feed-in tariff scheme guarantees the tariff for a set number of years (between 10 and 25 years, depending on technology) and is index-linked to inflation throughout. However, the level of tariff for newly installed and future solar photovoltaic and wind projects reduces year on year in anticipation of technology cost reductions.

To supplement the existing support schemes, the UK government is consulting on the design and implementation of a feed-in tariff scheme that will support renewable heat, rather than electricity, from 2011. This is the proposed Renewable Heat Incentive Scheme, which would cover all scales of technologies such as ground source heat pumps, biogas and solar thermal. The scheme has the dual aims of providing a reasonable level of compensation and support for a wide range of technologies and investor types.

The scheme will be open to individuals, community groups and businesses. In all cases the owner of the plant will be responsible for funding the upfront capital costs and recouping payments under the scheme over the useful life of the plant.

The government intends to release payments over a number of years according to the heat produced. Payments will be conditional on the owner maintaining and using the heating system appropriately. The tariffs are intended to be fixed for the life of a project (giving greater investor certainty) and differ according to technology. The tariffs have been calculated on the basis of reference installations in each technology band. This means that returns will likely fluctuate from the target of 12% (6% for solar thermal, due to its more established presence and low installation challenges).

Investors' perspective: The key issues for investors concern clarity and certainty in relation to the scheme's key characteristics and duration. On this basis, fixed tariffs are typically preferred as the level of remuneration is constant throughout the scheme's lifetime (subject to performance). Because of this additional certainty, projects have a lower risk profile and the hurdle rate for investments is lower. This in turns lowers the cost to the consumer. A fixed scheme also removes any obligation on the part of the scheme operator to get involved in the complexities of system balancing, which reduces the complexities and costs of the project. This is particularly important for micro-generation schemes that produce low output levels per year.

Consumers' perspective: Feed-in tariffs that have been set too high have proved to be very expensive for consumers. However, the certainty that comes from feed-in tariffs over tradable certificate schemes means that the economic rent built into the level of the incentive should be lower. It is therefore the regulators' job to ensure that they evaluate current and future costs of the relevant technologies as accurately as possible. To assist with that process, feed-in tariffs should be offered in phases. This will enable regular review of the level and capacity limits needed for each phase.

Regulators' perspective: A key area of focus for regulators considering introducing or extending a feed-in tariff scheme is to ensure that they have an accurate forecast of all-in prices for the technologies that they want to incentivise. Failing to do so results in excess returns to the investors in these projects at the expense of consumers.

(b) *Obligation schemes*

Over 20 countries and regions operate incentive schemes based on an obligation to deploy a certain capacity of renewable energy. These include Australia, Japan, Norway, the United Kingdom and some US states.

Obligation schemes typically place an obligation on electricity suppliers[8] to surrender what are sometimes referred to as 'green certificates' that prove that the suppliers have generated energy to a certain level from renewable sources or purchased certificates from a generator of renewable energy.

The UK government's central support mechanism for renewable energy is the renewables obligation. It is designed to provide an output-based incentive for investment in all eligible forms of renewable electricity in the form of tradable renewable obligation certificates. These are awarded per megawatt-hour of electricity produced and have a market value.

The obligation falls on electricity suppliers and requires them to source a specific and increasing proportion of their electricity from renewable sources or else pay a financial penalty. The obligation was initially set at 3% for 2002/2003 and is 9.7% for the current obligation period. It is suppliers' responsibility to generate from their own projects, or else purchase from other renewable energy generators, a sufficient number of renewable obligation certificates to match their renewable obligation in each year.

If generators fail to meet their obligation, they can pay a buy-out price; these funds are pooled and recycled to suppliers on the basis of how many renewable obligation certificates they surrendered. The suppliers with the smallest portfolio of renewables therefore give money to those with the largest portfolios.

Renewable energy generators can sell their renewable obligation certificates to

---

8   As a contrary example, the Australian Mandatory Renewable Energy Target places an obligation on wholesale electricity purchasers. See www.climatechange.gov.au/renewabletarget/mret.html.

electricity supply companies that use them to demonstrate compliance with the renewable obligation. This enables generators to receive a premium on top of the sale of the electricity.

The value of a certificate is therefore driven by the level of the buy-out price each year, plus the market's expectation of the recycled value in that year (which is a function of the supply and demand balance against targets). The value is further determined by the method of sale. Long-term fixed contracts give generators certainty. The price that they receive is therefore lower (potentially) than the price that they could get through selling their certificates periodically.

The value of a traded renewables obligation certificate has continued to increase in recent years to over £50 per megawatt-hour, as shown in the chart overleaf.

To increase price certainty, the UK government has attempted to minimise the risk of excessive future price drops through the use of headroom mechanisms. These attempt to ensure that the market supply of renewables capacity can be delivered in a way which avoids the risk of the price of renewables obligation certificates crashing in the event that capacity greatly exceeds expectations.

The level of support received for each technology depends (since April 2009) on which banding that technology falls within – the idea being that more mature and expensive technologies need less support to incentivise investment than newer, more expensive and risky technologies. Prior to the introduction of banding, the renewables obligation was successful in contributing to the growth of primarily wind and landfill gas, as the cheapest of the eligible technologies.

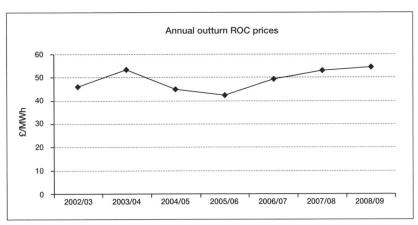

Source: Office of the Gas and Electricity Markets

Investors' perspective: The benefit from the investors' and generators' perspective of a market-based tradable certificate mechanism is that the price can go up when the supply of new renewable energy capacity in a particular year is low and this revenue will increase the equity returns to the project.

The risk is, of course, the downside of the same point; the market price is not a fixed price, with fluctuations resulting from, for example, the oversupply of certificates. To mitigate this risk, investors can try to secure long-term offtake contracts for the sale of their certificates at a fixed or floored price to a creditworthy counterparty.

Overall, however, it is generally considered that obligation schemes do not deliver the same results as feed-in tariffs.

Consumers' perspective: Consumers ultimately pay for the additional costs of having renewable energy in the generation mix. Where energy bills are low, these additional costs are more easily borne. However, when energy prices are high, these additional costs are less palatable. As the additional costs are currently spread across all customers, the low level of renewable penetration in the United Kingdom has not had a significant impact on end-user energy prices. However, this will change as the percentage of renewables in the energy mix increases. In terms of efficiency of spend and effectiveness of such schemes, there are debates as to whether it is better to rely on a feed-in tariff scheme or a tradable certificate scheme. In principle, the feed-in tariff scheme is designed to lower the risk premium to the investor. It should thus require a lower cost of capital than a tradable certificate scheme, which in turn should reduce the cost to consumers.

Regulators' perspective: In order to deliver growth aspirations without excessive costs to consumers, regulators have had to develop a number of improvements that assist with buffering the effects of the market on these market-based mechanisms. Such improvements include the use of a 'headroom' concept, whereby the obligation in each year is set above the forecast for renewables output. Another improvement is the banding of support, so that a broader portfolio of technologies is brought on board – particularly those that need greater levels of support – while those that are closer to grid parity receive lower levels of benefit.

3.5   Deterrent schemes

Governments can also incentivise investment in low-carbon energy solutions and more energy-efficient behaviours through deterrents such as additional taxes or 'cap and trade' schemes in relation to non-renewable energy usage and generation.

(a)   *Taxes on non-renewable generation*

Taxes can be placed on the consumption of non-renewable energy in order to encourage the business and public sectors to improve energy efficiency and reduce emissions of greenhouse gases. This gives organisations a price-based signal on their non-renewable energy usage.

In the United Kingdom, this deterrent mechanism has been applied through the Climate Change Levy Scheme, which applies a tax on energy used in the industrial, commercial and public sectors. The tax is designed to increase energy efficiency within the sectors covered and hence reduce overall emissions.

Imposing a tax on energy consumption creates a risk that energy-intensive sectors will lose out to international competitors that do not operate in such strict regulatory environments. As a way of mitigating this risk, levy participants should therefore be able to reduce the impact of the levy through appropriately climate-friendly investments and operational behaviours. The levy does exactly that. Climate change agreements were introduced in order to provide participants with an opportunity to benefit from an 80% discount on the levy if challenging targets were agreed and met for improving energy efficiency or reducing greenhouse gas emissions. In addition, any energy generated from renewable sources is eligible for renewables levy exemption certificates from the Office of the Gas and Electricity Markets. Electricity suppliers can then surrender these certificates in relation to a specific supply contract in order to claim exemption from the climate change levy.

The government also introduced the Enhanced Capital Allowances Scheme for investments in energy-saving technologies and products at the same time as the climate change levy. This tax-incentive scheme helps businesses to reduce their energy use and adopt low-carbon technologies. The levies received are used partially to offset the tax revenues forgone due to the application of the enhanced capital allowances.

(b)   *'Cap and trade' schemes*
Cap and trade schemes place a carbon emissions cap on specified carbon-intensive market participants – for example, large installations in the industrial sectors (eg, power generation, oil refineries and steelworks). The participants in the scheme then have to monitor and report on emissions each year, and return to the regulator an amount of emissions allowances equal to their emissions in that year.

Participants that fail to surrender sufficient allowances to cover their annual reported emissions must pay a penalty for each tonne of carbon dioxide-equivalent in excess of their surrendered allowances.

The EU Emissions Trading Scheme is the largest multinational emissions trading scheme in the world. Under the scheme, the governments of the EU member states:
- agree on national emissions caps that are approved by the European Commission;
- allocate allowances to their industrial operators;
- track and validate actual emissions in accordance with the relevant assigned amount; and
- gather the allowances that are submitted after the end of each year.

While such schemes are not primarily focused on incentivising investment in renewable energy sources, they represent an important step towards a less carbon-intensive economy by increasing the long-term costs for fossil-fuel users. This leads to a strong indirect incentive for the substitution of renewable energy sources in the participant sectors, particularly in electricity and heat production.

In addition, where a cap and trade scheme is linked to other emissions reduction schemes that directly reward renewable energy projects, such as the linkage between the EU Emissions Trading Scheme and the UN Clean Development Mechanism and Joint Implementation Schemes, there is also a direct incentive for organisations in the cap and trade scheme to invest in renewables projects in order to obtain carbon credits effectively to offset their own emissions.

The Clean Development Mechanism and Joint Implementation Schemes enable participants in the EU Emissions Trading scheme or independent carbon traders to purchase carbon credits from projects outside the European Union and then trade or surrender them within the European Union in the place of allowances. The Clean Development Mechanism encourages sustainable developments and lower greenhouse gas emissions in developing countries (eg, Brazil and China) without setting an emissions reduction target. The scheme covers projects that deliver renewable energy, energy efficiency and fuel switching. For each tonne of carbon dioxide avoided as a result of the project in question (as measured against an assumed baseline), a Clean Development Mechanism certificate is awarded, equivalent to one tonne of carbon dioxide or one allowance under the EU Emissions Trading scheme.

The Joint Implementation scheme works similarly to the Clean Development Mechanism. However, it supports projects in countries that have an emissions reduction requirement, but whose economy is thought to be in transition (eg, Croatia, Poland and Russia). The economic basis for including developing countries in emissions reduction schemes is that emissions cuts are thought to be less expensive in these countries.

In April 2010 the United Kingdom introduced a new cap and trade scheme, the Carbon Reduction Commitment, which:

- is aimed at incentivising energy efficiency in large public and private sector organisations; and
- applies to emissions that do not fall within the climate change levy or the EU Emissions Trading scheme.

All UK organisations that consumed electricity above a threshold level in 2008 and that used half-hourly meters to record their consumption will be obliged to participate in the scheme. They will also have to monitor their emissions and purchase carbon allowances, initially sold by the government, for each tonne of carbon dioxide that they emit. The government estimates that around 5,000 organisations will be required to participate fully and another 15,000 will be required to make information disclosures only. The scheme is intended to be revenue-neutral to the government and to provide an incentive to reduce carbon dioxide emissions by distributing revenue from the worst to the best performers in the scheme.

The basic principle is that the better an organisation performs in terms of reducing its emissions, the higher it will be ranked in an annual league table and, therefore, the higher the value of recycled revenues it will receive. During the introductory phase of the scheme, allowances will be sold at a fixed price of £12 per tonne of carbon dioxide. Following this period, allowances will be traded and therefore subject to price fluctuations.

The scheme will place an additional administrative burden on participants and will result in additional operating costs for those organisations that are not successful in reducing their carbon emissions and outperforming their peers in the Carbon Reduction Commitment league table. The quantum of this additional cost will depend on an organisation's league table position and the price that it must pay per tonne of carbon emitted, which will fluctuate.

There are also remaining uncertainties to be resolved regarding the interaction between the Carbon Reduction Commitment and the Renewables Obligation scheme in circumstances where an organisation invests in renewable energy projects. In addition, the price uncertainty that arises out of the trading of certificates after the introductory phase leaves participants with additional financial risk if these participants are unable to implement energy efficiency measures as quickly or effectively as their peers.

4.   Conclusion

The success of the deployment of renewable energy in different markets depends on the political support for renewable energy within that market. Strong political will is usually backed up by a strong financial support mechanism. In the short to medium term, many renewable technologies will continue to need some form of financial incentive to encourage investment.

The evidence to date suggests that some support mechanisms have had a greater impact than others in encouraging the deployment of renewable energy. A successful support mechanism should deliver a sufficient amount of renewable energy to meet set targets in an affordable way. Such a mechanism is more likely to meet the sometimes conflicting interests of different stakeholders.

What is clear is that the market is still evolving and there is no one-size-fits-all solution for energy support mechanisms. A different type of mechanism may be required for different stages of technology development or more appropriate for managing certain types of risk. However, in general, a successful support mechanism will offer transparency, ease of implementation and administration, and income certainty and level. Mechanisms should also be available for a sufficient timeframe to meet investors' requirements, but not allow excess returns at the expense of consumers. Meanwhile, regulators need to ensure that they strike a balance between overly intervening in the market and ensuring that support levels are at optimum levels. The ongoing interaction between these key stakeholders will be essential to ensure that each others' interests are appropriately balanced and to meet our renewable energy targets.[9]

---

9   This chapter draws on the personal experience of the authors and other members of the PricewaterhouseCoopers global network, as well as a variety of research sources, including the following: Trilemma, Redpoint, "Implementation of the EU 2020 Renewables Target in the UK Electricity Sector: RO Reform", a report for the Department of Energy and Climate Change (June 2009); www.iea.org/textbase/pm/?mode=re (a useful database summarising the various support mechanisms around the world); and Poyry Energy Consulting and Element Energy, "*Qualitative issues in the design of the GB Feed-In Tariffs*" (June 2009, URN 09/D/698).

Secondary UK legislation includes:
• the Renewables Obligation Order 2002/914;
• the Renewables Obligation (Amendment) Order 2002/2004/924;
• the Renewables Obligation Order 2004/2005/926;
• the Renewables Obligation Order 2005/2006/1004;
• the Renewables Obligation (Amendment) Order 2006/2007/1078;
• the Renewables Obligation Order 2007/2009/785; and
• the Renewables Obligation (Amendment) Order 2010.

Issues for financiers

Nicholas Sinden
Siobhan Smyth
HSBC Bank plc

1. Overview

1.1 Introduction

This chapter aims to provide an overview of project financiers' opinions of the renewables sector and the methods that these financiers use to structure lending to various types of renewables project.

Accordingly, this chapter looks specifically at what attracts senior debt providers to the renewables sector and what puts them off, and why. It also discusses two successful deals in some detail. The first is the Acciona Alvarado solar thermal transaction in Spain, which closed in June 2008. The second is the Fred Olsen UK wind portfolio, which closed in November 2008 – at the height of the credit crisis.

As at the end of 2009, various macro factors linked to the global financial crisis had negatively affected demand for renewables in the near term. However, another strong period of growth is expected for the sector, as indicated by:

- the credit markets recovering;
- financing becoming more readily available; and
- the cost of renewables approaching grid parity (ie, the cost of renewables is close to that of fossil fuels) as oil prices increase and renewables costs drop.

In addition, governments remain motivated to provide green stimulus packages (ie, incentives to reduce carbon emissions) and ensure supply security. Consequently, the authors broadly expect ongoing governmental support for the renewables sector.

1.2 Volume of financing

The growth of the renewable energy sector has kept many specialist bankers busy for the past few years. This is because the sector's rapid growth – especially in Germany and Spain, initially, and later in the United States – has required a significant mobilisation of capital. However, financing volumes for the sector dropped in 2009 as senior debt providers became more wary of long-term financing and developers became more selective in their investments in the sector.

Nevertheless, after the hiatus of early 2009 the sector is regaining some momentum. Besides, given the required build-out of the sector, the authors are confident that financing volumes will have returned to previous levels by the end of 2010.

The graph on the following page shows the amount of project financing in

Europe and the United States between 2005 and 2009. The levels of lending in 2009 did not reach those of 2008. Wind farms were the main beneficiaries of project lending. The figures below include the refinancing of existing projects.

1.3 Growth prospects

The targets for renewable energy usage are well known. The European Union requires that 20% of its energy come from renewable sources by 2020. EU member states have been set individual targets based on their individual characteristics. A significant number of countries are close to reaching these targets with, in the authors' opinion, the United Kingdom a noticeable laggard.

One of the fundamental challenges faced by the renewables sector is the cost of the underlying technologies compared to that of other sources of energy. As indicated below, for the most part renewable energy is more expensive, even when factoring in carbon emission costs.

In some markets, onshore wind power is close to reaching grid parity with wholesale electricity prices. This will continue to drive growth in onshore wind installations.

In addition, given the higher efficiencies and lower costs achieved in the solar sector, HSBC Research expects these technologies to achieve grid parity by 2013 or 2014, as shown in the graphs below. (The black lines show the price at which parity with retail tariffs will be achieved and the grey bar show the estimated year this will occur.)

Assuming that long-term wholesale European power prices will be around €60 per megawatt-hour (MWh), projects will remain partly dependent on the support of the regulatory frameworks (eg, feed-in tariffs) and the long-term commitment of

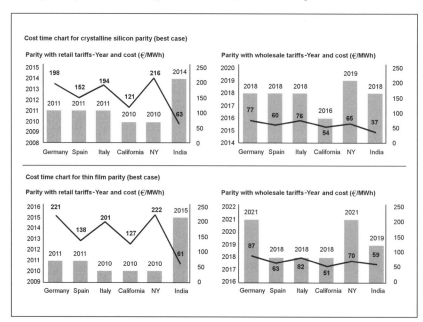

Technology	Plant size (MW)	Investment cost (€m)	Fixed cost (€/MWh)	Fuel cost (€/MWh)	Operation and maintenance (€/MWh)	Carbon emission (€/MWh)	Full cost (€/MWh)
Geothermal	50	1757	24	–	20	–	44
Coal thermal	1600	1100	13	21	5	9	48
Nuclear	1000	3000	36	5	7	–	48
Lignite	1000	1270	15	19	5	10	49
Combined-cycle gas turbine	750	700	9	34	4	5	52
Coal CFB	150	1400	18	22	5	9	54
Coal IGCC	480	1550	20	21	9	8	58
Wind onshore	100	1300	46	–	12	–	58
Open-cycle gas turbine	40	500	7	58	5	6	76
Biomass	25	2600	38	40	13	–	91
Wind offshore	200	2380	74	–	24	–	98
Solar CS	10	4700	191	–	18	–	209
Solar photovoltaic	10	2750	210	–	15	–	225

Source: HSBC Research

governments to green energy. Underpinning this commitment are the need for green stimulus, supply security and an evolving energy mix, and a growing energy demand globally.

Historically, developed countries (primarily, the United States and Europe) have led the development of renewable energy. Now emerging economies are demonstrating strong growth in the sector – particularly China, as a result of a growing demand for energy, a lack of indigenous fuel sources and the need to improve its energy mix.

2.   Sources of financing

Developers can follow a number of routes to fund their projects. The source and type of capital will depend on the stage of development of the project and the developer's own available cash resources. A common model is for smaller developers to carry a project through early development until it is ready to be built, and then sell the project and move onto the next one. This model struggled in 2008 and 2009 as capital became scarcer due to the credit climate, which in turn stalled the development of the renewables sector.

In general, bank debt remains the primary source of funding for new investments. Until the recent credit crisis, bank debt provided long-tenor financing for relatively low lending margins.

Below are the common debt instruments used in bringing a renewables project to market.

2.1   Corporate bridge

The quickest way to finance construction is from a company's balance sheet. The assets can be owned directly by the company or via a special purpose vehicle set up for the new development. In essence, the project is funded through either:

- equity; or
- if through a special purpose vehicle, corporate-guaranteed bridge facilities provided by senior debt providers directly or indirectly to the project to cover the investment needed until due diligence for the non-recourse debt has been completed.

These bridge facilities are generally guaranteed by the project sponsor's parent company. They offer the advantages of optimising liquidity, as hard cash is not paid in as equity on day one, and lowering the overall cost of financing. In the solar thermal case study described in section 7 below, bridge financing was put in place for the initial construction.

The intention will be to repay the bridge loan with the proceeds of the non-recourse facility (ie, a longer-term debt). Bridge facility providers are typically awarded a mandate for the non-recourse project finance take out.

Given the initial reliance on the developer's balance sheet, these facilities are typically provided only by senior debt providers to the larger players in the sector.

2.2   Senior limited/non-recourse debt

Senior non-recourse debt remains the primary source of capital for renewable energy projects.

'Senior non-recourse debt' means bank money that is lent directly to the project and repaid solely from the revenue produced by the project's assets.

In a default and enforcement situation, these senior debt providers will be able to enforce their rights and access the project security before any unsecured senior debt providers or equity providers.

Over the past few years the project finance market for senior debt for renewable energy projects has followed the trend of the wider power market by showing an appetite for highly liquid credit. This has been helped by a favourable attitude from banks towards maintaining a presence in this sector.

This lending appetite increased liquidity up until the end of 2007 and:

- pushed down margins and fees;
- increased gearing levels;
- boosted the appetite for merchant price risk in certain markets; and
- encouraged less conservative revenue assumptions – for instance, in the wind sector a move from using more statistically certain energy yield forecasts to less certain riskier estimates. This meant a move from the conservative 'P90' forecast wind energy yield number, which has a 90% probability of being exceeded, to a higher 'P50' measure (used in an equity case), which has a 50% probability of being exceeded and a 50% of being lower.

The uncertainty across financial markets following the credit crunch affected renewable energy project financings in the following ways:

- Overall lending liquidity for the sector decreased.
- Lending margins and fees for the sector increased, with senior debt providers requiring a premium to provide project finance.
- Tenors were shortened or cash sweep mechanisms were introduced to encourage early refinancing of the project. (A 'cash sweep' is where a portion of the free cash flow available after scheduled debt service is used to pre-pay outstanding debt if certain financial ratios are not met or if a certain point in the life of the loan has been reached.)
- Underwriting appetite was lacking, with transactions being closed on a club basis. (A 'club' is where banks will not take underwriting risk on a transaction and all senior debt providers are identified before the transaction is completed.)
- Financiers' approach to gearing levels, debt service coverage ratios and market price risk became stricter.

2.3   Multilateral and export credit agencies

Export credit agencies are government entities designed to support the export transactions of domestic manufacturing companies. The agencies' main form of lending consists of providing insurance or guarantees to the buyers of the equipment. Banks lend the money while the agencies provide a commercial and political risk guarantee for between 85% and 95% of the loan value. In addition, the agencies provide insurance cover or lend directly – sometimes on a non-recourse basis – to special purpose vehicles and use similar lending criteria as banks.

Multilateral agencies are organisations, such the International Finance Corporation (the private funding arm of the World Bank), the European Bank for Reconstruction and Development and the European Investment Bank, which have a specific development agenda. The agenda of the International Finance Corporation covers developing economies; that of the European Bank for Reconstruction and Development covers the developing economies within Europe; and that of the European Investment Bank covers economies in the pan-European and Mediterranean area.

During the credit crunch, an overall scarcity of capital meant that multilaterals and export credit agencies became increasingly important in the renewables markets. Generally, the multilaterals and agencies allow longer-term financing, regardless of the economic cycle, over a wide geographical area and range of international markets. In 2009 the Organisation for Economic Cooperation and Development laid down new guidelines for renewables to be used by export credit agencies. These guidelines allow for security and debt parameters to be more closely matched to the assets, with repayment periods of up to 18 years. Previously, agencies were considered an option for developing economies, but the recent downturn has seen them looking to help companies in markets closer to home. For instance, Eksport Kredit Fonden, the Danish agency, has assisted its domestic wind turbine developers in various mature European markets.

Multilaterals such as the European Investment Bank offer financing for renewables at significantly reduced pricing on a two to five-year bank-guaranteed basis. Several other initiatives direct money to renewables-related projects. Thus, a new UK investment scheme will benefit the UK renewables sector by providing loans of up to £4 billion. Both the European Bank for Reconstruction and Development and the International Finance Corporation view renewables favourably. They have stated publicly their intent to support sustainable projects in emerging economies.

2.4 Bond markets

A project bond is a long-term non-recourse form of senior debt which non-bank financial institutions, such as pension funds and insurance companies, will invest in. An investment bank will arrange the bond issue.

Despite the best efforts of financiers, bond financing for renewables remains rare. This is mainly due to the aversion of bond investors to taking construction risk on projects. However, the debt capital markets are suitable for refinancing bank debt once projects have established an operational track record.

Between 2004 and 2007, only three wind portfolio financings were launched in the capital markets (the German Breeze wind portfolio financings) – and these were based on the very stable German regime and an existing pool of assets. Monoline insurers, which enhanced the credit rating of the investment (ie, it could then be bought by institutional investors), collapsed in the credit crunch. The bond market's appetite for this type of asset has reduced greatly, since the assets' prices have increased to levels that are not competitive with the equivalent senior debt pricing.

However, the continued improvement in market conditions, coupled with structuring financing to attract an investment-grade rating, could make the bond market (either vanilla or structured) a source of long-term funding. As the markets emerge from the crisis, the debt capital markets should form the platform for a range of refinancing opportunities.

2.5 Equity and mezzanine financing

The above forms of financing will typically be in the form of senior debt with priority in an enforcement situation. Below the senior debt, investment can take different forms. A shareholder can put traditional share capital into a structure. Alternatively, other forms of debt subordinate to the senior debt can be placed into a structure attracting different types of investor.

Mezzanine and equity opportunities are increasingly available to investors as part of the overall financing package. This is increasingly seen as a desirable market for new investors prepared to pay a significant premium for positions in the post-development phase.

Typical investors include specific renewables funds and niche market developers, as well as major power players seeking to establish an asset base within a specific regulatory regime. Equity returns and payback periods vary significantly from country to country. They are dependent on the generosity of the incentive mechanism used, ranging from 7% a year to beyond 20% a year for equity returns and from five to 10 years for the payback period.

2.6   Acquisition financing

In light of current market conditions and the sheer scale of offshore wind projects in particular, developers are increasingly often partnering with, or being acquired by, large utilities eager to secure access to green certificates. This eagerness stems from a need to offset the carbon production from these utilities' traditional businesses. An example of this was UK utility Scottish and Southern Electricity's acquisition of Airtricity in 2008. The acquisition enabled Scottish and Southern Electricity to reach its targets for electricity generation from renewable technologies more quickly.

The current financial market conditions, however, are likely to result in greatly reduced market appetite for funding large-scale acquisitions. Thus, the market is likely to have to wait for confidence to be restored before seeing transactions on the scale of:

- the 2007 Trinergy acquisition, where International Power paid an enterprise value of €1.84 billion for an operating portfolio of 648MW; or
- the 2008 Airtricity acquisition, where Scottish and Southern Electricity paid an enterprise value of €1.45 billion for 308MW of installed capacity.

3.   The role of government

Governments play a key role in the development of renewable energy by providing a stable regulatory regime against which private investors can make long-term investment decisions.

The regulatory regimes for the provision of renewable energy typically take the forms described below.

3.1   Feed-in tariffs

Feed-in tariffs mean that every unit of energy produced is purchased and paid at a regulated all-in price. This has been the most common form of regulatory structure and arguably the most successful (although the countries that have the largest penetration of renewable energy generally also appear to be the most enthusiastic about green energy, with quick planning responses).

3.2   Green certificates

Green certificates typically provide renewable energy generators with two revenue streams:

- from the government for the electricity produced (which commands a wholesale power price); and
- from utilities that are under an obligation to purchase a certain amount of their energy from renewable sources, which these certificates prove.

These certificates usually have an explicit or implicit floor price. If the utility companies do not meet their targets, then the supply of certificates will not meet demand and the price will rise, creating an incentive for more supply. The Fred Olsen portfolio transaction described in section 7 is an example of a large transaction successfully banked on the basis of green certificates.

3.3   Tax

The United States has built its renewables sector on the basis of a series of tax breaks available upon construction of the asset.

The table opposite summarises the current regimes available for wind power, demonstrating the different combinations of support available.

Senior debt providers will look very closely at:

- the form of support;
- the country's track record for supporting renewables;
- the country's energy outlook; and
- its ability to pay for the cost of renewable energy.

The table opposite shows that the simplest regime (ie, feed-in tariffs) is associated with the countries with the largest development of renewable energy. However, regulatory regimes are changed regularly in nearly all countries, as this is an evolving market. Spain, for example, has regularly reduced its feed-in tariffs for wind power and latterly solar power as the costs fell and equity returns increased. In the authors' opinion, the United States stands out, at this time, as the main developed economy without an organised federal scheme for renewables. In general, governments have been pragmatic when imposing change such that senior debt providers are protected from adverse effects on their existing investments.

4.   Structuring a non-recourse deal

As previously indicated, the primary source of financing for renewables is non-recourse debt. A renewable project may be developed under a non-recourse structure for the following reasons:

- Debt constraints – in many renewables financing situations, the sponsor's balance sheet does not support the level of debt that the project requires, but the money can be raised against the asset being developed.
- Multiple sponsors – renewables financings often involve a local sponsor (with good understanding of the local regulatory climate or ownership of the site) and an international sponsor with technical expertise. The local sponsor usually has limited access to capital and requires a project financing route, which the international sponsor might not need if it were carrying out the project on its own.

	Feed-in tariffs/ pool + premium	Offtaker	Market risk	Green certificate	Grants	Priority despatch	Installed capacity in gigawatts (2008)	Tax incentives	Regulatory support
Australia	No	Market via power purchase agreements	Yes	No	No	No	1.3	No	Medium/weak. Strong growth potential, but requires political support
China	Yes	Local grid	No	No	No	No	12.2	No	Weak. 70% of turbine components to be domestically manufactured. Difficult market for new entrants
France	Yes	Electricité de France	No	No	No	No	3.4	Yes	Medium
Germany	Yes	Local grid	No	No	No	Yes	23.9	Yes	Strong
Greece	Yes	Public power corporation	No	No	Yes	Yes	0.99	Yes	Strong
Italy	No	Market	Yes	Yes	Yes	Yes	3.7	Yes	Medium
Spain	Yes (choice)	Local grid	No	No	No	Yes	16.8	No	Strong
United Kingdom	Feed-in tariffs for schemes under 5MW	Market	Yes	Yes for schemes over 5MW	Yes	No	3.2	No	Medium (banded incentive for offshore)
United States	No	Market	Usually power purchase agreement	No	Yes	Varies from state to state	25.1	Yes	Medium. Production tax credit has been extended by one year. Carbon credit scheme likely to be introduced, which will support growth
India	Yes	Local grid	No	No	Yes	Yes	9.6	Yes	Strong

Source: HSBC Research

- Jurisdiction – project finance is common as a source of financing in developing economies, where it provides a structure for multilateral agencies to lend into. The support of multilateral agencies mitigates the political risk for commercial senior debt providers.

The following chart illustrates the typical structure of a non-recourse deal, using a wind transaction as an example.

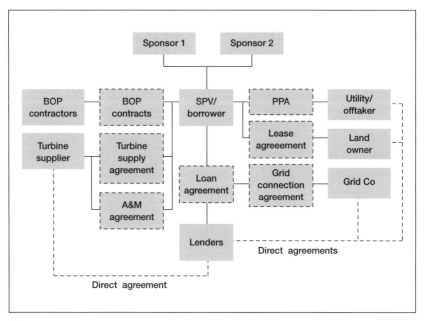

Notes:

SPV – the special purpose vehicle set up as the project company to own and operate the assets.

O&M – operations and maintenance

BOP – the balance of plant (civil engineering works and grid works etc).

PPA – a power purchase agreement between the SPV and an offtaker (typically a utility company).

While under a corporate financing a sponsor may be willing to leave things undocumented, a project financing requires significant due diligence, a more defined project structure and contractual protections. Accordingly, it can take between six and nine months for a renewable project financing to be put in place from the time the senior debt providers are initially approached. This time is taken up by technical and legal due diligence. This delay is reflected in high project development costs, which can make smaller-scale renewables projects (ie, where the total project value is under $25 million) unattractive. The sponsors' prior experience of project financing often affects how long the first project takes to arrange, as:

- the sponsors' expectations may need to be adjusted, which takes time; or
- the sponsors may not be prepared for the level of work required.

However, experience increases the speed and reduces the cost of future transactions.

4.1 Key project documents

Senior debt providers will spend a significant amount of time:
- reviewing the contracts and associated documents; and
- checking that the allocation of risk is appropriate between the project counterparts.

There is an expectation that documentation will contain certain clauses. Problems and delays in project financings often result from poorly defined contracts and associated documents that do not have the required risk allocation or market standard protections. For wind projects, the key documents concern the supply of turbines; while for solar photovoltaic projects, the key documents concern the supply of panels.

The key terms and conditions in the primary contracts on which senior debt providers will focus are summarised below.

4.2 Construction contracts

Projects can be built around:
- a single contract with a single point of contact and liability; or
- multiple contracts with multiple suppliers.

Multiple contracts can lead to interface risk between the various contractors and are therefore more cumbersome from a documentation perspective. They may also lead to more complex structures. In wind projects, in particular, multi-contract deals splitting turbine construction, civil works and grid connections are common. Senior debt providers generally understand the risks involved and focus on the warranties associated with the supply contracts. Lenders will primarily consider:
- the terms and breadth of scope of the warranty;
- any delay allowance and performance requirements;
- the liquidated damages and any associated caps; and
- the creditworthiness and experience levels of the entity supplying the warranty.

Similarly, for solar photovoltaic projects, multi-contract structures are common because of the simplicity of the build-out.

4.3 Power purchase agreement/regulatory framework

Renewables projects rely on a regulatory framework to support the revenue stream. Some regulatory frameworks specify both the offtaker and the price at which the power will be purchased. This is the case in Germany, where the local utility must purchase the power at the pre-agreed feed-in tariff. In other jurisdictions a power purchase agreement needs to be negotiated and entered into by the generator. The agreement will specify the terms and price of the offtake. The agreement will contain provisions governing any failure to produce power. It will also provide protection against changes in the law and set out liability caps and compensation payments if the agreement is terminated.

In addition, renewables projects have priority of despatch in some markets, which provides further comfort with respect to the revenue stream.

4.4 Operation and maintenance

The operation and maintenance contract will generally be medium to long term. It typically guarantees performance levels normally linked to the availability of the asset. If the project does not reach the guaranteed performance levels, then liquidated damages will be paid. Senior debt providers will be looking at the level of these liquidated damages and whether they will service the fixed costs and debt of the special purpose vehicle. Liquidated damages typically come with a cap – for instance, 30% of the original capital expenditure or the total of the fees payable over the term of the contract.

Senior debt providers will analyse this to assess how long the assets will last. For more technologically complex equipment, such as solar thermal, lenders may ask for:

- a longer operation and maintenance period with corresponding warranties; and
- a higher cap for the liquidated damages, which in turn is reflected in the final price of the contract.

Senior debt providers will also consider the credit risk that they are taking with the operation and maintenance provider: if the latter's credit standing is poor, the former may ask for guarantees or letters of credit to be placed as a condition of the contract.

4.5 Grid connections

Most power projects need grid connections, which are costly if the development is far away from main transmission lines. The cost and demand for a grid connection can delay projects coming online. Senior debt providers will require a firm commitment from the grid company that:

- it will provide a connection by a fixed date; and
- the debt providers will have a direct agreement with the grid company if the special purpose vehicle defaults on the debt.

5. Key lending issues for senior debt providers

This section considers the particular issues that senior debt providers will address in their credit analysis of a renewables project – in particular, technology issues.

5.1 Planning, permits and licences

By the time that project sponsors approach senior debt providers for a transaction, the relevant permits, such as generation licence and planning permissions, should be in place. Obtaining the relevant permissions can take a variable amount of time, as the UK experience has shown. This can therefore be one of the milestones of the project's timeline. Senior debt providers will typically not provide finance until all required permits and planning permissions are fully in place.

5.2   Technology

(a)   *Onshore wind*

Onshore wind has the longest track record of all renewable energy sources. There is strong competition between the wind turbine manufacturers, which is generally good for the purchaser. However, the period 2006/2007 saw demand outstrip supply and a corresponding movement in suppliers imposing their purchasing terms.

Onshore wind may be a mature sector, but it is not without risks. These can be understood or mitigated through proper due diligence and an appropriate risk allocation. The onshore wind sector has seen large transactions carried out. For example, the Fred Olsen wind portfolio discussed in section 7 cost £315 million.

Senior debt providers have been exposed to the following key risks in the wind sector:

- Wind assessment – lenders will consider in their due diligence at least one year's worth of wind data that:
  - is provided from a mast at an appropriate height; and
  - takes account of the relevant site's topography.

  The wind behaves very differently on the plains of Germany from in the highlands of Scotland, for example. This degree of caution is shown through the use of P90 wind yields to measure the debt. This is typically the case for equity. Senior debt providers may also consider a downside case at a P99 wind level through the life of the project to see whether the project breaks even.

- Turbine failure – experience across a wide range of turbines has demonstrated a fairly high incidence of short-term failures, especially with gearboxes across several manufacturers. The warranties for onshore turbines are typically for two years, with five-year availability guarantees through an operation and maintenance agreement with the supplier. (In any event, turbine suppliers have generally repaired turbine failures.) Accordingly, the warranty and operation and maintenance support for projects are regarded as critical parts of any project financing.

(b)   *Offshore wind*

The operating and construction risks associated with offshore wind energy are generally perceived as much higher than with onshore wind. This is because offshore wind is a relatively new sector and, consequently, few offshore wind farms are operating. From a financier's point of view, these types of project carry considerable construction risk, with very large civil, transmission and turbine contracts in place without a general engineering, procurement and construction wrap covering them. The level of contingency and recourse required from the shareholder to make up for this amount of risk can render any debt financing unattractive to the equity owner. This means that developers of offshore wind farms may well keep the construction risk on their balance sheet. The establishment of the required infrastructure to enable full offshore operation – in particular, the availability of specialised ships to service the number of wind farms envisaged – is uncertain at this point. This uncertainty leads to the provision of large levels of contingency in budget projections.

(c)   **Solar photovoltaic**

Solar power has developed considerably in Europe since the mid-2000s, thanks to extensive feed-in tariffs. These tariffs have been reduced as the sector now enjoys lower overall costs following technological developments that have in turn led to greater efficiency. However, the continual evolution of the technologies raises concerns over their reliability, which leads lenders to seek appropriate long-term warranties from the solar manufacturer. In addition, the photovoltaic and silicon sectors have recently experienced a boom and bust, leaving many solar manufacturers with a weak financial standing. This means that lenders consider manufacturers as high risk.

(d)   **Concentrated solar**

Solar thermal electricity generation harnesses the sun's energy to generate steam to drive conventional turbines. Concentrated solar power plants use a variety of techniques, including parabolic troughs and power systems, to focus the sunlight. Until recently, the market for solar thermal electricity generation appeared moribund. In May 2008 installed capacity worldwide was only 426MW, from just four plants in the United States and Spain. However, around 7 gigawatts of capacity are now in the pipeline globally (source: HSBC Research).

One key competitive advantage of concentrated solar power systems over solar thermal, which also makes them particularly attractive to senior debt providers, is their close resemblance to existing power plants in some important ways. For example, much of the equipment now used for conventional, centralised power plants running on fossil fuels can also be used for concentrated solar power plants. The solar technology can be integrated fairly easily into the existing electricity grid because of its similarities with thermal power plants in generating electricity. This makes concentrated solar power the most cost-effective option for large-scale solar electricity generation.

Although the levelised cost per megawatt-hour of solar thermal energy is still high in comparison to other renewable technologies (as seen in section 1.3 above), technological evolution provides a significant margin to reduce these costs, making solar technology more competitive overall. In a technically evolving sector, financiers will need reassurance that the revenue structure will not be affected by the arrival of lower-cost technologies in the future.

As with solar photovoltaic power, the sun resource is a premium input in concentrated solar power; a minimum of 1,800 hours of sunshine a year is normally required for concentrated solar power to be efficient.

(e)   **Hydro**

Hydro technology and design are long established, providing nearly one-fifth of the world's electricity. However, senior debt providers will consider carefully the issues associated with new hydro projects. These issues include new dams that will result in the displacement of large numbers of people and negatively affect the local fauna and flora.

Financiers will also need to consider carefully the hydrology (ie, the study of the movement, distribution and quantity of the water) in a given area and have comprehensive water studies carried out.

In the view of the authors, the risk associated with the movement, distribution and quantity of water may grow if global warming results in more severe droughts and floods. While new ways to exploit the power of water (eg, tidal energy) are being developed, most are still at the early stages of their lifecycle and, thus, are not yet considered financeable.

(f)   *Biomass, waste-to-energy and biofuels*
This area of renewables covers the products derived from sustainable resources of plants and animals described in the table overleaf

The expansion of bioenergy has often proceeded without strategic assessments of long-term sustainability, in terms of either climate change or competition with other uses of agricultural output – notably, food. Financiers' main concern with regard to biomass is the security of supply and the guarantee of feedstock under a long-term supply agreement. The environmental impact of large-scale biomass material production, especially when sourced from developing economies, is another source of disquiet.

In relation to biofuels, the success of Brazil's production and consumption of ethanol is well known. However, the commercial experience in Europe and the rest of the Americas has been less successful. While the European Union still maintains mandatory targets for the level of biofuels under Directive 2003/30/EC, the optimism

Biomass feedstock and their energy services		
Biomass feedstock	**Bioenergy produced**	**Energy services**
Agriculture and forestry residues	Wood pellets, briquettes, biodiesel	Heat, electricity, transport
Energy crops (biomass, sugar, oil)	Char/charcoal, fuel gas, bio-oil; bioethanol	Heat, electricity, transport
Biomass processing wastes	Biogas, bioethanol, solvents	Transport
Municipal waste	Refuse-derived fuel, biogas	Heat, electricity

of the 2000-to-2005 period has given way to difficulty in financing any deals. This is because senior debt providers struggle to come to terms with the combination of price risk on the inputs (agricultural products) and outputs (fuel-linked prices). Other factors have contributed to making these transactions slow and difficult to process. These include:

- the different definitions used in different countries as to what constitutes a 'biofuel';
- worries about the sustainability of biofuels; and
- the wavering of politicians in the food-versus-fuel debate (ie, the dilemma over whether to divert farmland or crops for biofuel production to the detriment of food supply).

5.3   Construction issues

(a)   Type of contract

As discussed above, financiers prefer a single engineering, procurement and construction contract to having to understand the individual risks involved in a series of contracts. Lenders have become comfortable with the level of risk associated with the construction of onshore wind farms where multi-contract structures are the norm. However, financiers will expect to see an experienced construction manager overseeing the activities of the different contractors.

Where the technological risk is large, senior debt providers will be looking for an effective wrap on the project – normally from the technology provider, which will need a creditworthy, robust balance sheet.

(b)   Warranties and bonding

The following bonding structure typically applies to construction contracts:

- The customer makes an advance payment upfront for, say, 10% of the contract value, which is offset against the initial milestones and work.
- The contractor posts a retention bond of between 10% and 20% of the contract value, which will be released subject to meeting agreed milestones, and performance and completion criteria.

In addition, the construction contract will contain warranties for how long the equipment should last without problems. This differs between technologies. For instance, onshore wind turbines typically enjoy a two-year warranty, while offshore wind turbines (a newer sector with a shorter track record) command longer warranty periods (ie, five years).

5.4   Regulatory framework

(a)   Regulatory and tariff risk

As a rule, senior debt providers prefer the risk profile of a feed-in tariff to that of a certificate scheme. This is because certificate schemes require that lenders analyse the electricity market (typically, with a merchant element). However, merchant power markets are well banked globally and most senior debt providers have expertise in analysing local markets.

(b)   Strength of counterparty

All senior debt providers will assess the credit risk represented by the counterparty that they are facing in the regulatory structure. In Western European countries, the counterparties are either top-rated utility companies acting as compulsory offtakers or state-owned transmission operators, all of which have a strong credit history. Where the offtaker is an independent utility, however, the power purchase agreement will include clauses governing credit support (ie, collaterals) in the event that the utility's credit rating is downgraded (these agreements will last up to 20 years).

In developing economies, where either the country or the offtaker has a poor credit rating, other support will be sought. For example, it is common to see a payment guarantee from the Treasury for the power despatched and debt repayment if the contract is terminated.

Setting up these types of agreement can be a lengthy process in countries that have no previous experience of such agreements.

5.5   Equator Principles

Most project finance lenders have signed up to a set of rules known as the Equator Principles. The principles consist of environmental and social guidelines based on the International Finance Corporation standards. They cover issues such as engagement with communities affected by new projects (a particular issue with hydro projects) and give specific targets on pollutant levels (eg, from thermal power plants). The principles need not apply strictly in developed countries, although many banks still ask their advisers to consider a rating. However, a specific environmental and social impact assessment will likely need to be carried out – even for a renewables project – unless developers can clearly show that the project will have no impact. Wind farms, for example, often raise concerns about noise levels and impact on wildlife (birds in particular).

6.   Key financing terms

This section highlights the key concepts that a lender will consider when negotiating a debt term sheet.

6.1   Portfolio v standalone assets

As developers often have more than one project in the pipeline, financings can be closed (for a single project) or open-ended (further projects can be brought in when they are ready). This is normally subject to criteria on technical and debt-sizing parameters. For senior debt providers, portfolio financings are attractive as risk is spread across various sites and, possibly, technologies. The Fred Olsen deal described in section 7 shows the benefits of bringing several assets together.

6.2   Greenfield or refinancing

Many financings in the renewables sector are for greenfield assets. However, a significant financing market has developed for refinancing existing assets. Refinanced assets may be looking for better pricing or bringing in other assets to create a portfolio. Also, the developer may be forced to refinance because some punitive terms (eg, cash sweeps) have kicked in on existing debt. The Fred Olsen case combines refinancing and greenfield financing in one transaction.

6.3 Tenor

The tenor that senior debt providers offer depends on the strength of the contracts, the project's lifespan and the bank market conditions at the time of financing.

One of the defining features of a project finance loan is the tenor achieved: 20 years for construction and repayment have been common in Europe, based on strong regulatory structures and country credit ratings. In the United Kingdom, where there is more merchant risk, tenors are often secured against offtake agreements signed for 15 years.

The recent credit crisis has changed this, with senior debt providers either unable to offer longer maturities or applying liquidity premium to their pricing where they were able to offer longer maturities in order to reflect the cost of locking up money longer term.

The structures that have emerged are based on shorter tenors of seven to 10 years, with balloons of between 30% and 50% (so-called 'mini-perms'). The loan terms provide that at this point there is either a hard refinancing or a cash sweep such that all cash is applied to repaying the loan balance. The intention is to provide an incentive to refinance early.

6.4 Security

Senior debt providers will take a full suite of security, to the extent possible under the local laws, including physical security over assets, key project contracts (to which they are not party) and bank accounts.

In addition, lenders will seek to put in place direct agreements with contractors so that they have a step-in period to solve problems and contracts can be assigned to them if a reorganisation is necessary.

Upon entering commercial agreements, but before bringing in senior debt providers, developers should ensure that contracts have appropriate provisions for, among other things, assignment and their lawyers have appropriate project finance expertise.

There appears to be no specific security issues for renewable assets. The exact basis for security arrangements varies from jurisdiction to jurisdiction. Senior debt providers will try to obtain the most extensive security possible. In Western Europe, security laws are comprehensive enough for financiers to feel comfortable to lend.

6.5 Gearing levels and debt service coverage ratios

The amount of debt that can be raised for a particular project is the product of the tenor and the debt service coverage ratio. The debt service coverage ratio is a measure of the ratio of cash flow available to service the debt (both principal and interest) to the actual debt service. For example, lenders may require a 20% coverage through the project life. For wind and solar projects in Europe estimated at P90 energy yields, requirements for debt service coverage ratio are between 1.15x and 1.2x.

Where there is a good energy resource, lenders may well accept a level of debt that is higher than what they are normally comfortable with. Where feed-in tariffs are in place, gearing levels typically reach between 80% and 85%. Countries that offer generous green certificate structures may see similar gearing levels. However, the debt service coverage ratio for base cases will need to be higher to offset the higher amount of risk.

6.6   Pricing

The margin that a bank will charge depends on the specific credit risk of the deal and the state of the credit markets at that time.

Obviously, projects that carry more risk, such as that linked to merchant revenue streams, will warrant a higher pricing.

6.7   Construction and completion regime

Completion risk varies between projects. New technologies demand that sponsors take more of the risk than lenders. For proven technology such as solar photovoltaic and wind, the completion risk can be passed on more easily to senior debt providers. These will be expecting to see performance bonds from each major contractor and warranties for an appropriate period. Lenders will use a technical adviser and lawyers to confirm the adequacy of these agreements and their conformity to market standards.

Where the agreements do not conform and renegotiation is impossible, the lenders will seek comfort from the sponsor in the form of completion guarantees or debt service undertakings.

7.   Case studies

This section illustrates, through two examples in the solar and wind sectors, what can be financed and structured.

7.1   Spanish project

Termosolar Alvarado is expected to be one of the first parabolic trough concentrated solar power plants to become operational on a commercial scale in Europe. The sole sponsor is Acciona Energía SA, the energy subsidiary of the Spanish infrastructure group Acciona SA.

In November 2007, following the successful closing of the 64MW Nevada Solar One power plant, construction began on a 50MW non-storage replica in Extremadura in Spain. This plant was constructed through a one-year corporate recourse loan, as Acciona had the financial resources available to follow this route. Financial close on the subsequent project finance facility, which took out the bridge loan, was reached in June 2008.

The favourable Spanish regulatory regime on renewables provides two options:
*   fixed offtake prices (via a fixed feed-in tariff); or
*   floating offtake prices with an attractive collar (high cap and floor similar to the fixed feed-in tariff) for 25 years, adjusted on a consumer price index basis.

After the flurry of solar photovoltaic transactions in Spain in 2007 and 2008, investors now favour concentrated solar power technology because of its higher efficiency and improved profitability under the tariff regime available (Royal Decree 661/2007).

The Termosolar Alvarado project benefits from Acciona's vertically integrated approach (sponsor, engineering, procurement and construction, and operation and maintenance).

The financing structure had gearing of 82:18, which was made possible by the excellent solar resource available at the site.

The loan repayment was based on an amortising profile with an average and minimum debt service coverage ratio of 1.3x and 1.1x over the life of the project under P50 and P90 scenarios, respectively.

Senior debt providers were attracted by the use of proven and tested parabolic trough technology, which is considered the most reliable of all concentrated solar power technologies available.

7.2   UK project

Fred Olsen Renewables Limited has established itself as one of the leading renewable electricity generators in the United Kingdom.

Since 2004 it has developed and financed three onshore wind farms in Scotland:

- Crystal Rig I (62.5MW);
- Paul's Hill (64.4MW); and
- Rothes (50.6MW).

Fred Olsen had financed these farms on a standalone basis. It had also begun the construction of Crystal Rig II (138MW) on its balance sheet.

Fred Olsen needed to finance the remainder of Crystal Rig II and saw this as an opportunity to simplify its existing financings. Along with its financial adviser, HSBC, Fred Olsen established its objectives as follows:

- to finance its new 138MW wind farm in Scotland;
- to provide a cross-collateralised portfolio structure that would reduce the level of default risk; and
- to provide a structure that would allow new assets to be brought into the portfolio in the future.

The deal allows Fred Olsen to expand its business further and free up the capital support that its parent company was providing.

Fred Olsen raised £303 million, which comprised:

- £115 million of existing senior debt and lease facilities;
- a £137 million term loan for the construction of Crystal Rig II, with a tenor of 18.2 years;
- a £17 million debt service reserve facility;
- a £30 million letter of credit facility for the construction and operation of Crystal Rig II; and
- a £5 million working capital facility.

The documentation allows for assets to be added to or removed from the portfolio and security structure, on the condition of meeting pre-agreed financial and technical criteria. This approach offers:

- reduced transaction costs;
- more certainty for the sponsor, as well as the knowledge that it will be working with a number of core lenders; and
- a larger pool of assets and a reduced equity risk, thanks to the cross-collateralisation.

For senior debt providers, the deal offers the advantage of security shared across a large portfolio of assets. If, in the future, Fred Olsen wants to add a new asset to the portfolio and that asset meets the pre-agreed criteria, then the lenders are unlikely to object to the addition (although they may do so). The lenders also keep the option not to lend to the new asset.

The project, unlike earlier wind financings, provides an open-ended portfolio structure that can accommodate projects with different types of power purchase agreement and levels of contracted offtake. In summary, the deal presents the following advantages:

- Fred Olsen has a long-term flexible financing in place for its current and future assets;
- The structure accommodates the evolving offtake and technological changes in the sector, unlike earlier deals; and
- Complex financings have been refinanced.

Fred Olsen's portfolio of projects
Before re-financing

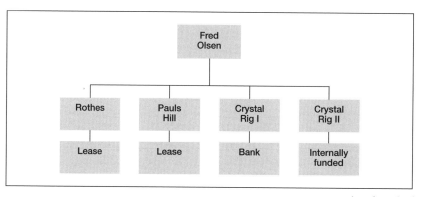

continued overleaf

After re-financing

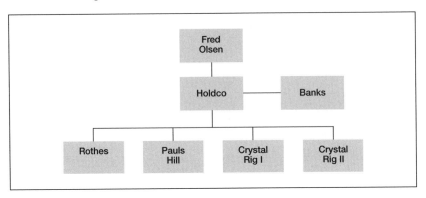

8.   **Conclusion**

This chapter has aimed to show how senior debt providers view renewables projects and which key aspects they look at when analysing a deal.

As mentioned above, lenders look favourably on the renewables sector – as both a business and ethical proposition. However, they will not give their money away unless projects offer some fundamental features.

Risk allocation in renewables projects

Tom Eldridge
Ike Nwafor
SNR Denton UK LLP

1. Introduction

One of the most important responsibilities of a transactional lawyer is to document risk allocation. This means drafting contracts evidencing a commercial transaction between parties operating within the applicable legislative framework. The contracts will contain the terms on which risk is shared, managed, accepted, avoided, limited and – ultimately – enforced by those parties.

Models of risk allocation develop over time. Thus, the risk model for development projects in traditional energy industries is an established one, while the model for delivery of renewable energy projects is evolving. The latter reflects policy and legislative developments. These policies are shaped by the sometimes competing considerations inherent to the security of energy supply and the Carbon Reduction Commitment.

The availability of subsidies, concessions and other forms of incentive will often shape the way in which risk is allocated. The lack of incentives or, equally as important, the lack of certainty regarding incentives will affect the way in which parties agree to accept risk. For what can be significant long-term financial (and resource) investments, the allocation of risk at the outset is a fundamental matter.

Several key principles of risk allocation apply to the delivery of renewable energy projects. This chapter addresses these. It shows how these principles apply to the risk model for renewables projects.

The chapter of this book entitled "Renewables – issues for financiers" examines risk issues arising in financing for renewables projects from a financier's perspective, including risks relating to:

- permitting;
- technology;
- contractual structure; and
- the regulatory framework for offtake arrangements.

This chapter looks at, among other things, these and other issues from the legal perspective, focusing partly on how risk allocation is documented contractually. In reading and considering these issues, this chapter looks at various renewable technologies and should therefore be read in conjunction with the chapters in Part 4 of this book, which analyse specific renewable technologies in more detail.

2. Parties to a transaction

The diagram set out below shows the key parties and the contracts for a typical renewables project.

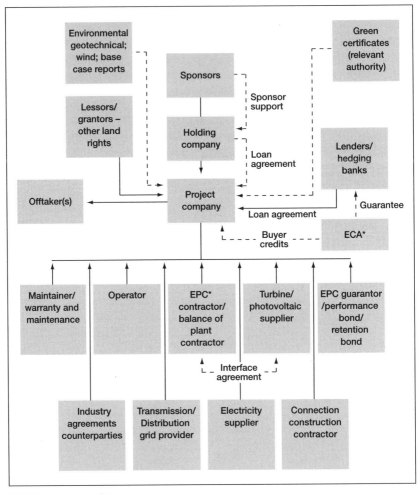

* ECA – export credit agency

** EPC – engineering, procurement and construction

3. Identification and allocation of risks – pass-down

Without a detailed risk analysis at the outset, the parties developing any form of energy project will not have a clear understanding of the obligations that they may assume. They will not be in a position to consider suitable exercises to mitigate risk at the appropriate time. For example:

- considerable delays can occur and expense can be incurred should problems arise when the project is underway; and
- arguments can ensue as to who is responsible.

The developer will be particularly concerned to ensure that it has identified and understood all risks that it will be assuming in connection with the project. The developer will want to be certain that it can manage and monitor these risks effectively. Where it cannot do so, it must pass down the risk onto another party that is better able to manage it. Where this is not possible for any reason, alternative management tools are required. These include:

- introducing liability caps;
- taking out insurance cover; and
- more radically, altering the structure of the project to extinguish the risk or at least reduce it.

The overriding principle of a renewables project – indeed, with any project – that is being project financed is that unmanaged risk cannot be left with the project company, which will generally be thinly capitalised. Developers will expect to pass down these unmanaged project risks through the project contracts, to ensure that they are assumed by the project company's contractors and suppliers.

Lenders to renewables projects will have similar concerns. Additionally:

- they will need to be satisfied that there are no regulatory constraints imposed on them by any of the authorities that regulate their activities or pursuant to laws applicable to them;
- they may have to report non-credit risks assumed by them in connection with their activities to their regulatory authorities; and
- the more risk they are expected to assume, the higher the margin that is required and the level of fees they might expect to receive from the project.

Only once the risks have been identified can the principal parties (the sponsors and the project lenders) decide who should bear which risks, on what terms and at what price.

4. General rules

The parties involved should observe some ground rules. These apply to projects in general, and renewables projects in particular, when determining which party should assume a particular risk. Thus:

- a detailed risk analysis should be undertaken at an early stage;
- risk allocation should be undertaken prior to detailed work on the project documentation starting;
- a particular risk should be assumed by the party best able to manage and control that risk. For example:
  - the risk of cost overruns in relation to the construction of the generating facility is best managed by the main construction contractor;
  - the risk of delay in relation to the production and delivery of turbines for a wind farm project or the photovoltaic panels for a solar project is best managed by the turbine manufacturer or photovoltaic manufacturer, respectively; and
  - interconnection risk and grid failure problems are best managed generally

by the offtaker/utility; and
- risks should not be left with the developer, especially where it is a special purpose vehicle. The point here is that where there is a disagreement between, for example, the feedstock supplier and the offtaker in a biomass/combined heat and power project over who should assume a particular risk, there may be a temptation to leave the risk in question with the developer. However, this is simply storing up problems for the future, as the developer will rarely be in a position to manage or control that risk, let alone pay for it.

5. Categories of project risk

Key risk areas for a project can generally be categorised as follows:
- construction and technology/completion-related risks;
- operation and maintenance-related risks;
- market risk;
- political risk;
- counterparty risk; and
- legal (including regulatory) and structural risks.

Participants will need to consider all of these risks and decide which party is best able to manage the risk. Once these risks have been identified, it is through the various contractual arrangements between the parties and insurance that these risks are – for the most part – apportioned and assumed.

5.1 Construction and technology/completion risks

The construction phase is seen as the key risk phase for project lenders (and indeed sponsors). This is the period when the project debt facilities are being drawn on by the developer, with the facilities funding the developer's payment obligations under the construction contracts and any other project-related payments that are required to be funded during this time. These might include:
- payments for necessary land acquisitions;
- consents and permits; and
- in some cases, first deliveries of supplies necessary for the completion and commissioning tests.

While the project debt facilities are being drawn, the project will generate no cash flows if it is a greenfield project or, perhaps, limited cash flows in the case of brownfield expansion projects. In all project financings, the bankability test assumes that the facility, once operational, will be generating sufficient cash flow to repay the principal amount and pay interest. Therefore, the ultimate risk to the project lenders is that the debt is fully drawn, but the project either never reaches completion or, if it does, cannot operate at the base-case levels assumed when the debt amount was sized.

It follows, therefore, that if the project contracts do not achieve satisfactory pass-down to the correct project parties of the types of construction/completion risk described in this section, it is likely that the relevant parties will look to the developer, sponsors and shareholders for some kind of completion support. This changes fundamentally the risk assessment of the project for the developer, with risk being allocated to, and assumed against, a developer or sponsor's (or group of sponsors') balance sheet rather than the project itself, as represented by the project counterparties and the contracts under which they assume their obligations.

Construction of the plant and the central interface issues relating to feedstock supply, turbine/photovoltaic panel installation and grid connection represent key risks for any project. The concern will be whether project completion can be achieved on time, within budget and in accordance with the applicable specifications and performance criteria.

Consequently, project completion represents a significant milestone for any project. The risk allocation model for a renewables project is never more clearly demonstrated than on completion. Project completion means that the plant is capable of dispatching electricity for a certain period of time at certain output and efficiency rates. In the context of a wind project, this means, in broad terms, that:

- turbines have been delivered and installed;
- the balance of plant works have been completed;
- the transmission lines have been installed;
- the plant is operational; and
- connection to the grid/offtaker has occurred.

Each project participant assumes and must discharge, in connection with each of the other participants, the key risks allocated to it. As a result of achieving completion:

- the contractor and the key suppliers will get paid the outstanding balance of the contract price (less any agreed retention amounts);
- the feedstock suppliers will be paid for first deliveries;
- the offtaker will receive electricity;
- the project lenders will begin to receive payment of the principal amount of the project debt repaid; and
- the project sponsors will be released from any contingent completion support obligations and will start to see a return on their equity investments.

In assessing completion risk, project lenders in particular will be looking at:

- technology risk, including factors such as whether the technology involved is proven;
- the overall contract structure; and
- the identity of the parties involved and their track record.

It is generally perceived that technology risk for solar/photovoltaic projects, onshore wind and biomass/combined heat and power plants is limited, subject to due diligence and technical reports on the relative feasibility of the relevant technology. Projects have been delivered and funded by commercial banks based on such technology. The technology for tidal and wave generation has less of an operating track record and is therefore seen as a greater risk. The main technology risk for offshore wind is not seen as any different from onshore wind generation. Having said that, construction risk for such projects is viewed as significant due to the complexity and cost involved in fixing and installing the turbines and transmission lines in deep, offshore waters. However, project lenders and sponsors will seek to analyse, model and stress-test the predicted wind or irradiation rates for wind or solar projects, for example. This analysis will form part of the basis for the expected generation output of the project and, therefore, the expected project revenue. The ability of a wind project, for example, to deliver forecast energy yields is critical to its economic viability.

The technology risk for a renewables project can in some cases overlap with regulatory risk. For example, a key risk for project lenders and other investors to overcome for a UK combined heat and power project is whether the project requires a licence or other authorisation to carry out the activities of generation, distribution and/or supply. The specifications of the plant will therefore determine whether it falls within the regulatory requirements for a licence. Project lenders will seek upfront due diligence reports on the ability of the plant to comply with the regulatory requirement for a licence. Further, if the proposed fuel source is biomass, then project lenders and investors will want to be confident that the biomass comes from a sustainable long-term supply source. The project lenders and investors will also seek to ensure that the biomass qualifies for accreditation for the purpose of government subsidies and/or green certificates. If the proposed fuel source is waste, then project lenders will want to be assured that the project sponsors have all necessary permits in place (if required). As mentioned, risks relating to regulatory constraints arising from the technology of a project can also be categorised as legal/regulatory risk (as highlighted in section 5.6 below).

The project lenders will also review the overall contract structure and, from a construction perspective, the nature of the construction contract(s). The construction contract(s) between the developer and the construction contractor(s) will set out, often in great detail, the tests that must be passed to achieve completion. These will usually include a mechanical completion or cold commissioning test and a full commissioning test which would include reliability testing, performance testing and, in facilities such as wind farms, noise testing. These commissioning tests will also involve a commissioning period, which will require that the facility operate at certain required levels for a continuous period of time. In addition, project lenders' requirements for commissioning tests, for some projects, include that both feedstock/supply arrangements and offtake contracts/grid connections are in place (although for wind farm projects, a turbine can be considered as commissioned when it is capable of generating electricity). Accordingly, these arrangements and connections will also form part of the overall project completion tests.

The project financing documents will contain a number of construction-focused provisions, including:

- a project long stop date;
- the project lenders' technical adviser advising on drawdown requirements being achieved;
- step-in rights relating to the subcontracts; and

- monitoring rights.

In addition to completion requirements in the construction contract(s), project lenders will often require that financial tests be satisfied. This would include:
- the satisfaction of future cash-flow ratios based on an assumed revenue throughout the term of the project; and
- the funding in full of debt service reserve accounts and/or maintenance reserve accounts, in either case ensuring that there is sufficient cash locked up to fund anticipated finance or maintenance costs in the future if project cash flows are lower than expected.

The experience of the construction contractor will also be an important contribution; of particular importance will be a track record of completing projects on time and within budget, and the parent company support that is available.

Below are the other key construction/completion risk areas likely to be of concern to the project lenders. The risks and suggested mitigants highlighted above and below are common to all power generation projects, not renewables projects exclusively.

(a)   *Type of contract*

Is the construction contract a turnkey contract with a prime contractor or are the design and works being undertaken by a series of consultants and contractors with project management being undertaken by the project company? The obvious examples here are wind and solar projects where the turbines or panels will be supplied to the contractor. Project lenders have a strong preference for turnkey arrangements, as this avoids gaps appearing in the contract structure and disputes between the contractors as to where particular responsibilities lie. However, for some renewables projects, such as wind projects, it is not unusual to have a split contract package. Indeed, in an offshore wind project, it would be unusual to have anything else other than split package. In a wind farm, the contracts would therefore comprise a turbine supply contract and a balance of plant contract. The turbine supply contract relates to the supply and supervision of supply of the turbines, and may also relate to the erection and mechanical completion of the turbines. The balance of plant contract relates to the other aspects of the construction required to complete the project. In the case of an onshore wind farm, this would include turbine foundations, inter-array cabling and the on-site substation. In addition, there may be a separate contract for connection of the generating equipment into the electricity distribution network, which will include construction and installation of connection facilities. Potential gaps in liability can be mitigated by the project parties entering into an interface agreement to regulate liability between them and/or a third-party arbiter agreement for a third party to resolve disputes between them, but such contracts are often strongly resisted, particularly by turbine suppliers. The prevalence of split contract packages in wind farm projects is often a reflection of the prohibitive price, in most cases unacceptable to developers, which specialist turbine suppliers would charge if they assumed the entire construction risk for a project.

(b) *Price*

Project lenders have a clear preference for fixed-price, lump-sum contracts. They will wish to be confident that:

- the facility can be completed within the funding which has been committed; and
- cost and time overruns are the responsibility of the contractor.

If there are to be any changes to the contract price, then project lenders will want to have a say on this – particularly if these changes are the result of new specifications on the part of the project company.

(c) *Delay*

Project lenders will want to ensure that there is a fixed date for completion with minimum ground entitling the contractor to extend the completion date. If there are any delays, the project lenders will want to see liquidated damages payable by the contractor in an amount sufficient to cover the project company's costs of servicing the loans and operating costs during the period of the delay. In practice, however, it is seldom possible to obtain full coverage in terms of liquidated damages for an unlimited period. What is generally negotiated is a fixed *per diem* amount subject to a cap. Under English law, a distinction is drawn between:

- liquidated damages, which are a genuine pre-estimate of the damage likely to result from the breach of contract (failure to complete on time) and which are enforceable; and
- clauses that penalise the contractor unfairly and are unenforceable.

A provision will generally be considered to be a penalty if the amount payable under the contract is extravagant and unconscionable compared with the greatest loss that could (at the time the contract was entered into) be conceived to flow from the breach. The wording used in the contract will be irrelevant; it is the substance of the clause that will be taken into consideration.[1] One factor that may distinguish the contract provision as a penalty is if compensation is by way of a single lump sum that is payable on the occurrence of several events, some of which would give rise to trivial damages and others which would give rise to substantial damages. A project lender should, therefore, make sure that the liquidated damages are set at a carefully calculated and realistic rate in order to satisfy itself that the provisions are enforceable. Further, the risk of any delays that are not the fault of either the developer or the contractor/supplier will need to be addressed in any other project contracts to which the developer is a party. For example, if there is a delay as a result of a *force majeure* event that delays construction, then the *force majeure* provisions in the offtake contract should take account of such delays and relieve the developer of liability.

---

1   *Dunlop Pneumatic Tyre Co Ltd v New Garage and Motor Co Ltd* [1915] AC 79; *Clydebank Engineering and Shipbuilding Co Ltd v Don Jose Ramos Yzquierdo y Castaneda* [1905] AC 6; *Philips Hong Kong Ltd v A-G of Hong Kong* (1993) 61 BLR 41, PC. On construction of contracts, see *Investors Compensation Scheme Ltd v West Bromwich Building Society* [1997] UKHL 28.

(d)   **Force majeure**

These provisions must be reasonable and the contractor should take all reasonable steps to circumvent the event resulting in *force majeure*. The *force majeure* event must not arise as a result of action or inaction on the part of the contractor, members of the contractor or people under the contractor's control. Further, where insurance cover is available for *force majeure* events, then the project lenders will want to see such insurance cover taken out. If insurance is not available, the project lenders will want others (the sponsors) to assume responsibility.

(e)   *Bonding requirements*

Project lenders will be concerned that appropriate performance and defects liability bonding is in place and from institutions with acceptable credit ratings. The amount of such bonds will depend on:

- the credit rating of the contractor;
- any guarantees offered by the parent company; and
- the payment profile during the contract.

It is not uncommon for performance bonds to be in the range of 10% to 20% of the contract price. Project lenders will expect the benefit of these bonds to be assigned to them.

(f)   *Supervision*

Project lenders often require that:

- their own engineer be entitled to inspect the construction works, receive reports and attend tests throughout the construction period; and
- where stage payments are used, their engineer verify that the appropriate works have been completed at the relevant time before the relevant drawdown under the facilities agreement is made.

Further, in relation to supervision, it is often the case that disbursements of the loan for construction-related payments (and indeed other significant items) will be channelled directly through the facility agent to the relevant payees. Project lenders will not usually be prepared to release drawdowns direct to the project company and take the risk that these moneys are not disbursed by the project company. Similar to other project financings, project lenders will require the standard account control mechanisms in place for ring-fencing project cash flow and specifying the order and manner of disbursement of such cash flow.

5.2   Operation and maintenance risks

Operation and maintenance, key to the successful functioning of the plant (and consequently revenue generation), are also fundamental. The nature of the risk will vary with the type of project – for example, maintaining an offshore wind farm is technologically more complex, more expensive and more likely to be subject to unplanned outages compared to an onshore wind farm. Of particular concern to the project lenders will be the following:

- Who is the operator? Does it have a proven track record of operating and maintaining similar projects efficiently?
- If the project company is also operating and maintaining the facility, does it have the necessary staff and skills? Who are the senior management personnel and, again, do they have a proven track record?
- How are operating costs to be managed? Who is responsible for increases in operating costs?
- Is the operator also responsible for maintaining the facility or is there a separate maintenance contractor? Do these include full lifecycle costs?

The project lenders will also be concerned that *force majeure* provisions in the operating agreement tie in with the other project agreements to which the project company is a party. In particular, any delays in production as a result of operating problems should not be the risk of the project company.

Further, as with construction contracts, the project lenders will be concerned that any breach of the operating agreement or the maintenance and warranty contract by the operator or the maintenance contractor will give rise to liquidated damages payable to the project company at an acceptable level. Accordingly, project lenders will want to know whether revenue adjustments will be applied against the operator.

Not every project will have an independent operator. Sometimes the project company itself will also be the operator. While this reduces the risk of an additional party default, clearly the project lenders will want to be satisfied that the project company has the experience and resources required to operate the project facility properly and to manage this risk. To this end, project lenders may require sponsors (which are expected to have the relevant expertise and resources) to enter into agreements with the project company whereby the sponsors agree to provide management, personnel and technical assistance. The project lenders will seek to ensure that such agreements are on an arm's-length basis and are not a potential route to obtaining an additional equity return from the project.

From the outset, developers and project lenders will assess operating risk using technical due diligence reports and/or project feasibility studies. For example, for wind projects, turbine failure during the operating period can be mitigated by taking a robust set of warranties for, among other things:

- defects;
- inadequate performance (power performance characteristics);
- technical availability;
- noise levels; and
- health and safety compliance.

The warranted performance criteria will reflect the base-case assumptions for generating project revenue. As always, the value of the warranties is subject to the credit risk of the warrantor.

5.3   Market risk
There are two principal elements here. The first is a risk that there is not a sufficient

market for the electricity. The second is a risk that the price at which the electricity can be sold is insufficient to service the project debt. Many projects are structured in such a way that long-term offtake contracts are entered into with third parties on terms such that this risk is wholly or partially covered. For example, the power purchase contract entered into may include a pricing mechanism containing a component that covers the operating costs of the project company, as well as its debt servicing requirements (usually called a 'capacity charge' or 'availability charge'). Subject to assuming the credit risk of the offtaker in these circumstances, a properly structured contract in these terms should remove most of the market risk for the project lenders. An offtake agreement linked to the market price usually means that the developer needs to pledge a specified level of the price of electricity.

Another variation used in power purchase agreements for managing market risk is the take-or-pay contract. This is a contract entered into between the project company and a third-party buyer whereby the third party agrees:

- to take minimum capacity delivery over a period of years; and
- to pay for that capacity whether or not it actually takes delivery of it.

In such contracts the project lenders will try to ensure that these contracts contain 'hell or high water' clauses, so that the buyer is obliged to pay for the products notwithstanding any default by the project company or otherwise. The incentive for the buyer to enter into such contracts will be the desire to obtain certainty of supply in circumstances and at a price that otherwise may be unavailable to it. Alternatively, the buyer may be a shareholder or otherwise related to the project company such that it is prepared to assume certain project risks. The availability of green certificates in conjunction with the power is often an incentive for certain utilities or energy industry participants to enter into such contracts. Take-or-pay contracts can be extremely valuable and can amount, in effect, to a virtual financial guarantee of the loan during the operating period. This is provided that the project lenders are comfortable with the creditworthiness of the buyer.

Take-or-pay contracts have been used in a great many project financings and in many different guises. Many power purchase agreements (including those used for merchant power renewables projects) are effectively structured on a take-or-pay basis. However, as referred to below, the thriving renewables sector in a number of countries has been attributed to market risk being taken out by state-backed feed-in tariffs and, to a lesser degree, green certificate regimes.

Where there are no guaranteed offtake arrangements such as those described above, then the project lenders are likely to have the following concerns:

- Is there a ready market for the electricity?
- In the event that a long-term power purchase contract is unavailable, can the project be structured on the basis that the developer will sell power on the spot markets?

This latter structure is likely to be restricted to projects in developed countries offering specific (favourable) characteristics. Use of robust hedging arrangements may go some way towards mitigating market risk for merchant power projects.

Finally, many countries have specific fiscal incentives (eg, feed-in tariffs and green certificate regimes) that are necessary to support renewables projects financially. For further information on this, please read the chapter of this book entitled "Renewable energy support mechanisms: an overview".

5.4   Political risk

In any cross-border financing, investors take a political risk. This encompasses:

- a fundamental change in the existing political order in the country; and
- the imposition of new taxes, new laws, exchange transfer restrictions or nationalisation.

The common link between all of these events is that they may jeopardise the prospects of equity returns, debt repayment and asset recovery on enforcement.

Where any project has been project financed, the political risks are more acute. This is because, among other things:

- the project itself may require substantive governmental subsidies (eg, feed-in tariffs), concessions, incentives, licences or permits to be in place and maintained (as noted in section 5.3 above); and
- the project may be crucial to the country's infrastructure or security and, accordingly, be more vulnerable to the threat of expropriation or requisition.

In renewables projects, these reasons are especially pertinent. The levels of investment reached in the renewables sector over the years are a consequence of the high levels of government subsidies and other incentives that have been made available to developers. Without that support, those developers would not have had the confidence to commit the financial and other resources in search of the long-term, high-value returns.

The term 'political risk' is widely used in relation to project finance and can mean both:

- the danger of political and financial instability within a given country; and
- the danger that government action (or inaction), such as the withdrawal of a subsidy, will have a negative impact either on the continued existence of the project or on the cash-flow generating capacity of a project.

There is no single way in which an investor can eliminate all political risks in connection with a particular project. One of the most effective ways of managing and reducing political risks, however, is to lend through, or in conjunction with, multilateral agencies such as the World Bank, the European Bank for Reconstruction and Development and other regional development banks such as the African Development Bank and the Asian Development Bank. There is a view that where one or more of these agencies is involved in a project, the risk of interference from the host government or its agencies is reduced. This is because the host government is unlikely to want to offend any of these agencies for fear of cutting off a valuable source of credit in the future.

Other ways of mitigating against political risks include:

- obtaining private market insurance, although this can be expensive and subject to exclusions. Further, the term for which such insurance is available will rarely be long enough;
- obtaining insurance from national export credit agencies such as the Export Credits Guarantee Department in the United Kingdom, the Compagnie Française pour le Commerce Extérieur in France and Hermes Kreditversicherungs-AG in Germany. Export credit support comes in a variety of forms and can include guarantees and/or direct loans and interest subsidies. However, such support will usually be given only in connection with the export of goods and/or services by a supplier to the project. As with lending in conjunction with multilateral agencies, co-financings involving national export credit agencies probably enjoy protected status;
- obtaining assurances from the relevant government departments in the host country, especially as regards the availability of feed-in tariffs and other subsidies, consents and permits;
- persuading the Central Bank or Ministry of Finance of the host government may be persuaded to guarantee the availability of hard currency for export in connection with the project;
- as a last resort (although this exercise should be undertaken in any event), thoroughly reviewing the legal and regulatory regime in the country where the project is to be located to ensure that:
  - all laws and regulations are strictly complied with; and
  - all the correct procedures are followed with a view to reducing the scope for challenge at a future date.

In some countries – particularly developing countries, which are keen to attract foreign investment – host governments (or their agencies) may be prepared to provide firm assurances on some of the above matters to foreign investors and their project lenders. Obviously, such assurances are still subject to a performance risk on the host government concerned, but at a minimum they can make it very difficult, as well as embarrassing, for a government to walk away from an assurance given earlier in connection with a specific project and on the basis of which foreign investors and banks have participated in a project. In some cases, stabilisation agreements and other contractual arrangements can be put in place with a host government to document these assurances, with enforcement routes including the use of bilateral investment treaties, as applicable.

There is no doubt that political risk represents a central consideration in any investment decision. This is particularly the case in renewables projects where the capital investment in new and developing technology depends on an often new and evolving set of government subsidies and incentives, in the form of feed-in tariffs and other carbon-based benefits.

Political risk has been, for many years and almost exclusively, associated with developing countries. A mixture of 'who dares wins' thinking, careful project structuring, resilient and faithful financial backers and often high premiums in the political risk insurance markets has seen successful projects delivered in these developing economies despite perceived political risk. Political risk events have also been seen as largely acts of aggression against foreign investors.

In 2010 the austerity packages that the governments of Western Europe are being forced to consider and implement may well turn on its head the perception that political risk is confined to aggressive acts of expropriation by governments in faraway developing countries. In Spain, for example, the government's proposed review of feed-in tariffs relating to the photovoltaic/wind industry will have far-reaching consequences and implications for both current and future investments in the renewables sector beyond Spain. A concern in Spain is the risk of a retrospective reduction of the feed-in tariffs for certain renewables projects, such as solar projects. Given the obvious adverse effect of such changes on project economics, industry participants are closely monitoring these developments. Related developments in other countries with feed-in tariff regimes, such as Germany and Italy, evidence the continuing trend for the gradual reduction of the level of subsidies for renewables projects, even though in some cases the changes are driven by green policy rather than fiscal constraints. For example, green policy, it appears, was the main driver of the recent abolition by the German legislature of subsidies for certain photovoltaic plants.

The renewables sector is by no means alone in bearing the full force of this type of political risk. For many years, investors in mineral and hydrocarbon rights have faced governments whose policies and approach may change and, indeed, have changed in response to a perception that the investor has benefited too much from favourable government support in the form of taxation, subsidies and other types of incentive. The challenge that presents itself to the economies of Western Europe and beyond is one of aligning a carbon policy with an incentives scheme that continues to attract long-term investment in renewable electricity generation.

5.5 Counterparty risk

This category of risk overlaps with a number of the categories above, but it is convenient to isolate it by way of example. What is meant by this is the risk that any counterparty with which the developer contracts in connection with a project may default under that contract. As identified above, this risk could be in connection with:

- the main construction contractor;
- a bank providing performance and/or defects liability bonding for a contractor;
- a supplier of turbines or photovoltaic panels;
- a purchaser of electricity;
- an insurer providing insurances in connection with the project;
- a third party providing undertakings or support in connection with the project; and
- a host of other relevant parties.

Inherent to certain renewables projects is the greater multiplicity of project parties and the fact that, compared with other sectors, turnkey construction contracts are less likely to be available as a means of managing the developer's counterparty risk.

If any of these parties defaults in the performance of its respective obligations, the project may run into difficulties. The first difficulty may be in connection with litigation over the contract in question, which may be expensive and time consuming to sort out. Even if the litigation is successful or an agreed settlement can be achieved avoiding litigation (whether using alternative dispute resolution procedures or otherwise), there is still the financial risk that the counterparty will not have the resources available to make the payment due. In many cases project lenders (and indeed sponsors) will recognise this risk. Where the project company is contracting with companies that are perceived to be financially weak, the project lenders will demand parent company guarantees (if available), or even bank guarantees or letters of credit, to support these obligations.

Finally, breach of many of these obligations in the project contracts will give rise to damages-based claims that are unliquidated claims. Thus, the basic common law damages rules will apply in common law-based jurisdictions, which can seriously affect the value of such claims.

Counterparty risk involves an element of legal risk. The availability of enforcement rights and remedies if a counterparty defaults and/or becomes insolvent will be dependent upon a mix of three variables:

- the governing law and courts of the project contracts;
- the location and domicile of the relevant project counterparty; and
- the type of enforcement regimes in those countries.

Taken together, these will need to provide the developer and the project lenders with sufficient confidence that the remedies available to them are adequate in support of default risk of passed-down obligations to the various counterparties. If they are not, then alternative protections such as insurance, often at significant expense, will be required.

5.6 Legal (including regulatory) and structural risks

Again, there is some overlap between both of these risks and some of the risks itemised above. Legal risk is the risk that the laws in the host jurisdiction (and any other relevant jurisdiction) will be interpreted and applied in a way that is inconsistent with the legal advice obtained from lawyers in the relevant jurisdiction at the outset of the project (change of law is considered in more detail in section 5.4).

Of specific concern will be the laws relating to title ownership and security and, in particular, the security taken by the project lenders over the assets of the project. Not all jurisdictions have available to them the concept of the floating charge (or equivalent) and the ability to appoint a receiver which is so convenient for project finance structures. In many jurisdictions, particularly in developing countries, many basic concepts of security are not well developed. Specifically, there may be no ability to take security over intangible assets, such as rights under contracts, and future assets in the form of receivables due under contracts, including green certificates. Consequently, local legal opinions may be of only limited comfort. In such cases, the project lenders may well have to discount the value of such security completely or in part at least. Reservations and qualifications in legal opinions, while universally unpopular with the beneficiary of the opinion, often give a clear understanding of the inherent legal risks in developing a renewables project in a certain jurisdiction. Initial careful due diligence on the legal regime in these cases is an absolute requirement.

On title ownership, the risk of challenge to the land rights needed by a developer for a wind farm or a solar project, for example, and the resulting risk of delay or cessation of the project, are typical legal risks. Due diligence will be required on issues such as environmental law constraints and any unusual restrictions on the developer's title. For example, for offshore wind farms, the nature of the developer's land interest in the coastal sea bed is likely to be an issue to be clarified through due diligence. Another example relates to solar projects which use roof-mounted solar panels: an issue that arises in some jurisdictions is the ability of the developer to effectively register and/or grant security over its property interest. For more information on issues relating to environmental risk and land rights affecting wind farm and solar projects, please see the wind and solar chapters in part 4 of this book.

Structural risk is the risk that all of the elements of a complex project structure do not fit together in the way envisaged, and the various legal advisers and other experts involved have not done their job properly.

Complex projects can involve hundreds of interlocking documents running to many tens of thousands of pages and the possibility of error and oversights cannot be completely ignored. The availability of professional negligence insurance policies will obviously provide some comfort for project lenders and investors here, but even the largest firms of professional advisers are unlikely to have sufficient cover for the largest projects.

6. Conclusion

The risks discussed in this chapter are for the most part the same risks which a sponsor or project lender would need to consider for any power project. However, in some cases, for renewables projects, the character, probability and consequences of such risks are different. For example, in relation to fuel supply risk, the key concerns for a gas-fired power plant are security of supply and price and ensuring that the power purchase arrangements contain back-to-back provisions dealing with a fuel supply failure. For a wind or solar project, the risk relates to finding the right location which has appropriate wind/solar characteristics and ensuring that the contracts (particularly the power purchase agreement) recognise the generation risk associated with adverse meteorological conditions. Technology risk is another example where the probability and consequences of the risk are higher for certain renewable technologies than they are for fossil fuel power projects. Further, certain project risks are unique to renewables projects and have therefore involved more detailed discussion in this chapter. For example, the regulatory risk associated with projects that rely on government subsidies (eg, feed-in tariffs) is of concern to sponsors and project lenders, which need a degree of certainty that these subsidies will remain in place for the life of the project.

A sponsor seeking to invest in a renewables project should focus on a comprehensive review of the risks associated with that project before approaching project lenders. Project lenders, on the other hand, should similarly focus and seek to understand the risks for that project – particularly those that would be truly unique to a particular renewables project. For example, the sheer multiplicity of contracts in an offshore wind farm project means that traditional 'wrapped' contracts are not a suitable mitigant for counterparty risk.

Looking ahead, it is the authors' view that for investment and lending activity in the sector to increase substantially, the sector requires standardisation of both risk allocation and management and the contracts and provisions in contracts which deal with risks and issues unique to renewable projects.

Structuring the project vehicle

Matt Bonass
Juliette Seddon
SNR Denton UK LLP

1. Overview

Joint ventures are becoming increasingly common in renewable energy projects, partly because of the size of the projects. For instance, a successful bidder in an offshore wind development (eg, Centrica in the United Kingdom) may want to sell down some of its interest in that project in order to share the capital costs.

The preceding chapters discuss the financing of renewable energy projects and the risk allocation in such projects. This chapter considers the different types of legal structure available to parties when putting together a joint venture and, predominately, whether the structure is incorporated or unincorporated.

Further, this chapter discusses the issues that arise on preparing to set up a joint venture and the factors to take into account in determining the most suitable structure. Lastly, we focus on the most common structure, a private limited company, with an overview of the issues to consider when preparing a joint venture agreement, together with common provisions in the documentation.

2. Available structures

As noted above, the basic choice to be made in structuring a renewable energy joint venture is whether:

- a separate legal entity will be incorporated as the joint venture vehicle; or
- the project will proceed with an unincorporated structure.

In practice, this means choosing between:

- incorporating a limited liability company (the most common structure for a renewables joint venture);
- opting for a partnership; or
- entering into a purely contractual arrangement.

2.1 Corporate structures

Structuring a project through a private limited company is by far the most common choice. Under English law, it is also possible to use a limited liability partnership, incorporated under the UK Limited Liability Partnerships Act 2000. This combines the advantages of limited liability with tax transparency. However, this chapter focuses on a private limited company.

2.2 Unincorporated structures

An unincorporated structure will take the form of either a partnership (in the United Kingdom, subject to the Partnership Act 1890), a limited partnership (in the United Kingdom, subject to the Limited Partnerships Act 1907) or a contractual arrangement.

A limited partnership is unlikely to be a suitable vehicle for a renewables project because the structure requires a general partner with unlimited liability (although the general partner itself can be a limited liability company). More problematic for a joint venture structured through a limited partnership is the prohibition on limited partners retaining limited liability while taking any part in the management of the partnership. This structure is therefore suitable only if a majority of the partners are passive investors in the joint venture.

In a contractual joint venture, the parties do not create a separate legal entity. Instead, their relationship is based simply on a contract between the parties. There is therefore no framework outside the contract (other than the general law of contract) which can be relied on to supplement the provisions of the contract. Accordingly, everything that the parties want to agree and write down in relation to the workings of the joint venture must go in the contract. This may be different from the position of a private limited company (see below).

The contract therefore needs to cover issues such as:
- what each party will contribute (whether in resources or funds);
- management and decision making;
- handling disputes; and
- termination.

This structure is less formal than a corporate joint venture. For example, there is no separate legal entity to manage and administer, and the venture is easier to terminate. However, the lack of formal structure means that some aspects of the joint venture are more difficult to address. Thus, transfers of interests can be problematic, as there is no discrete interest, such as a share, to transfer. Replacing one party with another in an unincorporated joint venture inevitably opens up the structure and workings of the joint venture to renegotiation between the remaining parties and the new party. That said, an unincorporated structure can incorporate elements of a private company through the provision of a transfer of interest regime, such as those commonly used in the oil and gas industry through joint operating agreements.

A contractual joint venture is commonly used for cooperative arrangements such as research and development collaborations and consortium project bids.

Thus, it is suitable for a number of joint ventures in the renewables sector.

3. Initial considerations

A variety of factors will determine the choice of structure for a joint venture. Unless one element clearly outweighs the others, all the factors discussed below should be considered.

3.1 Limited liability

Limited liability is attractive to joint venture participants. In fact, it is the overriding reason for incorporating a private limited company joint venture vehicle, rather than using an unincorporated structure. There are only limited circumstances where the shareholders of a company can be held legally liable for a company's debts (beyond any unpaid capital contribution). Some traditional energy companies may therefore choose to establish renewables joint ventures in separate entities in order to ring-fence any liabilities which relate to these ventures.

However, shareholders will not reap the full benefits of limited liability. If the joint venture company is a newly incorporated company, as is often the case, shareholders will invariably be called upon to guarantee the joint venture company's liabilities to third parties as the joint venture company is unlikely to be creditworthy in its own right.

In order to protect creditors of limited liability companies, there are statutory requirements on a company to maintain its capital, which means that rules regulating the extraction of the company's profits and assets go hand in hand with limited liability.

In a partnership, each partner is jointly and severally liable, without limit, for the wrongful acts and omissions of its partners, as each partner is deemed to be the agent of the others. This is the case unless:

- a partner is not authorised by the partnership to conduct certain business; and
- the party with which it is dealing is aware of that restriction.

3.2   Tax transparency
A company has separate legal personality and therefore is taxed on its profits and capital gains. Shareholders are then liable to tax on the dividends that they receive when they extract profits from the company. They are therefore generally exposed to what is known as double taxation.

In a partnership or contractual joint venture, each partner is taxed directly so the potential for double tax charges does not arise. The partners may also be able to enjoy direct tax benefits through pass-through taxation. In the United Kingdom, this is also the case in the context of a limited liability partnership, as noted above.

3.3   Publicity
An incorporated entity will generally have to make public filings of its constitutional documents and accounts. An unincorporated entity will not be under the same obligation. In a contractual joint venture, both parties to the contract may be required to disclose information on the joint venture as part of their general obligation to make public filings, but such information is likely to be less visible as it will be diluted by the other information disclosed. This may be an issue for the parties to consider, but in practice it is unlikely to be a decisive factor.

3.4   Degree of integration
A corporate structure represents the pooling of all trading activities, assets and liabilities in relation to the joint venture business in one vehicle. Unincorporated structures offer the participants a lesser degree of integration, which may better suit

the parties' aims. A partnership represents sharing of some aspects (eg, the liabilities of the other partners), but not of others (eg, tax). A contractual joint venture may offer collaboration in a particular area only or entail sharing of particular resources only, thereby maintaining distance between the parties in other respects.

3.5 Financing

Companies can give security over their assets by way of a floating charge in favour of a bank without hindering their ability to deal with those assets. Floating charges are invariably required in order to secure bank financing.

3.6 Accounting

If a party establishes an incorporated joint venture, it will need to consider whether it accepts that company being reported together with its group on a consolidated basis for accounting purposes. This may be the case if it owns a majority of that company's shares or controls a majority of the voting rights relevant to that company. This is less likely to be the case with an unincorporated association.

3.7 Size, duration, project, sector, stage of life

If the joint venture is to be small or of a specific duration, rather than an ongoing business, a non-corporate structure is often used. This is common in construction and exploration or technical or professional fields.

4. Set-up issues

Aside from selecting the appropriate structure for the joint venture, the joint venture participants will have to consider a number of factors when setting up the project vehicle, whether by way of an incorporated or unincorporated structure.

4.1 Initial documentation

Once the parties have decided to work together to form a joint venture and before they invest time and effort in detailed negotiations, they should put in place initial documentation which contains the protections discussed below. The initial documentation commonly takes the form of a letter between the parties known as a letter of intent, heads of terms or memorandum of understanding. It contains the outline structure of the joint venture and generally is non-binding, save for some provisions applicable to issues such as confidentiality of information and exclusivity for negotiations relating to the joint venture.

Mutual obligations not to disclose confidential information received from the other party should be put in place at an early stage and before financial or technical information is exchanged between the parties. The agreement needs to cover information given both in writing and orally. Under English law, there are principles of confidentiality under which either party can claim for misuse of its confidential information in the absence of a confidentiality agreement. Furthermore, in a number of EU jurisdictions, there is a specific duty of good faith which includes the concept of keeping information confidential. However, an agreement provides more robust protection because the parties can identify the kinds of information which they consider confidential and the manner in which such information can and cannot be used and disclosed.

Confidentiality provisions commonly include provisions about public announcements, requiring the parties to agree on the content of public announcements before they are made.

4.2   Exclusivity and non-solicitation

This is a binding provision in the heads of terms (or in a separate, binding confidentiality and exclusivity agreement). Under this provision, the parties agree mutual obligations:

- not to seek out third parties with which to negotiate a similar transaction; and
- neither to continue nor to enter into negotiations with any third party for a certain period of time.

In some jurisdictions, including England, any provision drafted as a positive obligation on the parties to negotiate with each other is unenforceable.

4.3   Due diligence

Each party should consider carrying out investigations into the other party and any assets that the latter will contribute to the joint venture in order to satisfy itself that the other party is an appropriate partner. Due diligence should cover commercial, financial and legal matters. The level of investigation will depend on the nature of the existing relationship, if any, between the parties and the nature of the joint venture. For example, it can be limited if:

- the parties are well known to each other;
- the joint venture is a small endeavour; and
- the joint venture is to be in place for only a short period of time.

4.4   Third-party clearances and consents

It is vital to establish early whether any third-party clearances or consents will be required to implement the joint venture. Obtaining these will need to be factored into the timetable for establishing the joint venture.

(a)   Competition clearance

Merger control may be relevant where the parties are combining existing activities in the joint venture or if the joint venture is a large equity joint venture. Merger control may simply require a notification to the authorities either before or after the joint venture is established. In some cases the parties may require approval from the relevant authorities before they can implement the joint venture, which has obvious timing implications.

Restrictive or anti-competitive agreements are regulated in many countries, including the United Kingdom. Accordingly, if the joint venture contains provisions under which the parties may be restricted from competing freely, then those provisions should be reviewed to ensure compliance with the relevant legislation.

(b) **Regulatory approvals**

If the joint venture operates in a regulated industry, the parties may require approval from the regulator. This will be another significant influence on the timetable for establishing the joint venture.

If the joint venture's activities are regulated, it will be important to check whether the legislation applicable to that industry or the licence required by the joint venture to carry out the relevant activity specifies a particular structure (eg, a company limited by shares or another type of corporate vehicle) and, in particular, whether and what type of corporate structure must be used. In the United Kingdom, any joint venture company operating in the water sector will need to be either a public or private company limited by shares. In contrast, any legal entity can hold a licence to operate in the electricity industry pursuant to the UK Electricity Act.

Regulatory regimes may also require that the assets relating to the regulated business be ring-fenced within the joint venture entity so that the company's assets are focused on the regulated activities. This may restrict the arrangements of the joint venture company, because of a direct prohibition or if some activities are subject to a consent regime.

(c) **Contractual consents**

Depending on their circumstances, the joint venture participants may need to consider the impact of establishing the joint venture on their existing arrangements. A joint venture participant may need the consent, under an existing contractual arrangement, of a lender, supplier or customer in order to enter into the joint venture. If that arrangement is significant enough, such consent may need to be sought before the joint venture is implemented.

4.5 Protection of IP rights

For any participant contributing IP rights, or if the function of the joint venture is to research and develop technology or know-how, the protection of IP rights will be a significant consideration. This may particularly be the case in the context of a joint venture to develop a waste-to-energy project or a fuel-cell development company. The parties should agree on how they will deal with IP rights so that the correct licences are granted to the joint venture and the correct rights are reserved to the parties.

The parties should also set up a framework for how the joint venture's IP rights will be protected when conducting its business and entering into contracts with third parties.

4.6 Guarantees

As discussed above, the joint venture parties are likely to be required by third parties to give guarantees or bonds in respect of the joint venture. In addition, the parties may want to consider whether they require guarantees or cross-indemnities in respect of their fellow participants' obligations to perform their obligations in relation to the joint venture. Of course, participants should be wary of requesting guarantees that they would not be prepared to give themselves, unless there is a significant difference in financial standing between the parties.

In the United Kingdom, a guarantee given in the context of a regulated industry may also be enforceable by the regulator.

5.   Incorporated joint venture

A private company limited by shares is the most common form of joint venture in the renewables sector. This section describes how such a company is usually structured.

5.1   Constitutional documentation

The joint venture company is incorporated pursuant to the relevant company law legislation in its territory of incorporation. In the United Kingdom, this is the Companies Act 2006. The rules of a company are contained in its constitutional documents. These may be referred to as 'by-laws' or, in the United Kingdom, 'articles of association'.

The joint venture participants also enter into a shareholders' or joint venture agreement which will further regulate the relationship between them. All of the features of the joint venture company discussed below are dealt with in one way or another in either or both of the shareholders' agreement or the constitutional documentation.

Given that the two documents may contain similar provisions in relation to certain matters, the shareholders' agreement needs to provide that in the event of any conflict between the terms of the shareholders' agreement and the constitutional documents, the shareholders' agreement shall prevail. Further, the parties are likely to be required to make public the constitutional documents. The terms of the shareholders' agreement can remain confidential between the parties. The parties will therefore want to place any sensitive information in the shareholders' agreement – for example, the way in which profits of the joint venture are shared.

5.2   Management and control

Shareholders exercise control over the joint venture company by voting at shareholder level and having rights to appoint directors to the company's board.

(a)   Shares

The joint venture participants need to decide how ownership of the joint venture company will be shared. Where there are only two participants to a joint venture, it is not uncommon to issue one class of shares held in equal numbers by each participant so that they have equal votes and rights to dividends. This structure is called 'deadlock', as neither party has the right to outvote the other. The flexibility of the share structure of a company means that a more sophisticated structure can be adopted, if appropriate, as share rights can be tailored. This means that different classes of shares with different rights can be issued to different participants. For example, preference shares that offer a preferential return but no voting rights (or voting rights with no rights to dividends) can be issued to an investor.

This can enable developers or energy companies to take an equity stake in projects, for example, or to take some form of preferential share which can give them enhanced rights to profit, information or control.

(b) Appointing directors

Board control is exercised by giving rights to the shareholders to appoint directors to the board of the joint venture company. In a simple structure with only two shareholders, these have rights to appoint equal numbers of directors in order to maintain the deadlock position.

(c) Decision making

Most decisions, especially those relating to the day-to-day running of the joint venture's business, are made at board level. If shareholders have equal weight on these issues, the quorum for a board meeting will require that the directors appointed by each shareholder be present so that one shareholder's directors cannot make decisions in the absence of the others.

The shareholders may not wish to delegate certain matters to the board and these matters will be voted on by shareholders. These may include the declaration of dividends, entry into material contracts, bank borrowings or the sale of a substantial part of the joint venture company's assets. However, one party may want decisions to be taken at board level, as this will require that the directors take decisions which promote the success of the joint venture company or act in the best interests of the joint venture company. This would be to avoid shareholders taking decisions that are in the best interest of individual shareholders.

Again, the quorum requirements will ensure the presence of any shareholder whose right to decide such matters has been agreed on. Finally, according to companies legislation such as the UK Companies Act, a company can do certain things only if the shareholders approve them. These matters are also reserved for shareholder meetings rather than board meetings.

(d) Voting

By the use of quorum requirements or by allocating different numbers of votes to different directors or to different classes of share, the agreed balance of power within a joint venture can be achieved on board and shareholder matters.

In addition to categorising decisions as either board or shareholder decisions, important decisions at either board or shareholder level may require the votes of certain directors or shareholders, or a unanimous vote. This effectively gives those directors or shareholders a veto right over the matter in question. One of the existing directors may additionally be nominated as chair of the board of directors and have a casting vote in the event that the board is otherwise in a deadlock position. Obviously, a party that has a majority shareholding in the joint venture company will not be keen to grant the minority extensive veto rights.

(e) Breaking deadlock and handling disputes

In a joint venture company which is deliberately structured so that the parties have equal voting rights or where a minority shareholder is given veto rights on certain matters, the shareholders' agreement usually includes a mechanism for resolving the potential deadlock which may arise when the directors or shareholders disagree on how to proceed in relation to a matter being voted upon. A dispute resolution mechanism is required in case the deadlock escalates into a fundamental breakdown in the relationship of the parties. This mechanism can ultimately include termination of the joint venture.

Some joint venture participants decide not to include provisions to resolve deadlock in their shareholders' agreement, preferring to negotiate a resolution at the time. Contractual provisions to resolve deadlock may well rarely be followed to the letter, if used at all. However, including a resolution process offers the advantage of providing encouragement to the parties to resolve their differences by setting negotiation positions.

Mechanisms for resolving disagreements over business or policy issues, which are not due to a breakdown in relations between the parties, commonly commence with an informal procedure that can be escalated into more formal proceedings if necessary. First, the disagreement is referred to the chairs or chief executives of each shareholder, who meet informally to resolve the issue (under a so-called 'gin and tonic clause'). If this is unsuccessful, the next step may be to engage a mediator to facilitate resolution of the issue between the parties. It is tempting to provide that an expert or arbitrator should be appointed to decide the dispute. However, this is suitable only for a dispute over a technical or legal matter – in a business or policy dispute, a third party is unlikely to provide a 'right' answer.

Deadlock could also arise from one party's boycott of meetings so that a quorum is never achieved and nothing can be decided. To prevent this, the constitutional documentation and shareholders' agreement commonly provide that if a meeting is adjourned due to lack of quorum, at any subsequent meeting those shareholders or directors (in the case of a board meeting) present constitute a quorum. The quorum may also be required only at the beginning of a meeting, so that if a party walks out, the meeting may validly continue. Both measures effectively give control to the parties present.

The parties may want to provide for their relationship as joint venture participants to come to an end when a deadlock becomes unbreakable. The shareholders' agreement may already include adequate termination provisions, such as the ability of either party to terminate on notice or to exercise a 'put option' to sell its shares to the other shareholder.

Where such provisions have not been included, special termination measures are still available. They include:

- the voluntary winding-up of the joint venture company;
- the sale of the joint venture company as a whole; and
- a buy/sell procedure between the parties, known as a 'shoot-out', which determines which party will buy, which will sell and at what price.

These options may appear draconian, but are rarely used in practice. However, they are useful in that they encourage a negotiated settlement.

(f) Transactions with shareholders

The nature of a joint venture as two or more parties coming together for a common business makes it likely that the company will contract with its shareholders, whether to receive or to provide certain services, for the secondment of employees or for any other reason. It is therefore important for the shareholders' agreement to deal with how the joint venture company does business under contracts which it has entered into with its shareholders. For example, a shareholder in a joint venture company established for a wind farm project may also supply wind turbines to the joint venture, either directly or through a company within its group.

Joint venture participants are often comfortable with the shareholder with which a contract is being made or with the directors of the joint venture company appointed by that shareholder being permitted to vote on approving the contract or changes thereto, as long as the directors or shareholder in question declare their or its interest. It is at times of disagreement over the contract between the joint venture and the shareholder counterparty that it is thought appropriate:

- to exclude the directors and shareholder in question from meetings where the joint venture's position is discussed; and
- to allow them neither to vote nor to see relevant papers from those meetings.

It is also possible to exclude the relevant directors or shareholders from any meetings which discuss a future contract between the joint venture company and a shareholder if that is the preference of the joint venture participants.

The joint venture directors are likely to owe fiduciary duties to that company. This is the case under English company law. One of the duties owed is to avoid situations in which they could have a conflict of interest and/or duty. The nature of a joint venture makes it likely that its directors are also directors or employees of the shareholding company. This does not mean that there is an outright prohibition on individuals being directors or employees of both a shareholder and a joint venture company. Under the UK Companies Act at least, shareholders can authorise such a situation of potential conflict. However, the joint venture documentation (in particular, the constitutional documents) will need to contain this authorisation.

(g) Business plan

Whether the joint venture is being set up for a specific project or an ongoing business, agreeing on a business plan at the time that the constitutional documents and shareholders' agreement are put in place can be another important tool for shareholders to control the business of the joint venture. For instance, the joint venture directors may have day-to-day authority to act within the scope of the business plan and be required to ask the joint venture shareholders to make decisions only when these are outside the scope of the business plan.

5.3 Funding

(a) *Equity*

A company has flexibility in how it is funded. Equity is just one way in which shareholders can fund a joint venture company. Funding by way of equity entails rights arising in favour of shareholders. Return on an equity investment is also subject to rules under company law which prevent capital from being returned and restricts the extraction of profits.

Equity subscriptions can also be made by way of contribution of cash, or of assets or other non-cash contributions. The latter is more complicated than a cash contribution, as the parties have to negotiate and agree on the most appropriate way to value the non-cash contribution, which is likely to involve professional valuations. Equity subscriptions are becoming more popular in renewables projects, probably because of a lack of available debt finance.

(b) *Shareholder loans*

Shareholder loans to the joint venture company often supplement equity contributions. In determining the debt-to-equity ratio for the joint venture company, the factors to be taken into account include:

- ease of repayment (loans can be repaid more easily than share capital);
- how much equity the shareholders want on the balance sheet (in order to raise external financing or to demonstrate a strong equity base to the market); and
- tax (see section 3.2 above).

If appropriate, shareholders can make loans to the joint venture on very basic terms (eg, an interest-free loan with repayment on demand). The terms of the loans can be included in the shareholders' agreement or in a separate shareholder loan agreement.

Quasi-equity, such as a loan note, offers a more structured way of funding a joint venture company. This is useful if there are multiple shareholders lending to the joint venture company. In these circumstances, the same terms can apply to each shareholder and notes can be issued on the same terms at different times.

(c) *External financing*

A joint venture, like any other company, can also seek external financing. The previous chapters have looked at third-party financing for renewable energy projects, particularly in the context of debt finance. Lenders will need to ensure that the fact that the project vehicle is a joint venture will not pose an additional risk for their lending. They will therefore want to ensure, among other things, that:

- there are effective provisions for the resolution of deadlock;
- the joint venture participants are committed to retaining their equity interests for a minimum time; and
- the relationships between the joint venture and its participants, such as material contracts and shareholder loans, are reasonable.

(d) **Guarantees**

Despite the joint venture participants having chosen a company structure in order to benefit from limited liability, this concept will in practice be eroded because they will be unlikely to be able to escape giving guarantees to third parties in respect of the joint venture company's obligations unless the company is creditworthy. External lenders will require their lending to be guaranteed. Counterparties to material contracts may also require guarantees not only of the joint venture company's payment obligations, but also potentially of the joint venture company's ability to satisfy the contract. Lastly, any shareholder in a joint venture company with a funding obligation may require a guarantee from its parent entity of the obligation to fund.

(e) **Future financing**

To avoid disputes and regardless of whether future funding requirements are known, it is a good idea to agree at the outset on:

- how funding will be provided to the joint venture company in the future; and
- whether the shareholders themselves will have any future obligations to fund.

(f) **Profit distribution**

The rules on distribution of profits by companies under company law generally means that only certain reserves can be distributed to shareholders. Thus, a profit does not necessarily translate into a dividend. Once the joint venture has profits available for distribution, its participants will need to decide:

- how frequently to distribute the profits;
- in what proportions to distribute them; and
- what proportion of the profits, if any, should be retained and used by the joint venture company.

5.4   Transfer of shares

(a) **Restrictions on transfers of shares**

As the joint venture participants will have chosen to enter into a joint venture specifically with each other, it is possible that the parties will not want the ability for any of them to transfer their shares freely at any time. However, it may be expressly envisaged that the parties will sell down a part of their shares to other participants. This is likely to be the case in the context of some of the larger renewable energy projects – for example, those projects auctioned under the Crown Estates Round 3 Licensing Round.

A 'transfer' should mean not only a transfer from one registered shareholder to a new registered shareholder, but also a transfer of the beneficial interest in shares – for instance, by declaring a trust over the shares and giving security over them. The question for the parties is then: how restrictive do they want to be?

(b)   ***Transfer of whole or part of an interest in a joint venture company***
The parties need to decide whether to permit any shareholder to transfer some of its shares or whether it must sell them all. It is common to provide that a party which wishes to sell must sell its entire shareholding and not just some of its shares. This preserves the balance of the structure that was originally set up.

(c)   ***Lock-up period***
If the joint venture is being set up for a specific project, it may be appropriate not to allow any transfers until the project has reached a certain stage of development and therefore have a lock-up for an initial period. This could also enable the project to break even.

(d)   ***Permitted transferees***
It may be appropriate to allow transfers to any third party, provided that they meet criteria that can include:
- being a member of that shareholder's group;
- having sufficient financial standing or technical know-how to perform its role in the joint venture; or
- not being engaged in a business which competes with that of the joint venture or of another participant in the joint venture.

The shareholders' agreement will therefore include categories of permitted transferees. If a permitted transferee ceases to be a permitted transferee, perhaps because it ceases to be part of a shareholder's group, there will usually be a provision to transfer back the shares held by that permitted transferee to the original holder of the shares or another member of its group.
If regulatory approvals are relevant to a transfer of shares in the joint venture company, a permitted transferee will need to be a party approved by the regulator.

(e)   ***Pre-emption rights***
The parties may wish to include pre-emption rights so that if a party wishes to sell its shares, the other shareholders have a right of first refusal to buy those shares. If the other shareholders do not wish to buy the shares offered, then the selling shareholder is free to sell its shares to a third party at the same price as that offered to the other shareholders. The selling shareholder is often free at that point to sell its shares to any third party. However, in some situations it may be appropriate to combine a pre-emption right with criteria that a third party must meet, so that the remaining shareholders still have some control over who the new shareholder will be, even if they had no desire to buy the shares themselves.

(f)   ***Transfers subject to consent***
An even tougher, yet simple restriction on transfer is not to allow a transfer without the consent of the other shareholders. This consent can be given or withheld entirely at the discretion of the other shareholders. However, it is common to have a restriction on such a clause to require that consent not be unreasonably withheld or delayed.

(g)   **Prohibition on transfers**

The shareholders may also agree that transfers of shares to third parties are not permitted at all. In this case, the parties will rely on the termination provisions in the shareholders' agreement to terminate the joint venture. This is common in joint ventures where the identity of the parties is paramount to the joint venture.

(h)   **Tag-along rights**

A tag-along right gives other shareholders a right to sell if a shareholder has negotiated a sale of its shares to a third party. The selling shareholder is required to ensure that the third-party purchaser also makes an offer to the other shareholders to acquire their shares on the same terms. If the offer is accepted, all shares in the joint venture company are sold on the same terms. This right is important for minority shareholders: it enables them to sell their minority interests at full value, rather than at a discount as can be the case for non-controlling interests. This right also removes the risk for minority shareholders to remain in the joint venture with a potentially unwelcome new majority shareholder.

(i)   **Drag-along rights**

A drag-along right allows a selling shareholder to force the other shareholders to sell to a third party with which it has negotiated a sale of its own shares. This right is especially advantageous to majority shareholders. This is because selling the whole company, rather than only some of the shares, is likely to make the sale more attractive to potential purchasers, and thus command a higher price per share.

A drag-along right will generally be granted only to a shareholder or shareholders which hold a substantial majority of shares.

(j)   **Change of control**

'Change of control' refers to a change in ownership of a joint venture shareholder. It is relevant to the discussion of transfer provisions for two reasons. First, the transfer provisions do not catch transfers of shares in a joint venture shareholder (as opposed to shares in the joint venture itself), which will change the ultimate ownership of that shareholder. If the ultimate control of the joint venture participants is important, such transfers will have to be regulated by other provisions in the shareholders' agreement. Second, if the shareholders want to prohibit changes in their respective ownership, the consequence of breach of that prohibition can be a forced transfer of shares by the shareholder that is undergoing a change of control. In this case, the transfer provisions can be used to determine the terms of such sale.

Where the joint venture's contractual counterparties require that the joint venture agreement include provisions governing any change of control, the terms of the joint venture should ideally mirror those of the third-party contracts.

(k)   **Loans and guarantees**

The shareholders' agreement should also provide what happens to a shareholder's loans when that shareholder sells its shares. The joint venture could simply be required to repay the loans when a shareholder exits the joint venture. However, this may not be practical, as the joint venture may need to rely on this source of finance continuing or may not have the funds to repay at the relevant time. An alternative is to require the purchaser of the shares also:

- to purchase the loans; or
- to agree to provide equivalent finance on equivalent terms if tax reasons or cost efficiency dictates that it is better for the joint venture to repay the loans than for them to be assigned.

The exiting shareholder will want to be released from any guarantees that it has given in respect of the joint venture from the time of the sale. It is therefore common to provide that the incoming shareholder must offer alternative guarantees as a condition of the sale.

(l) Deed of adherence

On any transfer to a party that is not already a shareholder, the new shareholder must be required to execute a deed of adherence before the transfer takes place which makes it a party to the shareholders' agreement.

In practice, it is likely that the entry of a new shareholder into the joint venture will stimulate the renegotiation of the joint venture arrangements, which will most likely lead to the shareholders' agreement being amended or replaced. This renegotiation may take place even though:

- the existing shareholders' agreement may have a perfectly workable mechanism for substituting one shareholder for another; and
- the other provisions of the agreement may have worked well.

5.5 Termination and exit

When setting up a joint venture, the participants should think about termination or exit. They should provide for such an eventuality in the shareholders' agreement, even if they are reluctant to discuss this at the beginning of their relationship.

The parties need to decide what will trigger an exit by a shareholder or termination of the entire joint venture. These triggers are discussed below. In each case, the parties will also need to decide on mechanisms by which exit or termination will happen.

(a) Mechanisms for exit

Exit mechanisms include winding up the joint venture and the purchase of a shareholder's shares by either another shareholder or the joint venture – perhaps by way of implementation of a buy/sell or shoot-out mechanism, as noted above. Detailed provisions on termination in shareholders' agreements are not necessarily followed because if a problem arises, the parties often negotiate a commercially agreed solution or exit. Termination and exit provisions are still useful as they provide a background against which the parties can negotiate, but they should be commercially workable and not overly complicated.

(b)   ***Consensual termination***
The parties may decide upfront to establish a joint venture for a fixed term with automatic termination unless each party agrees to renew. This is relatively uncommon.

(c)   ***Termination by notice***
This gives a party the ability to give notice that it wants to exit the joint venture. No reason or cause is required.

(d)   ***Termination for cause***
The parties can agree that certain triggers will cause the entire joint venture to terminate or, if a shareholder is at fault, the non-defaulting shareholders can remove the defaulting shareholder. Triggers could include:
- the joint venture not meeting specified targets;
- material breach of the shareholders' agreement by a shareholder;
- the insolvency of a shareholder; or
- change of control of a shareholder.

(e)   ***Sale or initial public offering***
The ultimate goal of the parties in setting up the joint venture may be:
- to sell it once it has reached a certain stage of development; or
- to float it on a stock market by way of an initial public offering (IPO).

The parties should therefore incorporate triggers in the shareholders' agreement whereby they agree:
- to sell automatically at a certain time or at a certain stage in the development; or
- that any shareholder can initiate a sale or IPO at the time certain targets are reached.

While included in the shareholders' agreement, these may not be capable of binding the parties in practice. However, they do set out the intentions of the parties and provide a background against which the parties can negotiate.

6.   Conclusion
Joint venture arrangements are likely to become increasingly common in renewables projects as the size of such projects increases. Numerous issues must be considered when structuring a joint venture. First, various factors need to be weighted before deciding whether the project company should be incorporated or unincorporated. Second, if the parties to the joint venture opt for an incorporated structure, they must also decide how the joint venture will be managed and operated.

The key considerations all relate to money and control. They include the following questions:
- How will the joint venture be financed initially and in the future?
- How will the capital expenditure of the company and other incomings and outgoings be controlled?

- How will any profits be distributed back to the shareholders?
- Who will control the board and shareholder decisions of the joint venture?
- What protections will minority shareholders have at board and shareholder level?
- What control will shareholders have to sell their interest in the joint venture?

Onshore wind projects: technical issues

David Groves
Wind Prospect Group Ltd

1. Introduction

Wind power is rapidly emerging as a key technology towards a low carbon, resource efficient Green Economy.[1]

In 2009 the world's wind generating capacity grew by 31.7%, adding a record 38,343 megawatts (MW) to reach 158,505MW of installed worldwide.[2] In 2009 electricity generated by wind turbines equated to almost 5% of the European Union's electricity consumption.[3]

Wind power is no longer seen as an 'alternative' form of energy and is becoming increasingly global in its use. In 2009 the United States' addition of almost 10,000MW brought its total wind energy capacity to 35,064MW.[4] This allowed the United States to hold onto the top spot in overall capacity, ahead of Germany, China, Spain and India. China ranked first in new capacity added, with 13,803MW, and moved into third place in terms of total capacity, just 673MW behind Germany. Rounding out the top five countries in new additions were Spain, with 2,459MW; Germany, with 1,917MW; and India, with 1,271MW.[5]

Europe, by far the leading region in cumulative wind capacity, added 10,163MW in 2009.[6] Traditionally, Germany, Spain and Denmark have been the major players in European wind development. However, the market is quickly becoming more diverse, with Italy, France and the United Kingdom leading a second wave of wind expansion. With their help, wind was the leading source of new power added in the European Union in 2009 for the second year running.[7] Wind also accounted for 40% of new electricity generating capacity in the United States.[8]

The development of a wind farm follows a similar process to the development of any large-scale power generation project, with the project advancing in phases to

1   Achim Steiner, United Nations under-secretary general and environment programme executive director, during a press conference in Copenhagen, quoted in Global Wind Energy Council news release, "Wind energy can meet 65% of tabled 2020 emissions reductions by industrialised countries", December 14 2009 (www.gwec.net).
2   Global Wind Energy Council, "Global Wind 2009 Report", April 21 2010, pp8-9.
3   Chris Rose, *Wind Directions*, "European wind energy continued its surging success in 2009", April 2010, pp6-8.
4   Global Wind Energy Council, "Global Wind 2009 Report", April 21 2010, p10.
5   *Ibid.*
6   Chris Rose, *Wind Directions*, "European wind energy continued its surging success in 2009", April 2010, pp6-8.
7   *Ibid.*
8   Global Wind Energy Council, "Global Wind 2009 Report", April 21 2010, p11.

manage risks and resources (ie, decisions are taken at key milestones in order to determine whether to progress to the next phase of development). The typical phases of a wind project are:

- initial site selection;
- project feasibility assessment;
- preparation and submission of the planning application;
- construction;
- operation; and
- decommissioning or repowering.

This chapter considers the key technical issues for each of these phases. Given the close interaction between the technical aspects of a wind project and commercial and legal matters, the author has, for ease of use, included in this chapter some of the concepts mentioned in the commercial and legal wind chapters. It is recommended that you read those chapters as well.

2.   Site selection

Finding suitable locations to site a wind energy project depends on many factors and differs significantly from country to country. Site identification is best approached as a process of elimination whereby unsuitable land areas are discounted and, thus, potentially developable areas identified. This exercise is usually performed using a geographical information system, combined with local knowledge. This is initially a desk-based review to establish whether sites satisfy technical and environmental criteria.

The first port of call is usually wind resource maps, which indicate regions that have higher predicted wind speeds. Higher-quality resource maps are modelled on the basis of terrain height maps (eg, higher elevations will have higher wind speeds); some even take into account terrain roughness (eg, a site near a wood will have a lower wind speed).

Initially, the areas with an average wind speed deemed unsuitably low will be removed from the map (this cut-off wind speed will vary from country to country and depends on various economic factors). Section 3.1 below considers the wind assessment process once a site has been selected.

Proximity to a grid connection is an important factor for consideration. A suitable grid connection can be a substation, a transmission line or part of a distribution network. Selecting a suitable connection will depend on:

- the distance to the connection point;
- the potential capacity of the site; and
- the capacity available in the grid.

Often a site-finding exercise will look to eliminate all areas that are beyond a certain distance from a connection point. This distance will vary depending on site generation capacity. Section 3.3 below considers the grid connection issue in more detail.

Due to the problems associated with potential noise, a setback distance is applied to all residences, usually ranging from 500 metres to 1.5 kilometres, depending on the size of the site and, sometimes, existing background noise at the location.

In more developed areas, because of the large quantity of residences, this exercise will result in very specific site boundaries.

Potential visibility of the proposed site and relevant landscape or heritage designations create further constraints that need to be addressed. These include, for the United Kingdom:

- areas designated for their natural landscape value (eg, national parks[9] and areas of outstanding natural beauty);[10]
- areas designated for ecological reasons (eg, sites of special scientific interest[11] and special protection areas);[12] and
- areas of cultural heritage importance (eg, World Heritage sites[13] and scheduled ancient monuments).[14]

Due to the potential impacts on radar systems, areas around airports (whether civilian or military) and other radar installations are also discounted. However, this could change thanks to ongoing research into technical fixes to radar interference problems. Section 3.2 below considers the radar issue in more detail.

The above process generally results in the identification of an indicative developable area. Similar techniques are often applied to prioritise and rate multiple potential developable areas in order to determine the best site(s).

3. Technical aspects of onshore wind farm development[15]

Once a potentially viable wind farm site has been identified, detailed studies and surveys are required in order to:

- examine various aspects of the site;
- confirm the viability of the project; and
- secure the necessary permissions that are required to design, install, commission, own, operate and maintain the wind farm.

Technical studies that should be undertaken during this process include those set out below.

---

9 Public parks and open spaces protected under the UK National Parks and Access to the Countryside Act 1949.

10 Landscapes designated under the UK National Parks and Access to the Countryside Act 1949. The Countryside and Rights of Way Act 2000 introduced new measures further to protect these sites.

11 Locations protected due to their scientific interest, in accordance with the UK Wildlife and Countryside Act 1981 (as amended).

12 Land and wetlands in the European Union protected due to their importance for birdlife, in accordance with the EU Birds Directive (79/409/EEC), which came into force in April 1979.

13 Places of outstanding universal value for cultural or natural features selected by the United Nations Educational, Scientific and Cultural Organisation.
Buildings and other man-made features above or below ground protected in accordance with the UK Ancient Monuments and Archaeological Areas Act 1979.

14 The technical issues faced by offshore wind farms vary significantly from those faced by onshore wind farms. This chapter covers the latter only.

15 www.enercon.de/en/enerconsturmregelung.htm.

3.1 Wind resource assessment

To gauge the financial viability of a potential wind farm, the site's wind resource must be assessed accurately. How the proposed wind turbines respond to the site's wind resource determines the likely energy yield from the site and, therefore, how much power can eventually be exported onto the grid.

The energy yield is calculated by taking into account:

- the wind data measurement;
- the wind resource and energy yield analysis; and
- the loss and uncertainty assessment.

Ensuring the accuracy and quality of the data at each stage is essential to create confidence in the energy yield predictions.

(a) Wind data measurement

A measurement mast at the site captures the wind data. Masts are either tubular or lattice towers equipped with tools such as anemometers and wind vanes mounted at various heights.

The data is measured for as long a period as possible; a minimum of 12 months is preferable in order to understand the annual variation in wind resource and is essential when trying to obtain project finance. Subsequent analysis should take into account complete, continuous 12-month periods in order to remove seasonal effects that may be present in the dataset, which could unfairly bias the results.

Mast location is important and should be representative of the likely wind turbine locations. For large sites, more than one measurement mast may be appropriate, as ideally masts should not be more than 2 kilometres from any of the proposed wind turbine locations.

Measurement masts range from 10 metres to 100 metres in height, although even taller masts are becoming available. Measurements are rarely taken at the eventual wind turbine hub height (ie, the height above ground level of the centre of the wind turbine rotor), so the data will need to be extrapolated in the wind resource analysis section of the work. As a result, data is measured at various heights in order to determine the wind shear (ie, the change in wind speed with height) at the site. This data is later used to extrapolate wind speed information from the mast height to the hub height.

Remote sensing devices are becoming more common in the industry. These use either lasers (Lidar) or sound waves (Sodar) to measure the wind speed and direction at various heights. They are easy to install, rarely need planning permission (although it is always important to check this) and have a longer life than a mast. Typically, however, remote sensing units are used alongside masts to measure the wind and build a greater understanding of the wind regime across a site (especially for sites located in complex terrain), rather than being relied on as the sole wind measurement tool. They are likely to become the more commonly used measurement tool in the future.

(b) Wind resource analysis

The analysis begins with removing from the data spurious records and periods of icing where the response of the anemometer may be questionable. This process aims to reduce uncertainty over the measured data. Quality checks are undertaken periodically to ensure that the measurement equipment has been recording correctly.

The key wind statistics at this stage are mean wind speed, wind direction and turbulence intensity. Turbulence can be considered as the fluctuation of the actual wind speed from the mean wind speed; it tends to affect the loading of the wind turbines. Wind turbines are rated according to the level of turbulence that they can withstand without compromising their design life.

The challenge at this stage is to understand whether the measured site data is representative of the longer-term wind regimes in the region of the site. In order to determine this, the site-based data must be correlated by a longer-term reference source. Traditionally, this data is sourced from local meteorological stations run by a national weather agency. However, when this data is either unavailable or unsuitable, data based on a mesoscale model (ie, a numerical weather prediction model) for the region may be appropriate. The key aspect of data selection is that the correlation strength must be high, ensuring that the data is representative of the site.

A wide range of statistical methods, indexing and averaging techniques can be applied to determine the ratio between the short-term measured data and the long-term reference mean. The result of the correlation process will be a long-term wind distribution for the mast location on site. Once this distribution is created, the rest of the wind assessment process can start.

The transfer of wind speed from anemometer location to hub height location is undertaken in two stages. Stage one consists of a vertical extrapolation from mast height to hub height. This extrapolation can be undertaken in two ways:

- By using anemometers at different heights, the site wind shear can be determined. This can then be used to estimate the wind speed at the proposed hub height.
- If there is doubt over the quality of the shear estimation, it may be more appropriate to use a flow model to undertake the extrapolation.

A flow model considers the changes in height and surface roughness across the wind farm, and determines the likely changes in wind speed that will result. Different types of flow model can be used, depending on the complexity of the site with regard to slopes and forestry. The flow model is set up with height data and roughness data for the site. Typically, the height data comes in the form of a roughness map and either a contour map showing height data for the site in contour lines or a grid map showing height data as a series of grid points.

The roughness data represents the relative smoothness of the ground around the wind farm site and is determined based on the ground conditions at the site – typically ascertained from a site visit, aerial photos and survey maps. Wind turbine locations and measurement mast locations are added, along with the long-term wind distribution. The flow model is used to determine the wind variation across the site.

Stage two of the wind speed transfer is the horizontal extrapolation of the wind speeds from mast location to wind turbine location. This is always undertaken using a flow model. Indeed, if using a computational fluid dynamics-type model, the vertical extrapolation stage is not required, as this flow model works in a different manner altogether.

Once the wind distributions at hub height have been established, the next stage is to analyse the performance of the chosen wind turbine. For the purposes of energy yield assessment, wind turbines are characterised by a power curve and a thrust curve. The power curve shows the power production from the wind turbine over a range of wind speeds.

The thrust curve is used in determining the wake losses and allows the calculation of the wind turbine's wake size and spread. The wake reduces the wind speed behind the wind turbine, this reduction being the result of the energy being taken out of the wind by the rotor. The performance of downstream wind turbines is thus affected by the wake. Accordingly, wind turbines need to be spaced correctly to minimise the loss of production due to wakes (known as 'wake loss').

The wind speed distributions, together with the thrust and power curves, allow the wind farm software to determine the gross energy yield predicted for the site. At this stage the analysis has taken into account the wind speed, terrain and wind turbine wakes in its calculation of energy yield.

(c)   *Losses and uncertainties assessment*

The final stage of the assessment involves determining the difference between the theoretical amount of power that the wind turbine can convert from the available wind resource and the power exported at the metering point.

A series of losses must be taken into account. These can be categorised as downtime and performance losses.

Downtime losses are typically calculated by taking into account the following elements.

- Wind turbine availability – this is a measure of the proportion of total time that the wind turbine is capable of operating, subject only to suitable wind conditions. Leaving aside periods of maintenance, the availability of a wind farm over a year typically exceeds 95% – although there are exceptions. Reductions in availability include unscheduled repairs and maintenance. Wind farms are usually subject to an availability warranty from the wind turbine supplier for the first years of operation, which enables the wind farm owner to recover lost revenues in the event of low availability (see Section 5 below). The author's experience shows that the allowance made for unavailability losses when calculating the potential energy yield is typically set at between 3% and 5%.

- Wind turbine maintenance – wind turbines will undergo scheduled servicing once or twice a year. The total downtime from such servicing is around 1% of the time, based on projects with which the author is familiar.

- Wind farm infrastructure maintenance and grid outages – other equipment within the wind farm will be subject to periodic inspection and/or maintenance. The local grid network can also be expected to be unavailable occasionally. Resultant downtime is typically low: it is usually estimated to be less than 0.5% of the time, in the author's experience.

- Curtailment – the conditions of building permits for some wind farm projects require their operation to be constrained for various reasons, including the effects of shadow flicker and noise emissions on local residents. The constraints applied will vary between projects and may involve the pausing or de-rating (operation at a lower than optimum output in order to control noise levels) of individual wind turbines under specific meteorological conditions or at defined times of the day. The impact on yield has to be evaluated on a project-specific basis.

Performance losses typically include the following elements:
- Electrical losses – as the electricity generated by the wind turbines is transferred to the metering point, electrical losses will be incurred, notably in transformers and power cables. The exact value depends on the size of the site and the electrical design. Normal estimates are of a loss of between 2% and 3% in the author's experience, although for projects where the metering point is close to the wind turbines, lower losses can be experienced.
- Icing/blade degradation (dependent on the location and atmospheric conditions at the site) – if icing occurs, the wind turbines will usually be shut down to prevent damage to the wind turbines themselves and the risk of ice being thrown from the blades. Areas where dirt and flies can accumulate on the wind turbine blades can reduce the aerodynamic efficiency. Periodic cleaning may be required. In the author's experience, losses are typically estimated to be around 1% of the power that could be converted, but this is site-dependent.
- High-wind hysteresis – typically, wind turbines cut out at wind speeds of 25 metres per second and do not cut back in again until the wind has dropped to 22 metres per second or below – although some wind turbines can be adapted to operate at higher wind speeds (eg, Enercon wind turbines if fitted with the 'Storm Control' module).[16] So while the site may be reading high winds of 23 or 24 metres per second, if the wind turbine has already shut down, this valuable wind resource is missed. A loss of around 0.5% can be expected based on the author's calculations.

Losses are combined using a standard equation. From the author's experience of conducting wind resource assessments and due diligence reviews, the overall losses for any wind farm can be anywhere between 6% and 12%, depending on the site and the analyst's interpretation.

Once the losses are applied to the energy yield prediction, the net energy yield, or P50, is produced. This is the best estimate of energy yield for the site based on all the information available. 'P' stands for 'probability of exceedence' – 'P50' meaning that this energy yield value is likely to be exceeded 50% of the time; for the other 50% of the time, the energy will be under the P50.

---

16 Directive 97/11/EC on the assessment of the effects of certain public and private projects on the environment.

Other P values are determined through an uncertainty analysis. This is a measure of the confidence that the analyst has in:

- the information provided;
- the equipment used;
- the methodology available; and
- a range of other factors.

Each analysis team has a slightly different methodology for uncertainty assessment, but it is an important stage of the assessment to try to determine the spread. The most commonly used value for financing is the P90. This is lower than the P50 forecast and represents the energy yield estimated to be exceeded 90% of the time, so there is more certainty over achieving this yield.

3.2   Site design and environmental impact assessment

Detailed studies and surveys are commissioned to evaluate various aspects of the site, including:

- the constructability (taking into account ground conditions, gradients, hydrology, flood risk and access);
- the ecology;
- the archaeology;
- the landscape; and
- the cultural heritage.

These studies and the overall wind farm design will be informed by the findings from scoping studies and other consultations with statutory and non-statutory organisations, including non-governmental organisations.

(a)   *Environmental impact assessment*

Many studies are required in order to secure environmental and building permits. In the European Union, the use of environmental impact assessments is obligatory for the majority of wind farm developments, although member countries may determine whether a wind energy project should be subject to an environmental impact assessment on a case-by-case or threshold basis.[17] On a case-by-case basis, the government or its appointed body decides whether an environmental impact assessment is required based on the characteristics of the individual project. On a threshold basis, one or more criteria (eg, proposed installed capacity or number of wind turbines) is established above which an environmental impact assessment is mandatory. Even where a full environmental impact assessment is not required, it is the author's experience that some studies will still be needed in order to secure planning consent.

An environmental impact assessment is an iterative process that is undertaken throughout the design of the wind farm development. It involves:

---

17   www.decc.gov.uk/en/content/cms/what_we_do/uk_supply/energy_mix/renewable/planning/
on_off_wind/aero_military/aero_military.aspx.

- identifying the potentially significant environmental impacts;
- determining what those impacts are; and
- identifying potential mitigation measures as appropriate.

The purpose of an environmental impact assessment is not to result in a project that has no impact of any kind, but to understand and describe what those impacts are in a logical manner. It is a decision-making tool – that is, the information presented still needs to be weighed against other relevant issues, such as planning policy. Consequently, a wind farm development that results in negative environmental impacts (eg, habitat loss) will not necessarily be refused planning permission. However, those impacts must be outweighed by other factors (eg, the need for renewable energy) in order for the development to proceed. In the author's experience, mitigation measures can vary significantly, from enhancing local habitat for sensitive birdlife to reducing the number of wind turbines within the development.

A logical consequence of the iterative assessment process is that technical aspects of the wind farm project, notably the proposed locations of wind turbines, will change throughout the development process. Once planning permission has been granted, however, the fundamental design of the wind farm is usually fixed – even though there may be some (limited) latitude for micro-siting the wind turbines (ie, changing the position of wind turbines within a limited radius around the original location) or amending the wind turbine geometry (ie, tower height and rotor diameter). Some aspects of environmental impact assessment are considered below.

(b)   *Visual and landscape assessment*

The main techniques used are:

- mapping of zones of visual influence to indicate where the wind farm will be visible from;
- wireframe analysis to show the turbine locations from particular views; and
- photomontages or computer-generated images overlaid on photographs of the site to demonstrate how the wind farm would look when it is built.

Sophisticated industry-specific software is used to generate the requisite images. A cumulative landscape assessment may also be required in areas with existing wind farms or where other farms are proposed to be built in a similar timeframe.

(c)   *Noise assessment*

A noise assessment is undertaken in order to assess the potential impact of the development on sensitive receptors in the surroundings, such as local residents. Background noise levels are measured and noise emissions from the wind farm are predicted using modelling software.

(d)   *Ecological assessment*

The wind farm's impact on local flora and fauna must be considered. Once the ecological surveys are complete, often at particular seasons during the year, the development is kept clear of areas that are known to support rare species that would potentially be at risk from the turbines or the construction activity.

(e) ***Interference with telecommunications infrastructure***

Mobile telephone network operators are consulted because a wind farm could interfere with the electromagnetic links between the various masts that could cross a site. All identified links are given an agreed setback distance. Utilities that use telemetry links are also consulted.

Operators of gas and oil pipelines and power lines are consulted to ensure that the development is kept an adequate distance from any local infrastructure. Buffer zones or corridors are commonly applied to pipes and power lines, although wind farm roads may pass over or beneath such infrastructure, subject to design limits.

Operation of wind turbines can cause interference with television reception by scattering signals; this will be evaluated during the development process. The wind farm owner will usually bear the cost of any mitigation measures implemented once the wind farm has been built. The problem can usually be solved by simple technical modifications. The provision of satellite television receivers to some households may even be the most efficient solution.

(f) ***Traffic management and construction***

A buffer is applied to busy roads. Increased vehicle loads and movements on public roads also need to be considered (see section 4.2).

On site, each wind turbine requires a track to access the base of the tower, the route of which will be designed in consultation with the land owner and wind turbine suppliers in order to minimise the impact on farming activities and ensure that the wind turbine components can be delivered safely.

Site entrances are kept to a minimum. A construction compound is usually sited near the main site entrance during the construction period.

(g) ***Radar***

Many wind farm developments have been abandoned or seriously delayed by the impact that the projects could have on military or civilian radar systems. In the civilian field, wind turbines can affect radar systems used:

- to handle aircraft in and around airports; and
- for en-route navigation and air traffic control purposes.

Military radar system operators and government defence departments regularly object to wind farm proposals that they consider to be a threat to the integrity of national defence systems.

The basic problem is that traditional radar systems detect objects using a sweeping motion, looking for moving objects via the Doppler effect (ie, the change in frequency of a wave for an observer moving relative to the source of the wave). The rotation of wind turbine blades can give similar readings to aircraft, causing confusion for the radar system operator and compromising the usefulness of the radar system.

Currently the only successful means of overcoming the radar problem is to locate the wind turbines carefully to avoid the problem in the first place. In addition, wind farm developers should conduct negotiations with military and civilian radar system operators in order to convince them that a specific turbine will not cause any interference. The presence of a large number of both military radar installations and airports in the United Kingdom renders the radar interference problem particularly acute in that country. The UK government, key stakeholders (including the Ministry of Defence and civilian air industry bodies) and the British Wind Energy Association (now Renewables UK) formed a strategic working group in 2001 to identify solutions to the problem. Reporting in September 2008, the UK government identified that:

- approximately 1 gigawatt (GW) of wind projects were facing objections relating to air defence radar;
- 1.9GW were subject to airport traffic control objections; and
- 3.1GW faced planning objections relating to en-route air traffic control.[18]

A number of initiatives are being pursued, including:

- the use of transponders within wind farms to communicate with aircraft;
- the use of so-called 'stealth' technology in wind turbine blades to absorb radar signals and hide the wind farm from the radar system; and
- advanced software developments that will be able to filter out wind turbine-related radar reflections in order to allow radar installations to focus on their real targets.

(h)   *Geotechnical investigations*

As the proposed wind farm design evolves, the nature of the local ground conditions needs to be taken into account. Wind farms have been built at sites with a huge range of geotechnical properties, from desert conditions to peat bogs, from rocky terrain to former open-cast mines. Some wind turbines have even been built within industrial premises that may contain buried waste materials.

In all cases, it is important to draw an accurate picture of the ground conditions across the site in order to design the wind turbine foundations, hardstandings and road network. Intrusive geotechnical studies such as trial pits or boreholes can be very expensive. Accordingly, it is common to begin with desktop surveys of available data such as historical geological records and other information about the local area. In former coal-mining areas, coal-mining reports can be obtained to ascertain the presence of shafts and tunnels. As the project progresses, excavations or boreholes may be dug to allow structural engineers to evaluate the specific properties of the ground conditions at individual wind turbine locations. Samples of the earth will be sent for chemical testing as part of this process.

This information is fed into the design process to ensure that the wind turbine foundations and other infrastructure are designed appropriately.

---

18   geophysics.esci.keele.ac.uk/Research/seismology/windfarms/.

(i)   *Eskdalemuir seismic monitoring facility*

Wind farm developments in Scotland are subject to a specific constraint. The Eskdalemuir seismological recording station, between Moffat and Hawick in the Scottish Borders region, is used to monitor underground nuclear testing across the world as part of the United Kingdom's obligations under the Nuclear Test Ban Treaty. Low frequency vibrations from sources such as wind farms can be problematic for these sensitive seismic monitoring instruments. Accordingly, the UK Ministry of Defence placed for a while a blanket precautionary ban on wind farm development within 80 kilometres of the Eskdalemuir site.[19]

Following studies by Keele University, the Ministry of Defence relaxed its restrictions, allowing the installation of many wind turbines up to a maximum limit. However, in 2010 the ministry determined that the number of wind turbines (in operation or consented) in the region had reached the limit that could be accepted. Accordingly, it announced that it would object to any further planning applications for wind farm projects within a narrower 50-kilometre radius of the Eskdalemuir facility.[20] A working group comprising Renewables UK, the Ministry of Defence, the Scottish government and academic specialists is seeking technical solutions to allow installation of additional wind turbines in the region. Until a solution is found, further projects in what is a key area for wind development in the United Kingdom will be delayed.

3.3   Grid connection

The term 'grid connection' is used to describe the electrical system required to connect a wind farm project to the nearest appropriate section of electrical distribution or transmission network such that the electrical output of the wind farm can be exported.

There are significant differences in upfront and lifetime costs associated with whether a project connects to the distribution network or the transmission system.

The developer must submit a connection application to the network operator. The level of technical data required by the network operator varies and tends to increase over time. It therefore helps if the wind turbine has been selected and the basic electrical design for the project has been determined before submitting the application.

Network operators are usually obliged to process connection applications and return a connection offer within a set period of time of receiving a competent application and any associated fee. The timing and details of the offer process differ from country to country.

The connection offer will detail:

- the electrical infrastructure required to connect the wind farm to the existing network;
- the predicted costs of the scheme; and
- the forecast timescales for the works.

---

19   "Wind farms banned as MoD listening post demands hush to detect nuclear blasts", February 10 2010 (www.news.scotsman.com).

20   "The World's Biggest On-Shore Wind Farm", *Wind Today*, Second Quarter 2010, pp22-27.

An offer usually has a set validity period depending on the network operator's preference. Typically, the acceptance of an offer requires some form of payment to be made to the operator. The initial payment can range from a small deposit (eg, 10% of the total) to full payment of the predicted grid connection scheme costs, subject to the developer's credit rating and the particular operator's conditions.

Once the connection offer has been accepted, connection works can commence. The typical timescales are between one and three years, depending on:

- the complexity of the scheme;
- the number of land owner consents required; and
- the acquisition of planning permission where required.

3.4 Development post-permission grant

Planning approval is often granted subject to conditions that may limit the development or require the submission of additional information prior to starting the development.

In order for the planning decision notice to be issued, the applicant may also have to enter into a legal obligation to undertake works or make a payment in lieu of works.

Many conditions applied to planning consents for wind farms are technical in nature and require detailed design of infrastructure to be submitted to the permitting authority prior to starting construction. In some countries, such as Romania and Poland, all wind farm infrastructure must be designed in detail before permits are issued. By contrast, in other countries some aspects require only an outline design – for example, it may not be necessary to select the exact model of wind turbine prior to permits being issued, allowing the wind farm developer a greater choice of turbines within a given set of limits (eg, rotor diameter).

Where an application is approved and conditions are imposed, it is in the post-determination stage that the applicant must meet the requirements of the conditions and mitigate potential environmental damage.

Planning conditions include:

- noise control;
- traffic management;
- eventual decommissioning;
- limiting telecommunications interference;
- establishing an environmental management plan to avoid or limit damage to fauna and flora;
- control of the design and colour of the turbines;
- control of the detailed wind farm infrastructure design; and
- ongoing programmes to monitor anticipated and unforeseen effects.

Failure to proceed in accordance with the development as submitted or without discharging the appropriate planning conditions could lead to enforcement action by the determining authority. Enforcement action can be a convoluted process that can ultimately lead to court proceedings, fines, alterations or demolition.

4. Construction

The construction of a wind farm involves three distinct phases:

- pre-construction;
- construction; and
- turbine erection.

Prior to starting work on site, a variety of pre-construction activities must be completed. In view of the number of separate contractors involved in a wind energy project, it is considered best practice for all contracts in place to be managed by a single entity before work commences on site. Some investors will prefer to negotiate a single turnkey contract with a wind turbine supplier or other specialist engineering company to reach this objective. Others favour the appointment of multiple contractors and a project manager to manage interfaces, programme, commissioning and other aspects of the project. In the latter case, the project manager may be either a member of the sponsor's organisation or a project management consultant. He or she will usually be involved in preparing and negotiating the other contracts, bringing his or her experience to bear on technical and commercial matters. Naturally, it is advisable that the project manager have a strong track record in the field, as would be the case with other types of large construction project.

4.1 Pre-construction phase

The extent of pre-construction activities varies depending on the location of the wind farm. However, they can include those outlined below.

(a) Property rights and permits, and authorisations

Prior to work commencing on site, the developer must be granted the necessary property rights and obtain any permits and authorisations. Construction logistics may need to accommodate farming practices such as harvesting crops that will be set out in the property agreement. Other permits required before construction starts can include permission to use specific roads for construction traffic or rights to extract rock from within the site for use in road construction.

In some jurisdictions these consents can be secured alongside the primary wind farm consents; in others the consents must all be dealt with separately.

(b) Satisfying a wide variety of planning conditions

This can include preparing method statements to demonstrate that the wind farm will be built safely and respectfully of the local environment. The sponsor may have to discharge and the permitting authority sign off a number of planning conditions before any works at the project site can begin.

(c) Detailed site design

Detailed site design may be required to allow the civil and electrical works to be priced in a competitive tender process. The detailed design also provides drawings and documents required by the contractors to construct the site roads, wind turbine foundations and other site infrastructure.

(d) **Production of statutory and non-legal documents**
The contractors involved in building the wind farm will be required to comply with
a number of statutory and non-statutory procedures. These may include:

- health and safety frameworks;
- environmental method statements; and
- construction method statements.

The local planning authority approves the construction method statement –
frequently used in the United Kingdom, but also applicable elsewhere – which
presents broad details of the scope, timing and methods of the work. The statement
may also impose environmental, planning or physical limitations on the works that
satisfy some of the planning conditions. The local authority may authorise variations
to the construction method statement and planning boundary in agreement with
the land owner and statutory consultees.

(e) **Preparation of essential health and safety documentation**
Health and safety regulations must be followed. In many countries this rightly
requires a focus on health and safety management before construction commences.
For example, in the United Kingdom the Construction (Design and Management)
Regulations, last revised in 2007, require that designers consider the impact that
their designs will have on health and safety during the construction and operation
of the item that they are designing. The impact of the regulations is far-reaching.
Using experienced professionals is essential to ensure that the project fully complies
with the law.

4.2   Construction phase
Once on site, the construction phase typically takes between six months and 12
months, depending on:

- the wind farm size;
- the location;
- the ground conditions; and
- the equipment lead times.

Small wind farm projects in favourable conditions can be built relatively quickly;
larger projects or those located in areas subject to adverse weather can take much
longer, especially if civil engineering works straddle the winter months, when snow
and ice can suspend works. Very large projects are likely to take longer. The 781.5MW
Roscoe Wind Complex in Texas (claimed to be the world's largest onshore wind farm
by owner E.On[21]), took just over two years to complete – a relatively short time for a
project of that scale.
The construction phase can be split into a number of distinct areas:

- on-site civil works;
- on-site electrical works;

- off-site access route alterations;
- off-site cable route works; and
- wind turbine assembly and erection.

Wind turbine erection takes place once key areas of the site have been completed and are ready to accommodate the assembled turbine.

(a)   *On-site civil works*

The on-site civil works generally extend to:

- widening/upgrading existing tracks and creating new access tracks. This may include new culverts and bridges;
- improving existing and creating new drainage regimes; and
- constructing turbine foundations and crane hardstanding pads.

The roads, foundations and crane hardstanding pads will all be built to the turbine manufacturer's requirements. The footprint of the civil works must remain within the project's consented planning and property boundary.

Intrusive geotechnical investigation may be required to confirm ground conditions at key points within the project boundary.

Some on-site civil works require licensing by environmental agencies such as the Scottish Environmental Protection Agency in Scotland. The licences must be in place prior to the works beginning.

(b)   *On-site electrical works*

On-site electrical works are usually designed around a system voltage of 38 kilovolts (kV) and below, and generally extend to:

- erecting a new on-site switchgear building containing switching facilities and transformers;
- creating a new electrical network connecting the on-site switchgear building and the wind turbines;
- earthing;
- metering;
- monitoring voltage;
- providing compensation equipment; and
- creating a new telecommunications network to support remote turbine operations and monitoring.

In larger wind farms the on-site electrical works may include a new on-site substation allowing the wind farm to connect into the high-voltage grid network (eg, 132kV, 275kV or even 400kV in the United Kingdom and similar voltages elsewhere). Where this is required, the electrical works will generally be split into medium-voltage and high-voltage works, as few contractors can undertake high-voltage works. The design of the on-site electrical infrastructure is optimised in order to strike a balance between minimising power losses and limiting capital and operating expenditure.

(c) *Off-site access route alterations*

Wind turbine components are transported to site on trucks that are considerably larger than standard transportation vehicles. Rotor blades for the larger wind turbines in use onshore, such as the Siemens SWT-2.3-101, can extend to 49 metres,[22] while nacelles are typically transported on low-loader trailers with limited ground clearance. Alterations may be required to the surrounding road infrastructure to allow transport of the wind turbine to the site. The alterations are generally limited to minor changes at junctions or tight corners, but may extend to lengths of new track and/or bridges being constructed to facilitate the abnormal geometry or load being transported.

Alterations to the public highway along the access route require approval from the relevant statutory and/or local authority roads department. They may require separate planning applications, statutory and legal agreements and notices.

(d) **Grid connection works**

In most jurisdictions the network operator will restrict the works that may be undertaken by any party other than itself in order to maintain control of timing, quality and security of supply. Grid connection works are therefore separated into:

- works that may be undertaken by the sponsor's contractors (termed 'contestable works' in the United Kingdom); and
- 'non-contestable' works that may be undertaken only by the network operator or its authorised representative.

The sponsor usually has limited influence over the cost and timing of any non-contestable works. Therefore, it is usually in its interest to undertake as much as the work as possible itself.

Grid connection may be achieved on site – for example, where an existing overhead power line of suitable capacity passes through the site (often a key criterion in selecting the wind farm location in the first instance). In such circumstances a new switching station or substation will be built within the site boundary. The connection to the network can be made with relative ease, the network operator having only limited tasks to perform once the infrastructure is built.

However, where the grid connection is outside the planning boundary of the wind farm, a new overhead or underground grid connection route is required. This new grid connection route mayb extend to several kilometres and where underground, involve significant works on and off the public highway.

Overhead connections are usually cheaper and faster to build, and easier to maintain. However, as they can be contentious with local communities in many countries, it is common for the connection to be installed underground for part or all of the route. Regardless of whether the grid connection is on site or a new overhead or underground connection is required, the electricity network operator may also require the modification or replacement of key electrical equipment on the existing grid network. This work is almost always conducted either by the network operator or through an authorised non-contestable works contractor appointed by

---

22   "The World's Biggest On-Shore Wind Farm", *Wind Today*, Second Quarter 2010, pp22-27.

the network operator. Such works may be minor (eg, changing switching parameters) or may involve significant works (eg, replacement of transformers, new electrical switchgear or monitoring equipment). The required works are specific to the choice of the electrical equipment and the wind turbine.

(e) **Wind turbine assembly and erection**

The turbine foundation and crane hardstanding pad must be complete, signed off and ready to accommodate the assembled turbine before the supplier delivers the wind turbine to the site. A specialist contractor or the turbine manufacturer itself generally undertakes the assembly and erection of the turbine. This process is usually straightforward, with weather-related delays being the main issue of concern for sponsor and contractor alike.

Once erected, the turbine is connected to the local on-site electrical and telecommunications network and is tested before being commissioned. Following commissioning and takeover, the owner is responsible for satisfactory operation of the wind farm throughout its lifetime.

5.   Operation and maintenance

Following commissioning, the wind turbine manufacturer normally provides defects, power curve and noise warranties for between two and five years, in accordance with the terms of the wind turbine supply agreement. However, in the author's experience, latterly these have tended to be for two years only.

Beyond the terms of the turbine supply agreement, warranties for parts can be extended to give a total warranty term of up to 12 years under the terms of warranty and maintenance contracts with the wind turbine suppliers. These contracts usually include an availability warranty that also covers the first years of operation in parallel with the warranties afforded under the turbine supply agreement. In the author's experience, the warranted level of availability differs from project to project, but is typically between 95% and 97% over a full year. If the guaranteed availability level is not reached, then the turbine supplier must compensate the wind farm owner.

Further, under the terms of a typical warranty and maintenance contract, the wind turbine supplier is responsible for all operations and maintenance associated with the wind turbines, including scheduled maintenance and unscheduled repairs at a fixed (or partially fixed) price. This reduces many of the key risks associated with the project for the owner, although at a premium.

A typical wind turbine maintenance crew consists of two people for every 20 to 30 machines. Most wind farms, however, especially smaller ones, are not permanently staffed and require remote monitoring only. This is carried out either by the wind turbine supplier or, increasingly, by third-party operations companies. By contrast, some large wind farms involve around-the-clock operation – E.On employs a full-time team of 10 people at the Roscoe Wind Complex in Texas.[23]

---

23   In some markets (eg, the United Kingdom, Poland and Italy), revenue can be enhanced by the careful trading of certificates via short and long-term contracts.

5.1   Return on investment

Appropriate management of operating wind farms is critical to maximising returns on investment. Wind farm owners will seek to maximise profit in the short and long term, while ensuring that they meet financing criteria – notably, servicing debt if loans have been secured to fund the project. In recent years the owners of operational wind farms have placed more attention on the farms' management as owners have realised that a passive approach does not deliver the results that they require (even though wind turbines are designed to operate with limited intervention). Maximising return on investment therefore requires a strategic approach encompassing asset management, proactive maintenance, performance assessment and strong overall management.

In order to maximise their returns on a long-term basis, developers must optimise their revenues and reduce their costs. Developers need to consider various issues when formulating their operation and maintenance strategy, notably:

- the function between power price and production. For projects that are subject to feed-in tariffs, the power price is out of the owner's control.[24] By contrast, maximising production is not. Consideration must be given to the wind regime, scheduled maintenance, turbine availability and turbine performance to maximise production;
- the optimisation of equipment, consumables, service, warranty and management. For example, proactive monitoring and condition monitoring can be used to identify probable component failures before they occur, enabling repairs to be undertaken with a minimum of downtime. Strategic purchasing of spare parts can also minimise downtime for remote projects, as well as reducing transportation costs as the need to air-freight parts to the site is minimised; and
- condition monitoring. Wind farm equipment condition and performance are the key factors affecting the asset value, which can be determined only through inspection, current data from the projects and, importantly, long-term records of servicing, faults and repairs. Increasingly, condition monitoring is used to prolong the life of key components, ranging from visual inspection of gears using endoscopes to permanently active vibration monitoring and oil particulate and temperature analysis.

5.2   Wind farm management

Whether they are managed in-house or outsourced to a wind farm management company, a number of activities must be coordinated to ensure the smooth running of a wind farm project. Where these activities are outsourced, the project owner should consider the extent to which it requires a performance regime (often called 'key performance indicators', 'service level agreements' and/or 'guaranteed standards of performance') with a liability and remedy regime for material or persistent failures.

The typical responsibilities of the wind farm operator are outlined below.

---

24   www.gigha.org.uk/windmills/TheStoryoftheWindmills.php.

(a)   **Site inspections**
Visual inspection of wind turbines should be conducted on a regular basis, usually at least every three months, although monthly visits are common.

(b)   **Site management**
This involves arranging the maintenance of the entire wind farm, including access tracks, auxiliary buildings and electrical infrastructure.

(c)   **Wind turbine repairs and maintenance**
The operations manager is responsible for ensuring that scheduled wind turbine maintenance is carried out. Typically, this should occur during periods of low wind in order to minimise production losses.

Unscheduled repairs and maintenance must be carried out as swiftly as possible to maximise availability. Maximum response times for diagnosis and repair must be stipulated in the service contract. Most trip events or breakdowns of turbines occur in strong winds, meaning that every 1% of loss in availability could mean a significantly higher loss in production. Prompt assessment, organisation and supervision of the work is therefore of utmost importance.

To reduce the potential for extended downtime due to the lead time in obtaining spares, the operations manager may be asked to advise on and maintain an inventory of spares on site, particularly if the windfarm is located in a remote area.

(d)   **Stakeholder liaison**
The operations manager is the conduit for communications between:
- wind farm owner;
- turbine manufacturer;
- maintenance contractors;
- land owners;
- neighbours;
- statutory and local authorities; and
- emergency services.

It must have adequate resources and protocols in place to ensure that each line of communication is effectively managed.

(e)   **Response to alarms**
Operations managers usually need to guarantee response times to alarms 24 hours a day, seven days a week. Depending on the alarm, a local or remote reset may be necessary or technicians may be asked to attend the site in order to diagnose the cause of the problem.

(f)   **Health and safety management**
All those involved in wind farm operation and maintenance must be fully trained. In addition, visitors to the site usually undergo some form of basic induction prior to entering the wind farm. Equipment such as safety wires for ladder climbing,

emergency lighting and fire extinguishers need to be checked frequently.

(g) *Reporting*

Reports covering the following aspects are prepared monthly and annually:

- production versus budget;
- scheduled and unscheduled servicing;
- availability;
- capacity factors;
- alarms;
- downtime;
- expenditure; and
- any other information which would be of interest to the wind farm owner, its funders and insurers, and the relevant statutory and local authorities.

(h) *End of warranty inspection*

As the warranty term comes to an end, it is advisable for the owner to commission an inspection of the wind farm in order to identify any issues which need to be addressed before the warranty expires and assess the cost of doing so that falls back onto the owner.

The operations manager arranges this inspection. It may commission specialist inspections (eg, of gearboxes) if considered necessary.

(i) *Performance analysis*

It is recommended that wind farm owners undertake periodic performance assessment of the asset, focusing on availability and power curve performance. This work may need to be subcontracted to a specialist, particularly in the case of power curve performance assessment. This is typically undertaken based on data obtained from the wind turbine control and monitoring system, unless a significant issue necessitates a specific power curve test. Problems with individual wind turbines may be identified through the course of the analysis which, when rectified can enhance the performance of the wind turbine and lead to an increase in revenue.

An availability analysis will ensure that the availability of the wind turbines reported by the warrantor is accurate and may also lead to the identification of problems which require remediation, such as repeated faults with the same root cause.

6. Decommissioning and repowering wind farms

At the end of their operational life (usually around 25 years), wind farms may be either decommissioned or repowered.

Decommissioning wind farms will result in the complete removal of the old wind turbines and most of the associated infrastructure, as the older wind turbines are very small in comparison with the units of between 800 kilowatts (kW) and 3,000kW now typically erected in Europe. In most cases, the foundation will be removed within 1 metre of the ground surface, with the remainder being left in situ. Inter-turbine cabling will be either disconnected and left buried or removed and sold as scrap.

Earthing copper will also be reclaimed and sold for its scrap value. If the wind turbines can be sold as viable generators, then they will be:

- re-erected elsewhere (see below);
- used for spare parts for other wind farms; or
- sold as scrap.

Other infrastructure, such as switchgear and transformers, may have a residual useful life – standard electrical equipment may have a useful life of up to 40 years, twice the design life of a wind turbine – and may be re-used elsewhere. If roads and hardstandings are no longer required, they will be re-instated to the natural condition of the land on which they were created.

The repowering of wind farms can involve a number of different options, including the replacement of:

- the entire wind farm infrastructure, typically with fewer, larger wind turbines being installed;
- the turbines with similarly sized units, allowing greater re-use of the wind farm infrastructure; or
- worn components forming part of the turbines, but not the whole turbines. This could allow the refurbished turbines to keep generating for another 20 years while minimising waste.

Future technical developments may also mean that the ratio between the rotor diameter and the generator capacity may change. This in turn would mean that more power could be delivered from the same wind farm footprint. Some wind farm owners may still prefer to replace their turbines with fewer, larger machines, but road infrastructure and planning limitations may preclude the installation of physically larger units in many locations. Owners of wind farms can also be expected to seek to extend the operating life of turbines through careful maintenance, component replacement and regular testing.

As mentioned above, decommissioning of a wind farm need not mean scrapping the turbines. Among the first wind farms to be repowered was Haverigg, the second wind farm to be built in the United Kingdom (in 1992) with five 225kW turbines. When the owners of the wind farm repowered it with four 850kW units in 2005, three of the original turbines were sold to the community-owned Isle of Gigha (one of the Hebridean Islands in Scotland), where they continue to operate.[25]

Few wind farms built in the United Kingdom have been decommissioned to date (the first commercial wind farm at Delabole, Cornwall, having been built only in 1991). Those that have been demolished have made way for new wind farms using the latest wind turbine designs. It is the author's view that many, if not all, of the United Kingdom's existing wind farms will be repowered rather than simply decommissioned at the end of their useful life, given:

- the shortage of suitable locations for wind energy development in the United Kingdom;

---

25   Global Wind Energy Council, "Global Wind 2009 Report", April 21 2010, pp8-9.

- the inherent problems with the planning system;
- grid capacity constraints; and
- the increasing need for renewable energy generation.

From a global perspective, while the repowering volume is modest now, the author expects the current global operating capacity of approximately 158GW of wind turbines[26] to be repowered in some form within the next 20 to 30 years. Repowering on this scale should ensure a ready supply of suitably sized wind turbines and a sizeable market for redundant equipment (recycling, use for spare parts).

7.   Conclusion

Wind farm projects are making an increasingly significant contribution to electricity generation around the world, with around158GW of wind turbines installed by the end of 2009 and global installation rates increasing. Issues associated with all aspects of wind farm development, construction and operation will remain. However, wind energy companies can be expected to continue overcoming the challenges, while the industry can be expected to continue growing. The author's predictions for wind energy into the future are set out below.

In development terms, constraints in the leading markets will be likely to increase as the impact of cumulative capacity starts to be felt. This will affect:

- permitting (cumulative impacts on wildlife habitats and landscape);
- grid (higher grid penetration leading to capacity constraints and tougher grid connection requirements); and
- land use (with the prime wind areas already developed, developers will be forced to seek out less attractive sites for new projects, with the attendant problems).

Technical developments in place and those in development should be able to address many of the technical issues, notably the development of wind turbines optimised for lower average wind speeds and improvements in grid compatibility of generators and control technology. As wind energy takes off in ever more diverse parts of the world, development in regions with significantly weaker grid networks, such as sub-Saharan Africa, will result in design issues not yet faced in Western Europe or North America.

The wind turbines themselves are changing, with more manufacturers bringing ungeared generators to the market. Enercon pioneered the design of wind turbines without the use of problematic gearboxes. Others are following its lead, with Siemens and GE both announcing the development of wind turbines without gearboxes. Beyond the traditional wind turbine manufacturing powerhouses of Europe, India and the United States, Chinese manufacturers have expanded rapidly thanks to a huge domestic market. They can be expected to start selling wind turbines overseas in the near future. Additionally, admired South Korean engineering companies Samsung, Daewoo and Hyundai have entered the market for wind turbine

---

26   www.scotland.gov.uk/News/Releases/2010/01/06141510.

manufacturing. With a limited potential domestic market, the South Korean manufacturers will have their sights set on international sales and can be expected to make inroads into several key markets early in the current decade.

Construction of onshore wind turbines is unlikely to change greatly in the near future. However, an increasing focus on projects in new markets as diverse as South Africa, Poland and Chile will bring fresh challenges to the industry.

Grid infrastructure development and reinforcement will continue to unlock the potential for wind farms in remote locations, with the introduction of various technological advances. High-voltage, direct current links are being considered to cover very long distances (eg, in Australia), while high-temperature, low-sag conductors are being used to increase the power-carrying capacity of existing overhead power lines. In Scotland, the proposed Beauly-Denny interconnector was granted consent in January 2010. When built, the interconnector could allow the connection of up to 2.5GW of wind farms in northern Scotland.

Technical solutions to minimise impacts on radar systems and seismic monitoring installations will further unlock the wind energy potential of countries like the United Kingdom where these issues are still preventing the deployment of several gigawatts of onshore and offshore projects. Of the numerous studies underway, it is to be expected that some will bear fruit. Besides, political pressure to meet the renewable energy targets deadline of 2020 set by the European Commission will increase the pressure on engineers to solve these problems.

Once operational, wind farms need to be reliable in order to ensure that maximum returns are achieved. With increasing project sizes and a realisation that more must be done to maintain their assets, wind farm owners can be expected to invest more in condition monitoring and preventative maintenance, while greater and more frequent performance assessment will result in a better understanding of performance issues and help to reduce problems caused by rogue turbines. Greater scale of operating fleets also offers scope for independent maintenance companies to build businesses based on maintenance of wind turbines during and after the warranty period.

At the end of the wind farm's operational life, provided that there remains an incentive to generate electricity from wind farms, it is to be expected that wind farms will be repowered due to the inherent advantages of installing wind capacity at locations that are well understood. Component replacements and upgrades can be expected to be commonplace as owners seek to extend the life of operating plant.

In summary, the outlook for the onshore wind industry is bright. Based on experience, future technological challenges will be overcome thanks to an innovative engineering approach, in much the same way that the industry has grown and expanded since its inception in the early 1980s.

Onshore wind projects: commercial issues

Eric McCartney
Chapin International LLC

1. Introduction

The climate change revolution has been encouraging private investment in the wind energy sector around the world. Worries about climate change, support from world governments, rising oil prices and ever-present concerns over energy security continue to fuel the industry's growth and market fundamentals.

A wide variety of other factors, ranging from cost reductions to access to new high-wind resources and better grid regulations, have also driven the sector's unprecedented growth over more than 10 years. During this period the net generating capacity of wind power has grown by a mean rate of over 30% a year, which equates to the doubling of capacity every two and a half years.[1]

Wind energy, however, has to date been developed mainly in industrialised countries that have encouraged investment through both generous support schemes and a solid legal infrastructure for renewable energy. The commercial issues surrounding the development of wind projects are thus very different in industrialised markets from the issues in lesser-developed countries. Consequently, this chapter discusses the commercial issues of developing wind projects in a generic manner. Some issues are not relevant to the development of projects in developing countries and are noted accordingly.

Although the legal framework for a wind project may vary from country to country, the basic development process is the same. The objective of this chapter, therefore, is to outline the development of a typical project, but in a manner that will ensure equity and debt financing from a maximum number of sources.

The following commercial issues must be addressed when developing a wind project:

- site evaluation;
- securing the site;
- costs of development;
- wind measurement;
- ground study;
- transportation study;
- environmental evaluation;
- permits and authorisations;
- electricity offtake;

---

1   *Wind Energy Report 2009/2010,* World Wind Energy Association.

- existing generation mix;
- availability of grid and interconnection;
- tax incentives and exemptions;
- financing sources;
- currency and foreign exchange considerations;
- carbon credits; and
- project operation and maintenance.

Given the close interaction between the commercial aspects of a wind project and technical and legal matters, the author has, for ease of use, included in this chapter some of the concepts mentioned in the technical and legal wind chapters. It is recommended that you read those chapters as well.

This chapter does not include information on offshore wind projects.

2. Site evaluation

When selecting a site for a wind project, a number of basic rules should be taken into consideration. First, the project developers should identify any legal showstoppers as early as possible, ideally before significant costs are incurred. Unfortunately, some of the best wind sites may be impossible to develop for a variety of legal reasons (eg, their proposed location is a protected environmental or archaeological area, which makes it impossible to site turbines there in accordance with local law and regulations).

Assuming that there are no legal impediments, two key factors to consider – given the potential cost implications – are:

- the proximity of a potential site to the grid; and
- the feasibility to connect the project to the grid at the proposed interconnection site.

The farther a project is located from a viable point of interconnection, the higher the costs, risk and time implications associated with both interconnection and the securing of the necessary land rights. This, together with potential line losses, may make the project less commercially viable.

Having said that, the location of a transmission line within, or adjacent to, a site does not necessarily make interconnection easier or cheaper. Wind projects typically deliver electricity at 30 kilovolts (kV). In the absence of an existing substation, step-up transformers and a switchyard will be necessary. As a result, the costs associated with interconnection can vary significantly. For example, if a substation already exists, the cost of interconnection will likely vary between €2 million and €3 million. However, in a worse-case scenario, where the project is looking to interconnect to a 750kV line and a full substation needs to be built from scratch, the costs can reach between €15 million and €20 million. (This cannot be expressed in percentage terms as this cost is fixed. By contrast, the project's building costs depend on the size of the project.) These issues are considered in more detail in section 11 below.

Consequently, the project developer should determine whether the site allows for a sufficient number of turbines to be erected to make the project commercially viable.

3.   Securing the site

Once the initial evaluation of the site is done, the project developer must secure the necessary land rights. This is typically done by entering into a lease option agreement with the owner of the land in order for the developer to lease the land long term. The developer can usually exercise the option only after the successful installation and commissioning of the wind turbines and associated plant and equipment, with the terms of the lease (including the rental provisions) coming into effect only at that point. Until then, the landowner typically provides a licence to occupy the site in order to carry out the installation and commissioning work.

In some instances, land may be available for purchase and a similar option arrangement can be entered into. In most cases, however, the acquisition of land presents a significant upfront cost and should thus be carefully considered.

The securing of the necessary land rights – whether by lease or purchase – can be a cumbersome part of the development, as the land can be owned by hundreds of parties, depending on the project's location. The process becomes even more complex where the local municipality and/or federal government of the country concerned owns some or all of the land at issue. Securing the land rights in such case may require special approvals that involve public consultations and government approvals and can be particularly time consuming and expensive.

Another key issue is that the land must be leased or purchased on a commercially viable basis and, in the case of a lease, for a period of time at least equal to the useful life of the wind turbines and associated plant and equipment (ie, 20 years).

Leases often include an option for the lessee to extend or renew for an additional period or periods. This will enable a project to operate beyond its useful life or provide the project owner with the possibility of re-powering the site.

The commercial terms of a lease are important because they directly affect the cash flow of the project. To date, the cost of leasing land has been set in a number of different ways. The most popular approaches consist of paying the landowner annually:

- a fixed amount per turbine;
- a percentage of the revenues generated on a per-turbine basis; or
- a combination of both.

For the most part, providers of finance prefer the percentage method. This method more closely aligns the cost of operation of the project with the generation of project revenues and protects cash flows during low-wind years.

However, a particularly low rent set on a fixed basis can also offer long-term benefits to the project and have a positive impact on the cash flow available to meet debt service and pay dividends.

Thus, there is no right or wrong way to negotiate a lease, other than ensuring that the cost of the land remains reasonable with regard to the long-term viability of the project. A sensible benchmark is for the land to cost between 3% and 5% of the expected project revenues.

4.   Cost of development

Generally speaking, the base costs of developing a wind project are the same for a 10 megawatt (MW) project as for a 200MW project (although they can vary significantly between countries). Depending on the country, the base costs can exceed €1 million before any construction authorisations have been obtained. These costs include those associated with:

- the installation of a wind mast on site;
- legal advice;
- wind studies;
- ground studies;
- interconnection studies;
- bird studies;
- acoustic studies;
- archaeological investigations;
- basic engineering;
- micro-siting; and
- visual studies and photo montages.

As these costs can be significant, the initial site evaluation and the overall feasibility of developing the site must be proven before detailed development work starts. For example, development in European markets can be difficult, time consuming, costly and risky. In France alone, a project must obtain 21 different authorisations.[2] The author has been involved with project sponsors that have had a success rate below 50% in developing projects in France, the United Kingdom and Italy. This means that despite the potential rewards, the wind development business is and remains risky.

Where a country does not have a legal framework in place for the development of renewable energy projects, the relevant stakeholders must rely on bespoke legislation and/or private contractual arrangements, which in themselves can increase the time, costs and risks associated with the project.

In all cases, it is highly recommended that a project sponsor have a professional team of in-house and/or independent technical, commercial and legal advisers in place to ensure that the project is developed on a commercially viable basis.

5.   Wind measurement campaign

Most sites are known to be windy before they are developed. However, this does not lessen the importance of actively pursuing a wind measurement campaign on site early in the development process. Most banks will not finance a wind project with less than 12 months of on-site wind data. The banks may even request data for longer periods if there is no reference wind data to which the on-site wind data can be correlated and to enable a wind consultant to develop accurate long-term projections.

---

2   *Agence de l'Environnement et de la Maîtrise de l'Energie* (www2.ademe.fr).

Generally, it is best to install a wind mast on site as soon as it has been determined that the site can prospectively be developed. It is recommended that the wind mast be at least 50 metres high. The closer the wind mast is to the prospective hub height of the turbine, the less extrapolation will need to be done and the more accurate the projected wind statistics will be. At the time of writing, the estimated cost of a wind mast is between €35,000 and €80,000, depending on the height and type of mast (ie, tubular or lattice).

Once 12 months' worth of data has been collected, a reputable wind consultant should be engaged to study the data and provide the developer with a detailed long-term wind projection that includes:

- the projected equivalent operating hours of the project; and
- the total project gigawatt-hours.

This is typically done on the basis of probable energy yields known as the P50, the P75 and the P90. 'P' stands for 'probability of exceedence'. The P90, which is the most conservative number (there is a 90% chance that the project will produce at least a set number of hours in a given year), is the one typically used by banks to size the debt for the project. Developers, on the other hand, generally use the P50 when analysing their equity return and internal rate of return.

The wind, together with the price of electricity per kilowatt-hour, has to be sufficient using the P90 to produce the cash flow needed to cover debt service at levels which are, in the author's experience, on average at least 1.25x. However, this depends on a number of factors, including:

- the banks' willingness to lend to the project; and
- constraints which may be dictated by country economics and require a project to be structured more conservatively to qualify for debt financing.

6. Transportation study

Wind turbines and their related equipment – including the tubular towers and, most importantly, the turbine blades – can be very heavy, cumbersome and bulky to transport. Although it can be valuable to commission a professional transportation study where terrains are complex, this is not a requirement. That said, local authorities generally require that a transportation plan be completed and certain permits obtained before allowing a special escorted convoy. Wind turbines are usually transported at night to cause the least nuisance to local traffic.

The weight, height and length of the equipment must be taken into consideration when evaluating its transportation. In most instances, the key leg of transportation is done by sea. It is therefore important that the developer evaluates the port of delivery. The port must be a deep-water port and have the necessary equipment and lay-down area for the off loading of equipment.

This may seem obvious to some, but there have been instances of wind developers obtaining all necessary permits, only to discover that the port could not receive the type of vessel necessary to deliver the equipment. A transportation evaluation can be easily done with the assistance of a turbine supplier and should not be overlooked.

As noted, a transport plan from the port to the site should be done and may be required as part of the permitting process. A simple evaluation determines whether there are any low-lying bridges or particularly sharp bends in the road. In many instances the road will need to be modified or upgraded for the transport of equipment. Mountain passes and circuitous roads present certain challenges for the delivery of equipment which, although more complicated and time consuming, are generally not insurmountable.

Unfortunately, in the author's experience, the costs of the infrastructure upgrades linked to the transport of turbines are seldom shared with third parties. Most projects see this cost as a form of investment in social responsibility in the country or community where the project is located. Such upgrades normally represent a very small percentage of the total costs of the project, depending on the size of the project.

7.   Permits and authorisations

Each country has its own procedure and requirements when it comes to permits and authorisations. The developer should review the permitting process of the host country. This involves engaging with the relevant authorities to identify the permits and authorisations required and the process, cost and time involved in obtaining such permits and authorisations. The developer should then ensure that the project development programme takes this into account – for instance, which permits and authorisations must be obtained before a relevant activity is undertaken or, in some cases, before another permit or authorisation can be obtained.

In developed countries, where strict procedures and regulations are in place, the most important element of the permitting procedure is generally the environmental impact assessment, as outlined in more detail below. There is typically a list of rules and procedures to follow. A project usually needs to elicit considerable social and political support to get through the permitting process. As noted above, in some countries only 50% of the projects which begin this process obtain all the authorisations necessary to build a wind farm.

The process is even more challenging in countries where no wind projects have been built or no laws or regulations exist for the development and construction of wind projects.

The best policy in these cases is to follow procedures similar to those followed in countries where such a framework exists. This is because these regulations are typically more stringent than those in the country where the project is being developed. Obtaining the local assistance of individuals with knowledge of local politics and existing procedures can be invaluable.

In many countries, the generation of electricity is a prohibited activity unless the relevant party:

- holds a permit or authorisation (eg, a generation licence); or
- is exempt from the prohibition (eg, small-scale generators).

Besides the construction authorisation and operating permits, a number of various other permits may be required. As noted above, it is recommended that the developer thoroughly evaluate the permit process at the earliest stage possible.

Failure to obtain a permit or authorisation could:

- considerably delay a project;
- increase the costs; and/or
- result in civil or criminal action should the developer undertake an activity without such permit or authorisation.

8. Environmental evaluation

An environmental impact assessment is an essential part of the development process. It generally consists of an in-depth study of the effects of a wind farm on the local flora and fauna. This includes the study of noise, shadow and visual impact on the surrounding communities. It is a good idea to conduct a detail environmental assessment of a site, regardless of the project's location and whether such an assessment is required by law.

A demonstration that the project will not have a negative impact on the environment is a useful tool to gain the necessary political and local support.

In most developed countries, laws and procedures dictate the manner in which an environmental study is conducted. Where such laws and regulations do not exist, it does not follow that an environmental impact assessment need not be carried out – and to international standards too. Indeed, whenever project sponsors seek external financing, an environmental study is required and this must be done in accordance with the Equator Principles (a voluntary standard for managing the social and environmental risk in project financing). Based on the author's experience, time and money are often saved if the study is prepared in whole or in part by a local professional environmental consultant who is licensed by the government and familiar with the Equator Principles.

9. Ground study

An in-depth study of the soil and subsoil conditions, together with bedrock, is sometimes required as part of the environmental impact assessment. Whether it is required by the assessment or not, a ground study will need to be conducted before construction begins and the final engineer and micro-siting of the park can be completed.

The results of a ground study can have an important impact on the project from a commercial perspective. The results of a ground study confirm that the foundations will be stable and support a wind turbine; they may require that turbines be eliminated or displaced because of unstable ground conditions. Although most problems encountered during a ground study can be resolved, it is an analysis of the time, cost and/or risk involved that drives the decision of whether to resolve the problem. For example, piling may be an adequate solution to unstable ground conditions, but it can add significant costs to the foundations and the overall construction budget – particularly for large projects with 20 or more turbines.

10. Existing generation mix

Most developers take for granted that a wind project can be developed in any country, as long as there is wind and the project can be connected to the grid. This is for the most part true. However, the size of the total market for electricity generation and the mix of the generation will put constraints on the amount of wind energy which can be commercially integrated into a system without having an adverse affect on the stability of the transmission grid.

With a diversified and stable generation capacity of more than 115,000MW and a stable, integrated transmission grid, a country like France has room to integrate a large wind energy development programme.[3] By contrast, in a country like Senegal, which has an installed generation capacity of less than 600MW[4] with a correspondingly sized transmission and distribution network, it would be much more challenging to implement a large wind development programme. Of course, these two examples are at the extremes of the spectrum. Romania lies in between, with an estimated wind energy potential of between 10,000MW and 15,000MW.[5] Despite this amazing wind resource, it is estimated that unless further investment is made to the grid, less than 1,500MW of wind generation capacity can be integrated into the grid without affecting the overall stability of the system.

The developer should therefore undertake the necessary research and analysis of the country's generation capacity, generation mix and demand before investing in the development of a wind project. Research would include obtaining information about existing wind projects and those currently under development or in planning. Such information is generally easily accessible, as most energy statistics are usually of public record. If the conclusion is that the total capacity of a country is technically constrained (eg, the country can support no more than 500MW of new capacity), and other projects are already under development or in planning to satisfy that requirement, then it would be unwise to invest significant sums of money to develop additional projects.

11.   Grid availability and interconnection

As mentioned above, the proximity of a prospective project to a grid connection does not necessarily mean that the project will be able to connect to the grid at that point, or at all. Wind energy is intermittent energy, and transmission grids and the existing installed generation need to react appropriately to ensure the safety, reliability and stability of the grid. A grid may require substantial upgrades in order to integrate a wind project. Although most developed countries have modern transmission networks and adequate base-load generation that can react to the intermittency of a wind project, this may not always be the case. In the United States, for example, all power projects, including wind projects, must conduct a detailed system impact study.

In smaller, lesser-developed countries, grids may need significant upgrades. In addition, the installed generation may put limitations on the amount of wind power that can be integrated into a grid without affecting its stability (see section 10 above). Accordingly, developers should conduct an evaluation of the grid and determine any

---

3   *Wind Energy Report 2009/2010*, World Wind Energy Association.
4   Senelec Annual Report 2008.
5   *Wind Energy Report 2009/2010*, World Wind Energy Association.

restrictions, limitations or upgrades that may be required early in the development process. The results of this evaluation could very well save time, money and energy in the development of a project.

12. Offtake arrangements

A critical element of a viable wind farm development is the offtake arrangements for the electricity and ancillary services (eg, green certificates). The form of the arrangements primarily depends on:

- the nature of the offtaker (eg, a public or private sector entity);
- the structure of the electricity market (eg, a single offtaker, a pool or a bilateral trading market); and
- the existence of a specific tariff and/or green certificate regime for wind projects.

The most attractive investment proposal for both project sponsors and their financiers consists of a long-term power purchase agreement with a robust offtaker that will pay premium feed-in tariffs and that is also supported by a green certificate regime.

In many countries, this position may not be achievable. This may be because the feed-in-tariff and/or green certificate regimes are not yet fully developed or market-tested. This is the case in Ukraine, which in September 2008 approved legislation to encourage the development of renewable energy including a feed-in tariff.[6] As a result, several thousand megawatts of renewables generating capacity are in the pipeline in Ukraine, but at the time of writing no project had been built or financed, and no precedent had been set for the basis of a power purchase agreement.

This is evidence that the structure of the offtake arrangements is the crux to the successful development of a renewable energy strategy in any country – even those where legislation has been implemented in support of such strategies.

For projects in an emerging economy, special attention must be paid to the power purchase agreement and other documentation. This is because specific points to be included in a power purchase agreement are germane to developing markets and make it acceptable to financing parties. These points include:

- *force majeure*;
- exchange rates;
- pricing structures;
- government guarantees;
- payment guarantees;
- termination clauses; and
- contract arbitrage.

This is in addition to the structural issues typical to all power purchase agreements for wind power, such as the concept of take or pay and the availability of the grid.

---

6   *Wind Energy Report 2009/2010*, World Wind Energy Association.

13. **Tax incentives and exemptions**

Various tax incentives and exemptions (eg, potential tax holidays on corporate profits, exemption of value-added tax and import duties, and enhanced capital allowances) are often available for renewable energy projects. They should be carefully investigated and evaluated when developing a renewable energy project in any jurisdiction. Withholding taxes on dividends and interest paid to banks also need to be evaluated and structured in order to minimise the impact on the cash flows of the project.

If a country has no legislation containing such tax incentives and exemptions, the developer should try to obtain specific tax incentives and exemptions for the project from the relevant government. This is because if a project is only marginally profitable without such tax incentives or exemptions, it may be difficult to qualify for financing or attract interest from investors. In addition, the ability to claim such tax incentives and exemptions is sometimes conditional upon the developer satisfying certain conditions (eg, having obtained all permits or authorisations and construction having begun).

14. **Grants and subsidies**

Certain countries will provide subsidies or grants to finance the development and/or construction of a project. Other grants or subsidies may be available from development institutions to incentivise the development of renewable energy projects in emerging economies. The process to qualify for these grants, especially those available for development, is cumbersome and may require the completion of an in-depth feasibility study of the project.

15. **Financing sources**

Investors and financing sources are usually easy to access in developed markets such as Europe and the United States. However, investment and financing need to be tailored to the local marketplace. Investor return requirements and financing structure and costs vary from country to country. It has been the author's experience that investor returns in Europe and other developed markets are between 10% and 15% after tax. By contrast, investors in emerging markets enjoy returns in excess of 20%, to compensate for the additional risks associated with doing business in those countries. More importantly, while there are numerous investors interested in renewable projects in developed markets, investors in emerging markets are few and far between.

It is therefore crucial that a project be developed in a thorough manner and at a highly professional level in order to qualify for the maximum number of financing options.

16. **Currency and foreign exchange considerations**

Generally, long-term power purchase agreements, feed-in tariffs and/or green certificate programmes are not financeable unless they are in a hard currency (eg, dollar or euro). When the power purchase agreements, feed-in tariffs and/or green certificate programmes are in a local currency, they must at the very least be carefully structured and indexed to a hard currency in order to mitigate any foreign exchange risk for the project.

Other foreign exchange concerns are associated with:

- transfer risk;
- expatriation of dividends; and
- the availability of foreign currency in a host country.

These risks must be considered, evaluated and mitigated to the necessary extent to ensure that financing can be obtained for the project.

17.   Carbon credits

No writing on the development of wind projects would be complete without some mention of the trading of carbon credits. The concept of a mandatory global carbon market was first introduced by the Kyoto Protocol. The Kyoto Protocol is an international agreement adopted in Kyoto on December 11 1997[7] and linked to the United Nations Framework Convention on Climate Change adopted in New York on May 9 1992.[8] Under the protocol, 37 industrialised countries and the European Union legally committed to reduce gas emissions by an average of 5.2% below 1990 levels by 2012. (For further information, please read section 3.5(b) in the chapter of this book titled "Renewable energy support mechanisms: an overview".)

The trading of carbon credits now represents an integral part of the development of projects in emerging markets. However, they should in no way be construed as the primary source of financing of a wind project. That said, any wind project that is developed in a qualifying country cannot afford to ignore this key feature of the market.

In terms of development, the process of validation of a project by the United Nations Framework Convention on Climate Change is complex and, in the author's experience, can take over two years. It is therefore essential that the process be started well before the project is expected to reach commercial operation. A number of organisations have emerged which specialise in advising on and completing the process on behalf of a project for a percentage of certified emission reductions produced by the project.

18.   Project operation and maintenance

If a developer has made it to this point, congratulations are in order. A solid operation and maintenance plan for a project is essential to the long-term viability of a project. Keeping wind turbines properly maintained is key to maximising revenues. If a wind turbine is not available and ready to produce energy when the wind is blowing, substantial potential revenues can be lost.

In addition to the turbine suppliers, a number of independent companies can

---

7   For the full text of the protocol, see www.unfccc.int/resource/docs/convkp/kpeng.pdf. For the status of ratification,   see www.unfccc.int/files/kyoto_protocol/status_of_ratification/application/pdf/kp_ratification _20091203.pdf.

8   For the full text of the convention, see unfccc.int/resource/docs/convkp/conveng.pdf. For the status of ratification,   see   www.unfccc.int/files/essential_background/convention/status_of_ratification/ application/ pdf/unfccc_ratification_20091016.pdf.

operate and maintain a project. An operation and maintenance programme should nonetheless be evaluated on a case-by-case basis and tailored to the size and location of the project, together with any specific requirements of sponsors and investors.

Spare parts, maintenance crews and cranes will be readily available in countries where a significant number of wind projects have been developed. In some instances, a maintenance team may not be needed on site. Larger projects or those located in remote areas, however, do not share in this luxury: they will have special needs and be exposed to risks that require an operation and maintenance contract. In these cases, the usual policy is to have on site a qualified team which will manage the project 24 hours a day, 365 days a year.

It has been the author's experience that despite the variety of options available, many sponsors choose the manufacturer of the turbines to operate and maintain the wind farm, at least for the first five years of a project.

Naturally, the location of a wind park in an emerging market, where the project may be the only wind farm in the country, presents challenges that must be considered and evaluated. Access to spare parts, cranes and qualified staff are important considerations; remote operation and maintenance is not an option. Consequently, a site must be large enough (ie, 20 turbines or more) to justify a full-time maintenance team on-site.

Although most scheduled and preventative maintenance will not require a crane, the availability of a crane needs to be evaluated. Cranes for wind projects typically need to be at least 150 metres high to access the turbines, which are installed at increasingly lofty heights. That said, cranes are usually needed only to replace blades and overhaul large pieces of equipment (eg, replacement of gear boxes and/or generators).

The average useful life of a wind turbine is 20 years; the blades are designed to have the same useful life, as long as they are not exposed to extreme conditions. However, it is not uncommon for lightning to damage blades to the extent that a turbine is put out of commission for an extended period of time and a crane is required to conduct the necessary repairs.

Insurance can generally cover equipment and repair costs, as well as potential lost revenues while the turbine is out of commission.

19. Conclusion

As this chapter has shown, the development process of a wind project is a complicated one that requires a significant investment in time and money. In order to succeed, special attention must be paid to the idiosyncrasies of the development process. One small mistake or misjudgement could result in a significant increase in the cost of a project or a total failure to obtain the necessary authorisations and permits to build and operate the project. Conversely, the successful development of a project can create significant value and upside.

Legal framework for wind projects: a US perspective

Gregory Blasi
Joshua Hill
Jay Matson
Loeb & Loeb LLP

1.   Introduction

After lagging behind Germany and other European countries for 20 years, the United States is now the leader in installed wind power capacity.[1] In 2009 the United States had 35,064 megawatts (MW) of installed wind power capacity, accounting for 22.1% of the world's total.[2] Despite the impact of the global credit crisis on project finance, the US wind power sector has continued to experience rapid advances. For example, the United States installed 9,996MW in new wind power capacity in 2009, while Europe installed 10,526MW and Asia 15,442MW during the same period.[3]

The increased viability of wind power projects in the United States is due in the main part to:

- the country's wind energy resources;
- technological advances in wind turbine design and construction; and
- shifts in the economics of wind power generation.

In addition, energy policy and financial and tax incentives have spurred the development of wind power projects. All factors contributed to the US Department of Energy's finding that wind power could provide 20% of the country's electricity needs by 2030.[4]

1.1   The US wind energy capacity

In 2009 wind power provided 39% of all new electricity generating capacity installed in the United States.[5] However, a 2010 study from the US Department of Energy shows that the potential for wind energy in most US states goes well beyond the current installed wind power capacity. This study was conducted as a collaborative project between the Department of Energy's National Renewable Energy Laboratory and AWS Truewind LLC. It found that onshore US wind resources could generate around 37 million gigawatts annually.[6] This is nine times greater than the overall

---

1   *Global Wind Report 2009*, Global Wind Energy Council, p10.
2   *Id.*
3   *Id*, p9.
4   US Department of Energy, Energy Efficiency and Renewable Energy, "20% Wind Energy by 2030", July 2008 (www1.eere.energy.gov/windandhydro/pdfs/41869.pdf).
5   American Wind Energy Association, "US Wind Industry Annual Market Report Year Ending 2009".
6   National Renewable Energy Laboratory and AWS Truewind LLC Resource Study (www.windpoweringamerica.gov/wind_maps.asp).

current US electricity consumption.[7]

However, the expansion of new wind power generating capacity requires expanding the US transmission grid to allow access to superior wind resource regions. The 10 windiest US states are Texas, Kansas, Montana, Nebraska, South Dakota, North Dakota, Iowa, Wyoming, Oklahoma and New Mexico.[8] Many of the windiest parts of these states have no immediate or convenient access to transmission lines. By contrast, one the benefits of some potential US offshore sites is their proximity to transmission lines and large population centres.

The National Renewable Energy Laboratory's wind resource study referenced above does not take into account the development of offshore sites. As of May 2010, the United States had no operating offshore (oceanic) wind power generating projects. By way of comparison, there are 830 installed and operational offshore wind turbines across 39 wind farms in nine European countries, totalling 2,063MW.[9] However, the US offshore installed and operational capacity is bound to increase soon as many states in the northeastern United States enjoy a large and strong offshore wind resource, with very limited opportunities to develop wind power generating sites on land. Along these lines, in April 2010 US Secretary of the Interior Ken Salazar announced his approval of the first US offshore wind project to be constructed on Horseshoe Shoal in Nantucket Sound, off the coast of Massachusetts.[10]

1.2   Energy policy and financial incentives

Renewable portfolio standards (RPS) are state rules and regulations that require retail electricity suppliers to obtain certain portions or percentages of their supply from renewable energy sources by certain dates. As of May 2010 over 30 US states, plus the District of Columbia, had binding or voluntary RPS programmes in place.[11] Under most RPS programmes, retail electricity suppliers may purchase tradable credits that represent an equivalent amount of renewable energy production in lieu of purchasing power supply directly from a renewable energy producer. For each unit of power that an eligible producer (eg, a wind power generator) generates, a credit is issued. These credits (sometimes also referred to as 'renewable energy certificates') can then be sold either together with the power generated or separately to retail electricity suppliers.

These credits generally increase flexibility for retail electricity suppliers and reduce the cost of compliance with the purchase requirements. The RPS have greatly increased the demand for wind-generated power. Prospects for growth of wind power are good as additional states adopt standards and the states that already have such programmes in

---

7   American Wind Energy Association (www.awea.org/newsroom/releases/02-18-10_US_Wind_Resource_
        Larger.html).
8   National   Renewable   Energy   Laboratory   and   AWS   Truewind   LLC   Resource   Study
        (www.windpoweringamerica.gov/wind_maps.asp).
9   European Wind Energy Association (www.ewea.org/).
10   US Department of the Interior, press release (www.doi.gov/news/doinews/Secretary-Salazar-Announces-
        Approval-of-Cape-Wind-Energy-Project-on-Outer-Continental-Shelf-off-Massachusetts.cfm).
11   See Renewable Energy World (www.renewableenergyworld.com/rea/news/article/2010/05/where-the-
        wind-blows-and-sun-shines) and the US Department of Energy Database of State Incentives for
        Renewables and Efficiency (www.dsireusa.org/).

place increase their requirements for renewable energy resource supplies.

The US Congress has authorised various tax and financial incentives to encourage the development of renewable energy projects. The choice of which financial incentive programmes are best suited for a particular project can be a complicated matter and requires the consideration of a variety of logistical, financial and legal factors. In addition, most federal financial incentive programmes are authorised by Congress and are subject to funding and programme availability. New financial incentives at the federal level are being considered; developers of wind energy projects should be sure to consult with legal and financial advisers with regard to which, if any, programmes are available and suitable for the project being considered. Besides, many states have their own financial incentive programmes. While the federal programmes mentioned above are discussed in more detail below, developers of wind energy projects should also consider what state financial incentives may be available.

1.3 Technological advances

The technology for wind power generation has advanced rapidly over the past 10 years. For example, in 2006 wind turbines were on average 11% larger than in 2005.[12] However, research, design and development in this area have focused on more than achieving gains in wind turbine size. Advances in the design and composition of blades, rotors and tower structures have been combined with improvements in drivetrain design (gearbox, generator and power conversion) and control systems. As more technology innovations are put into use, the cost and performance challenges of wind power generation are becoming less daunting. Thus, in 2009 work was completed on the Horse Hollow project – the world's largest wind farm – near Roscoe, Texas.[13] This project is owned by Florida Power & Light through its subsidiary Nextera Energy. It has an installed capacity of 735.5MW.[14] With the assistance of technological innovations, large-scale projects such as Horse Hollow should become more common in the near future.

The recent advances in wind power generation provide a positive example of the benefits of combining active government policy, financial incentive programmes and technological momentum to create formidable increases in wind power generation capacity.

Although the potential for continued growth in installed wind power capacity in the United States is great, there are some challenges. Some of these relate to:

- upgrading, improving and expanding the US transmission system to enable wind power generated in outlying areas or offshore to be delivered to urban areas in need of supply;
- lowering the capital costs (through government-sponsored policies and programmes such as production tax credits, investment tax credits and cash grants); and

---

12 US Department of Energy, Energy Efficiency and Renewable Energy, "20% Wind Energy by 2030", July 2008 (http://www1.eere.energy.gov/windandhydro/pdfs/41869.pdf), p5.

13 www.windpowerworks.net/12_case_studies/horse_hollow_united_states.html.

14 www.nexteraenergyresources.com/content/where/portfolio/pdf/horsehollow.pdf.

- resolving siting and permitting issues (eg, regulatory approvals, wildlife concerns, tribal and historic lands restrictions and zoning issues).

2.   Siting and permitting wind energy projects

Although wind power is one of the cleanest and least environmentally harmful sources of energy, it does not enjoy preferential permitting treatment in the United States. In addition, while wind energy projects generally garner widespread public support, siting such projects can raise significant local concerns. However, almost all wind power development projects begin with a determination of whether a potential site has enough wind to support a wind farm.

Many project developers identify sites through the use of wind resource maps. These maps typically show the predicted mean annual wind speeds at certain heights above the ground. Areas with annual average wind speeds of around 6.5 metres per second and greater at an 80-metre height are generally considered to have suitable wind resource for wind power development.[15]

Wind resource at specific sites should be measured for at least a year to understand how winds vary over the four seasons. The wind resource assessment process can be complicated and time-consuming. Large-scale projects typically incorporate numerous wind measurement locations and varying heights, wind direction, shear and terrain locations. New wind modelling technology also provides wind power developers with a crucial understanding of complex terrain, turbulence and wake factors.[16]

While choosing a site with a good wind resource is fundamental to a project's chance of success, a variety of other factors can determine whether the project is viable. The siting of wind power projects involves a varied set of laws. The wind energy siting process tends to be a highly localised one, with each project's permitting requirements being specific to:

- the local permitting jurisdiction (municipality, county, state or federal);
- the characteristics of the site; and
- the technical scope of the project.

State and local authorities have different responsibilities with regard to permitting or approving wind power projects.

The diversity and complexity of local and state permitting and approval requirements in the United States cannot be overstated. In many states the primary site permitting and approval jurisdiction rests at the local level where projects do not trigger federal authority (eg, being located on land that is under the jurisdiction of a federal authority). This means that locally elected and appointed officials and local agency employees will make siting, permitting and approval decisions. For projects located in rural areas, the site permitting and approval jurisdiction is usually with the relevant county. Projects located in urban areas may be subject to county, city or another municipal jurisdiction. Local authorities are usually charged with the

---

15   National Renewable Energy Laboratory and AWS Truewind LLC Resource Study (www.windpoweringamerica.gov/wind_maps.asp).

16   US Department of Energy, Energy Efficiency and Renewable Energy (www.windpoweringamerica.gov/ne_siting.asp#wind_resource).

responsibility of ensuring that projects comply with zoning, development and building standards. These authorities may also be concerned with environmental standards, water quality and community development aspects.

Wind power projects under local siting and permitting jurisdiction must typically go through a conditional use permit process in which an application for a project is submitted by a developer to a local commission or other agency responsible for approving and issuing permits and approvals for site development. Developers may be required to work with a variety of local authorities, such as zoning boards and county commissioners. A local agency will approve and issue a permit for site development that is usually a 'conditional use permit'. Conditional use permits typically require that:

- an applicant show that the potential project will be compatible with surrounding and adjacent land uses; and
- the development project have met with the approval of state or federal agencies (eg, those concerned with environmental matters, water quality issues or impacts on wildlife).

Since the local site permitting and approval process requires the involvement of elected and appointed officials, securing community support is crucial. Projects that do not enjoy the support of a large component of the citizens and/or have garnered special resistance (no matter how small) can face significant struggles in the permitting and approval process. For large projects that require access and road easements, the support of affected land owners is also necessary. In any event, whether a potential wind power project is large or small, community support is crucial to the local permitting and approval process. Developers often engage in public information campaigns to ensure that the local public understands the employment benefits, the positive impacts on tax revenue and the importance of clean energy. Even with the recent surge in US support for clean energy projects, developers should seek to increase their chances of successfully completing the local permitting and approval process by consulting closely with local agencies, commissions, communities and special interest groups.

States may also control the site permitting and approval process through special decision-making bodies or through rules and regulations that establish jurisdiction over the permitting and approval of wind power projects. A minority of states have energy siting councils or boards that have jurisdiction over the siting of wind power projects, usually in conjunction with the state utility commission (the commission that is charged with regulating certain aspects of the state electricity industry, such as the supply of electricity service).[17] Wind power developers should be sure to understand whether, and the extent to which, their potential project is affected by state permitting and approval jurisdiction.

Wind energy projects may require the participation of a federal agency if the potential site is located on (or requires a right of way through) land administered by a federal agency, such as:

---

17   Wind Power Siting Regulations and Wildlife Guidelines in the United States (www.batsandwind.org/pdf/afwastsitsum.pdf).

- the US Bureau of Land Management;
- the US Forest Service;
- the US Corps of Engineers;
- the Minerals Management Service; or
- any other agency that manages land for the US federal government.[18]

Before any federal agency can approve a project on federal land, the agency must consider the environmental impact of its decision. This environmental review requirement is set forth in a federal procedural law known as the National Environmental Policy Act.[19] The act is significant because, depending on the level of review required, the determination can take more than a year. Developers should carefully consider the potential impact of the act's requirements on their project and their development timeline.

As with the structure of state and local permitting and approval jurisdiction, federal jurisdiction can be equally complicated. Depending on the location of the development site, various federal agencies may participate in the permitting and approval process. The following is an overview of the federal agencies that have mandates related to wind power development projects:[20]

- The Federal Aviation Administration conducts aeronautical studies on all structures taller than 60 metres for possible interference with air traffic and military radar. Developers are usually required to submit an application to this agency in connection with each wind turbine.
- The Bureau of Land Management has a Final Programmatic Environmental Impact Statement on Wind Energy Development on BLM-Administered Lands in the Western United States. This impact statement sets forth best management practices, sets standard requirements for projects and lays the environmental groundwork to expedite permitting for projects located on Bureau of Land Management land in 11 western states.
- The US Army Corps of Engineers is responsible for issuing development permits for projects that will affect wetlands.
- The US Fish & Wildlife Service has the authority to prosecute violations of certain rules and regulations protecting migratory birds.
- The Minerals Management Service oversees the permitting of offshore wind power projects on the outer continental shelf.
- The US Forest Service approves projects subject to its management jurisdiction.

As illustrated by the foregoing discussion, the federal government's role in the

---

18   See the following websites for more information regarding these US federal agencies:
      • www.blm.gov/wo/st/en.html (US Bureau of Land Management);
      • www.fs.fed.us/ (US Forest Service);
      • www.usace.army.mil/Pages/default.aspx (US Corps of Engineers); and
      • www.mms.gov/ (Minerals Management Service).
19   42 USC §4321 *et seq.*
20   US Department of Energy, Energy Efficiency and Renewable Energy, "20% Wind Energy by 2030", July 2008 (http://www1.eere.energy.gov/windandhydro/pdfs/41869.pdf), p5.

site permitting and approval process varies greatly depending upon project circumstances, particularly site location. Whether or not a potential development site is located on (or requires a right of way through) federal land, careful consideration must be given to the jurisdiction of federal agencies.

One of the largest and most controversial issues in the siting and approval of wind power projects is the impact on wildlife. Birds are particularly affected by wind projects. US federal law provides some protection to a large portion of avian species with habitats in the United States. The most important federal avian laws are the Endangered Species Act, the Migratory Bird Treaty Act and the Bald Eagle Protection Act.[21] These laws generally prohibit injuring or killing certain protected classes of bird. The concerns of permitting agencies regarding the impact on birds (and bats) can require that developers and/or permitting agencies conduct long and costly studies to determine the answers to wildlife impact questions.

While the Endangered Species Act provides a means for developers to obtain a permit for the incidental injury or killing of protected animals, the Migratory Bird Treaty Act and the Bald Eagle Protection Act do not provide for a similar permitting method. To avoid violating the wildlife protection laws, most developers give careful consideration to project layout and design aspects so as to minimise the injury or killing of protected wildlife. Developers also gather data to determine the project's harmful affects on wildlife. Typically, at least one year of data regarding wildlife use of the potential site is compiled to set a baseline for the project. After construction of the project, developers continue to gather wildlife impact data and compare new data to the baseline survey. Post-construction data on wildlife injuries and fatalities assist developers with making operational adjustments to minimise exposure to liability under applicable laws.

Many developers also attempt to reduce liability under wildlife protection laws and gain favour with permitting agencies by adopting wildlife protection plans. Such plans set out the policies that a developer will follow to mitigate the impact on wildlife. However, the adoption of a wildlife impact plan will not provide legal protection in the event that a protected species is killed or injured as a result of the project.

When making siting decisions, developers should determine whether the site area is inhabited by a protected species (at both state and federal levels). There are databases available that show species native to particular geographic areas. Such databases can assist in preliminary siting decisions with regard to the potential of project approval and liability under wildlife protection laws.[22]

There are also databases available that describe historical and cultural sites. Many Native American tribes, as well as local and state governments, have designated certain cultural resources as protected (eg, physical sites listed on the National Register of Places, certain fossils and human remains). It is important to evaluate thoroughly potential development sites to understand the impact of development

---

21   The Endangered Species Act can be found at 16 USC §1531 *et seq*; the Migratory Bird Treaty Act can be found at 16 USC 703 *et seq*; and the Bald Eagle Protection Act can be found at 16 USC §668 *et seq*.

22   For example, see the US Fish & Wildlife Service Endangered Species Program Species Report (www.fws.gov/ecos/ajax/tess_public/pub/stateOccurrence.jsp).

on such cultural resources. During the construction phase, mitigation measures may be necessary to avoid harming or disturbing these protected resources.

The consideration that must be given to the potential impacts on Native American cultural resources is important because Native Americans own large areas of land with valuable wind resources. These resources have led many developers to attempt to site wind power projects on or near Native American lands. However, siting wind power projects on Native American lands comes with unique jurisdictional challenges. Both federal and tribal law governs easements, leases and other uses of Native American lands. Some state laws may also apply.

The US government owns a large portion of Native American land in trust for Native Americans; federal Indian law governs the lease of such land. Under federal Indian law, tribes can lease tribal land for up to 25 years and may include an option to extend the lease for another 25 years.[23] In addition, some specific tribes may lease tribal land for up to 99 years. Most tribal land leases for land owned in trust by the US government must be approved by the Bureau of Indian Affairs and comply with the requirements set out by the National Environmental Policy Act.

Adding to the complexity, a few tribes can lease land owned in trust by the federal government for up to 75 years without approval by the Bureau of Indian Affairs. Other leasing restrictions apply to tribal corporations operating under charters issued by the secretary of the interior.[24]

Developers must work closely with Native American tribal governments during the site evaluation process. Tribal governments may agree to adopt new laws or amend existing laws to facilitate wind power projects. Of course, a careful review of the laws affecting development on the particular Native American lands under consideration should be conducted at the beginning of the project.

This overview of the siting issues with which developers must contend in developing US wind power projects is necessarily brief. The foregoing discussion aims to highlight:

• the complexity of the site permitting and approval processes; and
• the extent to which multiple jurisdictional authorities may be involved.

In addition, this is a rapidly evolving area of the law. Careful planning and extensive due diligence are required in the early phases of any project to ensure that the permitting and approval process is thoroughly understood.

3. Power purchase agreements

3.1 Characteristics

Generating units, including those powered by wind, will produce capacity, energy and certain ancillary services for sale in the marketplace. These products are often sold to an individual customer under a bilateral contract commonly referred to as a 'power purchase agreement'. The agreement describes the terms and conditions of

---

23   25 USC §415.
24   25 USC §477.

the sale. Although the Federal Energy Regulatory Commission (FERC) regulates wholesale sales of electricity, it has not adopted a standardised agreement that must be used. So-called 'master agreements' that can be used as a model exist, such as those developed by the Edison Electric Institute and the International Swaps and Derivatives Association, but the specific terms of any particular sale are driven by negotiation between the buyer and the seller.

While the power purchase agreement is important in setting the terms of sale, it often plays a second, but equally important, function as a basis for securing financing on reasonable terms or, in some cases, obtaining any financing at all. The agreement serves this function because it reduces the financial risk associated with a project. The agreement's length and the amount of energy being sold as a proportion of the project's total output each can have an effect on the level of risk mitigation. An agreement for the sale of 100% of the output of a wind project over 20 years, for example, will make a project more attractive than an agreement for 50% of the project's output over five years.

Among many others issues, the following are important aspects of a power purchase agreement.

Term: While the term is a fundamental provision in any contract, the start date is of particular importance for wind-related power purchase agreements. Since agreements are often necessary for a project to secure financing, they may be executed months or more before the wind project is scheduled to begin commercial operation. Accordingly, the agreement should explicitly state:

- when the sale/purchase begins – usually, on the commercial operation date; and
- that the duration of the sale is calculated based on that date, not the date that the agreement is executed.

Delivery point: The parties should determine the precise point at which the purchaser takes the power. The delivery point can occur anywhere. It may be:

- the project's collector station;
- the point at which the project interconnects to the integrated transmission system; or
- the point at which the customer takes the power off the system for use.

In the last case, however, the generator would bear the burden of arranging for and transporting the power from its project to the delivery point. With this will come some risks of service disruption that many wind projects do not wish to take.

Performance incentives and/or guarantees: Wind is an intermittent resource. How much or when the wind will blow and, consequently, how much power will be produced varies over time. Often, the purchaser pays for the power that is actually delivered. However, the wind project can be encouraged to meet certain output levels by including incentives or guarantees in the power purchase agreement. For example, the seller may be required to pay any incremental costs that the purchaser incurs to obtain power that the wind project was unable to provide above certain levels. Alternatively, the seller may be required to pay liquidated damages if it fails to provide a certain amount of energy over a specified period.

Environmental attributes: Renewable energy certificates are an important characteristic of renewable power, including wind. The certificates capture the value associated with the green attributes of renewable power. The power purchase agreement should clearly state whether the purchaser is buying power alone or the power and the certificates.

3.2 FERC oversight

Independent power producers selling power at wholesale, including wind projects, are normally subject to significant oversight by FERC. However, there are several ways to reduce that regulatory burden.

Traditionally, a generator must file for and obtain permission to sell its power from FERC.[25] Each power purchase agreement is filed with FERC for approval. However, as competitive markets grew more robust, the regulatory lag associated with obtaining approval from FERC proved an impediment to power sales. In lieu of filing each agreement, FERC now allows a seller to obtain 'market-based rate authority' under Section 205 of the Federal Power Act.[26] To obtain such authority, the seller must demonstrate that it and its affiliates have no horizontal or vertical market power. If authority is granted, FERC will continue to oversee the generator's sales, but rather than file each agreement, the seller will submit a summary of its sales each quarter in a filing known as an 'electronic quarterly report'.

If even these reports are more regulatory oversight than desired, a generator may seek an exemption known as a 'qualifying facility', as established in the Public Utility Regulatory Policy Act. Two categories of unit can qualify for the exemption:

- facilities whose primary energy source is renewable (eg, wind) with a capacity that is not greater than 80MW;[27] and
- co-generation facilities, which produce both electricity and thermal energy (eg, heat or steam).[28]

Whether the unit is a small power production facility or a co-generation facility, the generator must also:

- file a self-certification with FERC (which does not require any further action on the part of FERC); or
- submit an application to FERC and receive an order granting certification.[29]

If a generator is designated as a qualifying facility, it is exempt from FERC jurisdiction; it can sell power without filing the power purchase agreements or

---

25   16 USC §824a(b).
26   16 USC §824d.
27   . There is a limited exception that provides no maximum generating capacity for solar, wind or waste facilities that have been self-certified as a qualifying facility by December 31 1994 (18 CFR §292.204 (2009)).
28   18 CFR §292.203 (2009).

obtaining market-based rate authority.

Under the 2005 Public Utility Holding Company Act of 2005, FERC also oversees utility books and records. Generators that qualify as exempt wholesale generators do not have to grant FERC access to their books and records (though they still must seek rate approval from FERC by filing any power purchase agreement or obtaining market-based rate authority).[30] An 'exempt wholesale generator' is an entity that engages exclusively in the business of owning and/or operating an eligible facility and selling electric energy at wholesale.[31] An 'eligible facility' is a facility used:

- to generate energy for wholesale sales only; or
- for the generation of electric energy and leased to one or more public utility companies, provided that such lease is treated as a sale of electric energy at wholesale.[32]

This means that to qualify as an exempt wholesale generator, a generating facility must not sell any of its power at retail. In addition, the generator must file a notice of self-certification with FERC confirming that it satisfies the definition of an 'exempt wholesale generator', as described above.[33]

4.   Transmission and interconnection

Unless the wind project is interconnected to the transmission system, its power cannot be delivered to the customer. Traditionally, a wind project approaches its local utility and requests an interconnection. Utilities subject to the jurisdiction of FERC have adopted standardised procedures and agreements to interconnect generators. One set of procedures (the Large Generator Interconnection Procedures) applies to so-called 'large generators' (ie, above 20MW).[34] Another set of procedures (the Small Generator Interconnection Procedures), some of which are more streamlined, apply to generators that are 20MW and smaller.[35] Utilities can change the terms and conditions of the standardised procedures and agreements for large or small generator interconnection agreements, but only if the changes are consistent with, or superior to, the standardised provisions. Pursuant to the standard interconnection procedures, the generator (known as the 'interconnection customer') will:

---

29   See 18 CFR §292.207 (2009). On March 19 2010 FERC adopted a rule under which generators with less than 1MW of capacity are not required to file an application or self-certification in order to obtain the qualifying facility status. See Order 732, Revisions to Form, Procedures and Criteria for Certification of Qualifying Facility Status for a Small Power Production or Cogeneration Facility (Docket RM09-23-000, 130 FERC ¶ 61,214 at P 34 (2010)). Such generators are automatically deemed qualifying facilities, as long as they satisfy the other requirements.

30   18 CFR §366.3(a)(2) (2009).

31   42 USC §16451 (citing 15 USC §79z-5a) (2009).

32   Id.

33   18 CFR §366.7 (2009).

34   Standardization of Generator Interconnection Agreements and Procedures, Order 2003, FERC Stats & Regs P 31,146 (2003); order on rehearing, Order 2003-A, FERC Stats & Regs P 31,160; order on rehearing, Order 2003-B, FERC Stats & Regs P 31,171 (2004); order on rehearing, Order 2003-C, FERC Stats & Regs P 31,190 (2005); affirmed sub nomine National Association of Regulatory Utility Commissioners v Federal Energy Regulatory Commission, 475 F3d 1277, 374 US App DC 406 (DC Cir 2007).

35   Standardization of Small Generator Interconnection Agreements and Procedures, Order 2006, FERC Stats & Regs P 31,180; order on rehearing, Order 2006-A, FERC Stats & Reg P 31,196 (2005); order granting clarification, Order 2006-B, FERC Stats & Regs P 31,221 (2006).

- be responsible for the costs of the interconnection facilities; and
- have to pay upfront for any upgrades to the network that may be required due to its interconnection, though the payment is recoverable in the form of credit for transmission service over the utilities system.[36]

While most interconnections occur at a point on the transmission system, there are instances in which smaller units will interconnect with lower-voltage facilities that are considered distribution in nature. Under the Federal Power Act, local distribution service is not subject to the jurisdiction of FERC. Nonetheless, providing service over distribution facilities for the purpose of selling power at wholesale is subject to FERC's jurisdiction. Thus, at the time the interconnection is being made, if the generator intends to sell power wholesale, FERC's interconnection rules will apply. Otherwise, the interconnection process is subject to state regulatory requirements. When interconnecting at the distribution level, the developer must be careful to identify and address early on any issues related to delivering the power over the distribution system, delivery being separate and distinct from interconnection. Depending on the terms of the applicable tariff, which would be subject to state regulation, delivery charges (in addition to any interconnection charges) may apply.

Seeking interconnection service from a local utility can be undesirable because of the time it takes for the utility to complete the interconnection or its cost. A wind developer may be able to build interconnection facilities faster and/or for lower cost than the utility.[37] That approach is permitted. However, when a wind developer builds its own interconnection facilities, it runs the risk that it will be subject to regulation as a transmission entity by FERC.[38] Under the Federal Power Act, a third party may seek to take service over the wind developer's interconnection line. If this occurs, the wind developer becomes a transmission provider regulated by FERC, subject to an array of requirements, including, but not necessarily limited to, the adoption of an open access transmission tariff – the baseline tariff developed by FERC for the provision of transmission service.[39] In addition, the wind project may be required to serve others with capacity on the interconnection line that the wind developer wanted to use itself. FERC has allowed wind projects to retain sufficient

---

36   Order 2003 at PP 21-22 ("[T]he Commission has found that it is just and reasonable for the Interconnection Customer to pay for Interconnection Facilities but not for Network Upgrades....Interconnection Facilities will be paid for solely by the Interconnection Customer, and while Network Upgrades will be funded initially by the interconnection Customer...the Interconnection Customer would then be entitled to a cash equivalent refund (i.e. credit) equal to the total amount paid for the Network Upgrades....").

37   On June 17 2010 FERC issued a notice of proposed rulemaking, initiating a proceeding to revisit the rules on development of new transmission facilities and allocation of the associated costs. See Transmission Planning and Cost Allocation by Transmission Owning and Operating Public Utilities, 131 FERC ¶ 61,253 (2010).

38   The regulatory requirements associated with transmission would be in addition to the regulatory requirements applicable in connection with wholesale sales.

39   See *Aero Energy LLC*, 115 FERC ¶ 61,128 (2006); order granting modification, 116 FERC ¶ 61,149 (2006); final order directing interconnection and transmission service, 118 FERC ¶ 61,204 (2007); order denying rehearing, 120 FERC ¶ 61,188 (2007). The issues associated with a generator becoming a transmission provider regulated by the commission are commonly referred to as 'Sagebrush issues', named after the facilities at issue in the *Aero* decisions.

capacity to serve their own needs, but only to the extent that there are definitive plans to use the capacity.[40] Retaining capacity for some unidentifiable or theoretical future use is considered hoarding and will not be permitted by the commission.

5. Tax and financial incentives

The wind industry in the United States has developed, to a great extent, in response to tax incentives authorised under the 1986 Internal Revenue Code, as amended. Traditionally, the United States has used the production tax credit[41] and accelerated depreciation[42] to promote the wind industry. Over the years, Congress has allowed the production tax credit to lapse from time to time. In those years in which the production tax credit lapsed, the amount of new wind installations plummeted. For example, Congress last allowed the production tax credit to lapse at the end of 2003. In 2003 there were close to 2,000MW of new installed wind generation. In 2004, after the lapse of the production tax credit, new installations of wind generation dropped to approximately 500MW.[43] The production tax credit was reinstated in October 2004 and currently remains in effect through 2012.

Clearly, tax incentives are vital to the US wind industry. However, as a result of the financial crisis, the traditional production tax credit incentive is no longer the primary incentive available to wind developers. Prior to the onset of the financial crisis, a so-called 'tax equity financing model' developed which allowed wind developers to monetise tax credits that they could not use. The primary investors in the tax equity market were large financial institutions. The financial crisis dramatically lowered the number of institutions investing in tax equity and therefore reduced the amount of tax equity available to finance new wind projects. The 2009 American Recovery and Reinvestment Act of 2009[44] provided a new approach to encouraging investment in wind projects, which is discussed below.

There are three tax incentives available to wind developers. The first is the traditional production tax credit. Section 45 of the Internal Revenue Code provides for a credit against federal income tax based on the amount of electricity:

- produced by the taxpayer from various renewable sources, including wind, at a qualified facility in the United States; and
- sold by the taxpayer to unrelated persons during the tax year.

The production tax credit is available during the 10-year period that begins on the date the facility is originally placed in service.

There are many technical rules under the Internal Revenue Code that must be met in order for a taxpayer to claim the production tax credit. However, the production tax credit is available only with respect to qualified wind facilities that

---

40   *Milford Wind Corridor LLC*, 129 FERC ¶ 61,149 (2009).
41   IRC §45. The energy tax credit for qualified small wind energy property (ie, wind turbines with a nameplate capacity of no more than 100 kilowatts) is beyond the scope of this chapter. See IRC §48(a), (c)(4).
42   IRC §168.
43   American Wind Energy Association, "Wind Power Outlook 2005" (www.awea.org/pubs/documents/Outlook%202005.pdf).
44   PL 111-5, enacted on February 17 2009.

are originally placed in service before January 1 2013.[45] While the production tax credit has been extended previously and there are predictions that it will be extended further, under current law the production tax credit for wind generation is not available for electricity produced at wind facilities that are placed in service after 2012.

The second incentive was created by the American Recovery and Reinvestment Act. The act allows a taxpayer to elect to claim an investment tax credit equal to 30% of qualifying project costs in lieu of the production tax credit.[46] The investment tax credit is a one-time credit for the year in which the property is placed in service, based on the amount invested in the facility rather than on the amount of electricity that is produced and sold.

Similar to the production tax credit, there are many technical rules to be followed in order for a taxpayer to claim this credit. Like the production tax credit, this investment tax credit is available only to the owner of a qualified wind facility that is placed in service prior to January 1 2013. In addition, the amount of the investment tax credit is subject to recapture (or payback) if, during the five-year period after the wind facility is placed in service:

- the taxpayer sells or otherwise disposes of the property; or
- the property ceases to be investment tax credit property with respect to the taxpayer.

The amount of recapture depends on when, during the five-year period, the taxpayer no longer qualifies for the investment tax credit.[47] The longer that the taxpayer owns and uses the property for its qualifying use, the lower the recapture.

The third incentive was also created by the American Recovery and Reinvestment Act. Pursuant to Section 1603 of the act, wind developers may apply for cash grants from the Treasury equal to 30% of the eligible costs of the project in lieu of claiming any investment tax credit that otherwise would be available with respect to such property. This cash grant programme is made available to:

- eligible taxpayers with respect to wind projects that are placed in service by the end of 2010; and
- for those projects for which construction has commenced by the end of 2010 and are placed in service by January 1 2013.

Like the investment tax credit, all or a portion of the cash grant is subject to repayment of the tax payment if the taxpayer disposes of property or ceases to use it in a qualified manner within five years.[48] Although the amount of the cash grant is equal to the investment tax credit, the taxpayer does not have to have a tax liability to benefit from the cash grant. As a result, the cash grant programme has been an

---

45   IRC §45(d)(1).
46   IRC §48(a)(5), added by ARRA §1102(a).
47   IRC §50(a)(1).
48   ARRA §1603(f); US Treasury Department, Payments for Specified Energy Property in Lieu of Tax Credits under the American Recovery and Reinvestment Act, Program Guidance (July 2009/Revised March 2010) (Treasury Program Guidance), at 18-20.

enormous success: approximately $3.1 billion has already been paid to renewable energy developers.

Wind developers need to be aware of several issues in connection with the cash grant programme. Construction commences when physical work of a significant nature begins, whether by self-construction or construction by a third party pursuant to a binding written contract.[49] Generally, any physical work on the specified energy property is sufficient.[50] A safe harbour is provided which allows an applicant for the cash grant to treat physical work of a significant nature as having begun when the applicant pays or incurs more than 5% of the total cost of the project. This includes the cost of all units if multiple wind turbine structures are considered part of a single project.[51] Even if the taxpayer elects to treat a wind farm that consists of multiple turbines, towers and supporting pads as a single unit, the grant is available only with respect to the qualifying property that is placed in service before the credit termination date.[52]

Tax-exempt owners of wind projects are not eligible for the cash grant.[53] However, in a significant liberalisation of the rules, the Frequently Asked Questions and Answers of the American Recovery and Reinvestment Act indicate that the cash grant is available to the owner of the property leased to a tax-exempt entity.[54] Thus, a developer can lease a wind project to a government agency without losing the cash grant, as long as the lease qualifies as a true lease and is not considered a disguised sale. The drawback to this approach, however, is that assets that are leased to a tax-exempt entity are subject to a slower depreciation schedule than would otherwise be available.[55]

An additional benefit available to developers of wind projects is accelerated depreciation under the modified accelerated cost recovery system. Under that system, qualifying components of a wind project are generally eligible for depreciation deductions over a five-year period based on the double declining balance method (switching to straight-line when it produces a larger deduction), subject to the applicable half-year or mid-quarter convention.[56] In the case of wind property, the taxpayer's depreciation base is first reduced by one-half of the investment tax credit or cash grant allowed with respect to the property.[57]

At the time of writing, the cash grant and the energy tax credit programmes were, as stated previously, scheduled to expire. There are several legislative initiatives to

---

49   Treasury Program Guidance, at 6; US Treasury Department, Payments for Specified Energy Property in Lieu of Tax Credits under the American Recovery and Reinvestment Act, Frequently Asked Questions and Answers Regarding Beginning of Construction, Questions and Answers 1-14.
50   Frequently Asked Questions and Answers Regarding Beginning of Construction, Questions and Answers 4.
51   Treasury Program Guidance, at 7-8; Frequently Asked Questions and Answers Regarding Beginning of Construction, Questions and Answers 15-22.
52   Treasury Program Guidance, at 8.
53   ARRA §1603(g); Treasury Program Guidance, at 4; US Treasury Department, Payments for Specified Energy Property in Lieu of Tax Credits under the American Recovery and Reinvestment Act, Frequently Asked Questions and Answers, Questions and Answers 17.
54   American Recovery and Reinvestment Act, Frequently Asked Questions, Questions and Answers 21.
55   IRC §168(g)(1)(B), (2), (h).
56   IRC §168(b), (d), (e)(3)(B)(vi).
57   IRC §§50(c), 48(d)(3)(B).

extend one or both programmes for eligible projects.[58] There is also a proposal to supplement the cash grant programme with a refundable investment tax credit programme under which a taxpayer could elect to receive either the cash grant or a refund of the tax credit.[59] Any refund payable would be paid only after the tax return was filed for the year in which the project was placed in service. Wind developers are anxiously waiting to see whether either programme will be extended and, if so, whether they will be modified.

6.   Conclusion

The renewable sector, including wind, will continue to grow in the United States over the coming years. There will be many opportunities to develop projects around the country. This chapter introduces some of the issues that a developer will confront, including:

- siting, permitting and environmental restrictions;
- the role of power purchase agreements (especially as it relates to financing);
- the need to interconnect to the grid (which can have unintended regulatory consequences); and
- the need to time project development in order to capture tax benefits that are often in flux.

Navigating the process is complex and can be difficult. By focusing on each issue early in the process, developers increase the odds of successfully completing their projects on time and within budget.

---

58   See, for example, the Renewable Energy Incentive Act proposed by Senators Dianne Feinstein (D-Cal) and Jeff Merkley (D-Ore); the 2009 Clean Renewable Energy Advancement Tax Extension Jobs Act proposed by Senator Chuck Grassley (R-Iowa); and the Renewable Energy Expansion Act proposed by Congressman Earl Blumenauer (D-Ore).

59   See the Renewable Energy Expansion Act proposed by Congressman Earl Blumenauer (D-Ore).

Increasing local support for wind turbines: a Danish perspective

Anne Kirkegaard
Jan-Erik Svensson
Gorrissen Federspiel

1.   Introduction

The attention given to climate change and how to mitigate it is overwhelming. The world's poorer countries are craving immediate action from industrialised nations. One of the most pressing concerns is how to promote environmentally friendly energy with the support of local communities without jeopardising the security of the energy supply and risking a negative impact on the economic development of a still-fragile world economy.

Denmark aims to become one of the world's forerunners in the use of renewable energy sources. This objective stems from:

- the desire to reduce the country's dependence on the limited supply of domestic and foreign fossil fuels;
- the necessity of ensuring security of supply; and
- a binding obligation to reduce emissions of carbon dioxide.

In February 2009 the Danish Parliament adopted an act to further the use of renewable energy sources,[1] as part of a February 21 2008 political agreement on future Danish energy policy.

The purpose of the Renewable Energies Act is to promote the production of energy from renewable sources, taking into account climatic, environmental and economic considerations, in order to:

- reduce dependence on fossil fuels;
- ensure security of supply; and
- reduce emissions of carbon dioxide and other greenhouse gases.

The act, which entered into force in early 2009, increases the subsidy for new land-based wind turbines to Dkr0.25 (around €0.0333) per kilowatt-hour for 22,000 full-load hours. In addition, the act sets out measures to promote support from local communities for the establishment of new wind turbines.

Denmark's target is for renewable energy sources to cover 20% of the national energy consumption by 2011 and 30% by 2020. In comparison, the target proportion of renewable energy in the total energy consumption across the entire

---

1   Act 1392/2009 on the furthering of renewable energy sources, with subsequent amendments.

European Union is 20% by 2020.[2] In 2009 the proportion of renewable energy sources in the total EU consumption was about 8.5%.

Although Denmark uses a comparatively high proportion of renewable energy sources, its 2008 installed wind turbine capacity of approximately 3.16 gigawatts (GW) does not seem large.[3] Nevertheless, this accounts for 24% of the country's electricity generation capacity.[4] In 2008 around 19% of Danish electricity consumption was based on wind-generated electricity – the highest percentage in the world.

The Renewable Energies Act introduces, among other things, four measures to promote the establishment of wind turbines by increasing local communities' interest in such projects. The previous Danish legislation required that hearings be held before new wind turbine projects could be approved. These hearings often resulted in fierce local opposition. As local politicians enjoyed some discretion under the law with regard to the granting of approvals for new wind projects, opposition to these projects had – unsurprisingly – a significant impact on the number of approvals granted. The new act still provides for a hearing period. However, it is expected that the act's provisions will deflate local opposition.

This chapter provides an overview of the legal and economic initiatives undertaken by the Danish legislature in order to strengthen local interest in and support for the erection of further land-based wind turbines.

2. Wind turbines

In Denmark, the first serially manufactured wind turbines were erected in the 1970s. These turbines had an original generation capacity of 22 kilowatts (KW) and were gradually upgraded during the 1980s to 55KW, 75KW and 95KW.[5] The erection of new wind turbines enjoyed a high degree of local approval because most were owned by local guilds or associations.

This local support started to deteriorate in the 1990s, when most of the new turbines were owned by individuals and energy companies, including commercial wind energy entities, rather than guilds and associations.

Today's commercial wind turbines are very different from the first turbines erected. Both technical and design aspects have changed greatly. A modern wind turbine consists of up to 10,000 different components, sweeps an area of over 9,000 square metres[6] and reaches a total height of over 150 metres.[7] The surface of the blades is treated to avoid reflection from the sun and to minimise the noise from the rotating blades. In addition, the emission of noise is reduced by, for example, insulating the engine house and suspending the generator and gears in rubber elements.

---

2   The Renewable Energies Directive (2009/28/EC). Annexe of the directive identifies the targets for each EU member state.
3   In addition, the world's largest offshore wind park, Horns Rev II (with a capacity of 209 megawatts (MW)), was connected to the grid in the autumn of 2009. Another large offshore wind park, Rødsand II (207MW), is expected to be connected to the grid and fully operational in late 2010.
4   See Danish Energy Agency (www.ens.dk).
5   See Danish Energy Agency.
6   Vestas Wind System's V112-3MW turbine sweeps an area of 9,852 square metres (www.vestas.com).
7   Siemens Wind Power 3.6MW wind turbine and Vestas Wind System V112-3.0MW wind turbine.

The increasing costs of erecting wind turbines have led to a shift in ownership to limited liability commercial entities whose shares are held by individual investors or energy companies.

3. Promoting new turbines

As mentioned above, the Danish authorities consider the approval and involvement of local communities as conditions to the continued installation of wind turbines. Accordingly, the authorities have implemented various measures.

The most interesting measures set out by the Renewable Energies Act are:

- the right for neighbours to compensation for the decrease in value of real estate due to the erection of wind turbines close by;
- the obligation on the owner of the wind turbine to offer for sale a share of at least 20% of the newly erected wind turbine;
- the establishment of a green subsidy scheme to foster local initiatives that aim to encourage the acceptance of new wind turbines; and
- the establishment of a guarantee fund from which local wind turbine associations may obtain guarantees for loans taken out to finance preliminary studies of ground and suitable sites of erection, among other things.

These measures, which are likely to be introduced in other countries and thus be of interest to a wide audience, are considered in more detail below.

The Renewable Energies Act applies to the Danish land and territorial waters, as well as the exclusive economic zone (the sea area up to 200 miles from the coast). Accordingly, the act also covers offshore wind turbines. However, the act is usually not applicable to offshore wind turbines established following a public procurement procedure. The fixed sale price of the electricity generated by these turbines is part of the competitive price offered by the bidders in the public procurement procedure.[8]

Under the act, the Danish municipalities must reserve areas for the erection of wind turbines with an aggregate generation capacity of 75 megawatts (MW) in both 2010 and 2011. The municipalities may include provisions regarding the maximum number of turbines, their size and the distance between them. It remains to be seen where these areas will be reserved and whether this move will be met by local opposition.

3.1 Compensation

Pursuant to general Danish law, a land owner may claim compensation for nuisance to an estate which exceeds changes that should be accepted as being inherent to social evolution. The provisions of the Renewable Energies Act depart from this rule in two important ways. Under the act:

- full loss will be compensated; and
- the compensation applies only to the erection of new wind turbines, as further defined below.

---

8   See, for example, the Danish Energy Agency's notice of invitation to tender for Anholt Havmøllepark (a concession to establish a 400MW offshore wind park).

This means that a land owner whose estate neighbours installations that use other sources of renewable energy (eg, biogas) will be entitled to claim compensation under the general Danish law, but not under the Renewable Energies Act.

In addition, only the following people are entitled to claim compensation under the Renewable Energies Act:

- neighbours of new land wind turbines exceeding 25 metres measured from the ground to the tip of a wing at its highest position; and
- neighbours of new offshore wind turbines that are not subject to public procurement.

A special regulation applies to the erection of offshore wind turbines for testing purposes.[9]

The land owners covered by the term 'neighbours' are land owners able to prove that the value of their real estate will decrease by over 1% because of the erection of new wind turbines. In the event that a land owner has contributed to the decrease in value by, for instance, selling or renting a piece of land to the developer of the wind turbines in question, the compensation payable may be reduced in part or in full.

A prospective developer must, with reasonable notice, summon a public meeting to provide information materials required and approved by the transmission system operator Energinet.dk.[10] The material must, among other things, list all the properties situated within a distance equal to six times the height of each proposed wind turbine. Any land owner wishing to claim compensation for an alleged decrease in the value of its real estate must file a claim with Energinet.dk within four weeks of the public meeting.[11] Land owners whose property is situated outside the prescribed distance for compensation must pay a fee when filing a claim.

The act foresees that the amount of compensation to be paid may be:

- reached by agreement between the owner of the wind turbines and the land owner claiming compensation; or
- determined by an appraisal committee consisting of a judge and a real estate agent.

There are no legal requirements as to the contents of an agreement by which the owner of the wind turbines and the land owner reach common ground regarding the amount of compensation payable. However, it is assumed that this agreement will set out the amount payable, the time of and conditions precedent to payment, as well as a cut-off for further claims by the land owner.

In the second instance, the committee establishes the compensation amount, if any, by taking into account the local circumstances. These circumstances include:

- the nature of the area in question;
- real estate prices in the area;
- any other wind turbines or technical installations in the area;

---

9   Section 22(a) of the Renewable Energies Act.
10 Energinet.dk provides secretarial assistance to the appraisal committee established by the act.
11 Only if special circumstances apply may claims for compensation be filed up to six weeks after the relevant wind turbine has been connected to the electricity grid.

- the distance from the wind turbines to be erected to the real estate in question; and
- the height and inconveniences to be expected from the wind turbines.

At the time of writing the appraisal committee had settled four cases on compensation pertaining to the erection of wind turbines in the municipality of Vesthimmerland.[12] The committee rejected the claim for compensation in two cases.

In the two cases won by the claimants, the compensation amounted to Dkr75,000 (around €10,000) and Dkr200,000 (€26,670) respectively. The committee's written decisions explain that:

- the wind turbine closest to the relevant real estates would be around 800 metres and 600 metres away respectively; and
- the noise would be below the applicable thresholds.

In addition, the committee stated in the first case that the top part of the closest wind turbine would be visible from the claimant's terrace and parts of the garden, whereas in the second case the wind turbines would be "very visible" from the primary outdoor space, including the terrace.

In the second case the committee accepted that in the summer evenings a shade effect could be present. The shade effect was estimated to appear for five hours each day (22 hours over 104 days in a worst-case scenario). The value of the real estate was stated as Dkr1.2 million (€160,000). Compensation of around €26,670 is therefore quite substantial, taking into account the limited inconveniences described in the committee's decision.

These decisions triggered some debate regarding the level of future compensations. The appraisal committee found it necessary to state that the compensation granted to some extent mirrored compensation of Dkr185,000 (€24,670) and Dkr150,000 (€20,000) agreed between the wind turbine developer and two land owners affected by the same project. Hence, no indication as to the level of future compensation can be deduced from the compensation granted by the committee in the Vesthimmerland cases.

Subsequently, no fewer than 19 land owners sought compensation for the erection of wind turbines in Lem Kær, in the municipality of Ringkøbing-Skjern. Seven claimants were granted compensation. One was denied, even though the appraisal committee estimated a decrease in the value of the real estate of Dkr50,000 (€6,670). However, this did not exceed 1% of the value of the real estate and, hence, no compensation was payable under the Renewable Energies Act.

The seven successful claimants received amounts ranging from Dkr25,000 (€3,330) to Dkr200,000 (€26,670). The reasoning in each case was similar, with the decisions referring to the distance to the wind turbine closest to the real estate at issue, as well as the effect on the views from the primary living quarters in and around the real estate. The decisions also refer to other installations such as existing

---

12   Three of the decisions of the appraisal committee are dated July 9 2009, with the most recent refusal dated August 5 2009.

wind turbines, electricity poles, slurry tanks and silos. The lowest compensation was granted in a case where the closest wind turbine would be placed approximately 670 metres behind the agricultural buildings on the estate, but where the wind turbines would be highly visible from a part of the garden. In the case in which the highest compensation was granted, the wind turbine closest to the real estate would be placed 1.5 kilometres from the house, but in an area of high recreational value with a direct view from all the main parts of the house. The wind turbines are expected to have a particularly dominant position in the landscape.

It appears that in the latter case, the expected visual impact of the wind turbines from the real estate greatly influenced the committee's decision to grant compensation. However, the compensation amounted to 4% of the real estate's estimated value of Dkr5 million, which does not seem generous.

The latest case published on the Energinet.dk website and decided on February 4 2010 relates to the erection of wind turbines in Vester Barde, also in the municipality of Ringkøbing-Skjern. Although the appraisal committee found a decrease in the value of the claimant's property of Dkr150,000 (€20,000), it rejected the claim for compensation on the grounds that the decrease in value did not exceed 1% of the value of the real estate.

Land owners may appeal the decisions of the appraisal committee before the courts. In contrast, any compensation agreed between a land owner and a developer is final. By statute, such compensation may be neither assessed by the appraisal committee nor brought before the courts.

While the possibility of obtaining compensation is one of the measures introduced to promote local support for the erection of wind turbines, the prospect of having to pay high compensation may discourage developers from initiating wind projects in the first place. The fact that compensation is payable only after the wind turbine has been erected may be a mitigating circumstance. However, this depends on how the word 'erected' will be interpreted. For instance, what would happen in the event that a wind turbine were only partly erected? The building site would, in many cases, result in a disfigurement of the landscape similar to or even worse than an operating wind turbine. (It appears from the wording of the law, however, that any compensation falls due only upon completion of the wind turbine.)

It follows from the above that the appraisal committee should lay out clear guidelines identifying the nature and level of expected inconveniences required in order to obtain compensation. Until such guidelines are established, it is unlikely that any land owner will agree on a compensation amount directly with a developer, as such compensation would be final. This will especially be the case if more land owners have filed claims for compensation.

3.2   Obligation to sell

The Renewable Energies Act introduced another measure to promote local support for new wind turbines: it is the obligation on developers of land-based wind turbines exceeding 25 metres and offshore wind turbines built without procurement to offer a minimum of 20% of the ownership shares in the wind turbine(s) to persons above the age of 18 years domiciled, at the time of the offer for purchase, within:

- 4.5 kilometres of the site of erection; or
- the municipality where the wind turbine will be erected.

The first group of persons has priority.

Wind turbines offered for sale must be operated by a separate legal entity. There are no requirements as to the form of the owner's liability (ie, it is no requirement that the operating entity be a limited liability company).

The revenue from selling shares to the public shall cover a proportionate part of the costs of the project. Thus, the value of each share held by the developer must correspond to the value of each share offered to the public. The pricing of the shares shall generally be based on a generation of 1,000 kilowatt-hours per ownership share,[13] although a dispensation for a different calculation method may be obtained.

The developer must prepare a prospectus describing the project and containing information about, among other things:
- the articles of association of the operating company;
- a detailed investment and operating budget;
- financing of the project;
- the liability per ownership share;
- the number and price of the offered ownership shares; and
- the deadlines and conditions that an offer to purchase must comply with.

Further, the prospectus must contain a statement from a state-authorised public accountant confirming that, among other things:
- the obligation to operate as a separate legal entity is fulfilled;
- the pricing of the shares is correct; and
- the information provided on the economic circumstances is correct.

The prospectus must also contain information on the liability per share and be approved by Energinet.dk.

The offer of ownership shares must be publicly announced in the local press within the area in which the eligible persons are domiciled. The announcement must provide a deadline of at least four weeks to forward an offer of purchase. Any ownership shares not sold through the public offer may be freely disposed of. The equality of ownership shares is secured by the requirement that no shares sold through the public offer may enjoy less favourable terms than other ownership shares. Also, the Renewable Energies Act bans a majority owner in a public limited liability operating company from redeeming by force shares sold through a public offer.

Thus, the act imposes several measures to ensure that local participation in new wind turbines is real and permanent, and on equal terms to the majority owner. This is underlined by the fact that the articles of an operating entity without limited liability must state the extent to which the entity may incur debts.

---

13 The purpose of the pricing method is to guarantee the transparency and comparability of the terms. This method appears to have been employed by the wind turbine industry for several years.

It remains to be seen whether sufficiently attractive sites for new wind turbines may be identified and, hence, whether these new turbines may be partly offered for sale to local people. The Renewable Energies Act contains measures to encourage local guilds to examine local opportunities to erect new wind turbines.

3.3   Guarantee fund

The initial costs of identifying relevant sites may be substantial. Without a firm guarantee that the invested funds will result in the erection of wind turbines, such costs may deter the initiation of new projects. In order to stimulate local interest in erecting new wind turbines, the Renewable Energies Act has established a guarantee fund with a capital of Dkr10 million (€1.33 million).

Local guilds or associations with no fewer than 10 participants may obtain a guarantee for loans taken out to finance, among other things:

- a preliminary study for sites;
- technical and economic analyses; and
- applications to public authorities.

Energinet.dk administers the guarantee fund. Applicants must fulfil a number of conditions at the time of both the application for the guarantee and the provision of the guarantee. In addition to the requirement that the guilds have at least 10 participants, the conditions include that:

- the majority of the participants be domiciled either in the municipality in which the wind turbines will be erected or no more than 4.5 kilometres from the proposed site of erection;
- the locals have decisive influence in the guild; and
- the guild's completion of the project be viewed as realistic.

The guarantee is capped at Dkr500,000 (€66,670) per project[14] and covers the principal of a loan taken out under market terms. The guarantee will lapse when the wind turbines are connected to the grid or no later than three months after the wings have been mounted on the wind turbine.

In the event that a project to erect wind turbines is not completed, the guarantee provided is not reclaimed. However, this may be the case if an incomplete project is transferred to third parties in whole or in part.

3.4   Green subsidy scheme

The final measure introduced by the Renewable Energies Act in order to promote local support for new wind turbines is the so-called 'green subsidy scheme for the enhancement of local scenic and recreational values'. The subsidy is based on 22,000 full-load hours for each wind turbine in the municipality connected to the grid after February 21 2008. Each kilowatt-hour is subsidised by Dkr0.004 (€0.00053).

---

14   Due to the limited amount available for each project, it is expected that the guarantee scheme may be administered within the *de minimis* threshold for state aid determined by the EU *De Minimis* Regulation (1998/2006).

Electricity consumers will finance the subsidy by payment of a surcharge on the electricity price.[15]

Energinet.dk administers the green subsidy scheme. The municipalities may, by filing an application to Energinet.dk, apply for money from the subsidy scheme to cover the costs of:

- enhancing the scenic or recreational values in the municipality (eg, creating a nature path); or
- developing cultural or informational activities in local associations with the purpose of increasing the acceptance of using renewable energy sources by residents of the municipality.

Payment of the subsidy requires an application from the municipality documenting the costs incurred.

The four measures introduced by the Renewable Energies Act clearly point at achieving local support for the erection of new wind turbines. Some 5,000 wind turbines are connected to the Danish grid (land-based and offshore), with a generating capacity of around 3.4GW.[16] As mentioned in section 1, this means that a little over 19% of Danish electricity is generated by wind turbines.[17]

Considering the comparatively small size of Denmark (around 44,000 square kilometres), the plan by the country's authorities to install more wind turbines may appear ambitious (as mentioned previously, municipalities are under a legal obligation to find sites for the erection of an additional 150MW of generation capacity by the end of 2011). In this respect, it should also be noted that new wind turbines are rather tall – with a total height of around 150 metres – and require substantial investments to identify attractive sites and conduct technical analyses prior to the actual erection, and to connect to the grid.

The guarantee fund may, to some extent, diminish or mitigate economic problems in the start-up phase, whereas the green subsidy scheme may help by providing funding for the establishment of scenic and recreational facilities in the area in which the wind turbines are erected. Although the measures to increase local support for the erection of more wind turbines appear necessary, the question is whether they are sufficient.

The mere possibility of obtaining compensation based on rules that are more favourable to land owners than the general Danish regulation on compensation for nuisances to an estate could arguably deter developers from initiating new wind turbine projects. This is especially so as the price of the electricity generated is fixed by law, without any provision for increase following unforeseen circumstances.

Conversely, the mere possibility of obtaining compensation may arguably result in less local opposition to new wind turbines and, hence, lead to the desired increase in generating capacity based on such wind turbines.

The possibility of claiming compensation for the decrease in the value of real

15 The surcharge payable by consumers to enable the transmission system operator to fulfil its public service obligations is known as a public service obligations tariff
16 Danish Energy Agency (Energistyrelsen), October 2009 figure.
17 Danish Energy Agency, 2008 figure.

estate, combined with the possibility of obtaining ownership interest in the wind turbines, may boost interest in the projects. As explained in section 3.1 above, any claims for compensation will be known at an early stage of any project. Accordingly, developers can take these claims into account when finalising their business case and offering the ownership shares to the public (see section 3.2).

4.   Conclusion

The erection of several new onshore and offshore wind projects should add 600MW of generation capacity in Denmark.

This chapter has shown that the Renewable Energies Act aims to promote acceptance within local communities of energy production from renewable energy sources. Offshore wind parks erected following public procurement procedures are not covered by the act. On April 30 2009 the Danish Energy Agency called for bids to be submitted by April 7 2010 as part of the public procurement procedure for the installation of between 300MW and 400MW of additional wind generation capacity. The procurement documentation prescribed that the wind park must be connected to the Danish electricity grid by December 31 2012. Upon connection to the grid, the new wind park is expected to generate at least 4% of the electricity annually generated in Denmark, which corresponds to the annual consumption of 400,000 households. The new wind park will increase the Danish wind generating capacity by approximately 10%.

There is, however, an inherent challenge in increasing the wind-based generating capacity due to the natural volatility in the load factor. As stated above, about 19% of Danish electricity is generated by wind turbines and this generation is heavily subsidised. The high proportion of wind-generated power and the inherent volatility of the load factor result in a corresponding request for balancing power. In Denmark, such balancing power primarily stems from fossil fuel (coal, natural gas or oil) or, alternatively, transmission lines from neighbouring countries (Germany, Norway and Sweden).

The cost of balancing power is higher than that of ordinary power purchased at the power exchange because the former must be available at short notice. Further wind-based generating capacity may thus lead to:

- further increased costs for electricity consumers; and
- according to some market participants, a somewhat less effective electricity market.

The question arises of how much wind power one can rely on and still obtain power balance at a reasonable cost. Denmark may have reached that limit: in late 2009 heavy wind led to the export of subsidised electricity at Dkr0 and even negative prices after December 1. This happened despite coordination between the varying Danish wind power generators and the hydropower generators in Sweden and Norway. An essential part of the generated wind power is thus stored as non-produced hydropower, which is then later generated when there is less wind. This coordination takes place through the electricity market (primarily, commercial transactions on the Nord Pool Spot Market).

The EU Renewable Energies Directive (2009/28/EC) provides that renewable energy must be cost effective. The Danish Renewable Energies Act also clearly states that economics must be taken into consideration.

Exporting subsidised electricity at negative prices is clearly not a cost-effective way of using renewable energy. This highlights the issues that arise when the proportion of the total electricity obtained from wind power reaches a certain level. Denmark has already reached the target set by the Renewable Energies Directive for 2011 and is facing difficulties in balancing the grid despite a working coordination with hydropower generators in Sweden and Norway.

The question is how to deal with the inherent volatility in generation where a large part of the generating capacity is based on wind power. Energinet.dk has made several proposals.[18] These include expansion and reinforcement of the electricity grid so that electricity may be transmitted from the planned wind parks to consumers (whether industrial or residential) via the interconnected international electricity market. Another proposal is an integrated planning of the energy systems for electricity, heat and transmission to enable the market to use wind power in a flexible manner. New types of electricity consumption (eg, heat pumps and electric vehicles) must be organised in such a way that consumption patterns can be adapted to wind power production.

Another way of increasing the flexibility of energy consumption may be to change the duties and taxes imposed on energy consumption. Most taxes are fixed on the basis of volume consumed, with no regard for the electricity generated in the period during which the electricity is consumed. Of course, such amendments require both the installation of intelligent electricity meters[19] and a substantial change in consumer behaviour, but the old saying 'money talks' should not be underestimated. For example, if the roll-out of an infrastructure for electric vehicles is completed and electric vehicles become a realistic alternative to vehicles fuelled by oil or gas, recharging a large number of electric vehicles during the period of the day with low consumption but a high load factor may assist in consuming excess electricity generated. Such timely consumption may be assisted by low taxes on electricity consumed during such hours.

The integration of large volumes of wind power creates substantial challenges for the transmission system operators responsible for balancing the grids so that generation during any hour of the day equals consumption.

Wind power is green and therefore politically correct, but it entails substantial challenges, in particular for the EU electricity market. These challenges should be faced before it is too late.

---

18   See www.energinet.dk for further information.
19   The Danish government has decided that intelligent electricity meters should be installed for 45% of all electricity consumers. Today, 45,000 of the largest electricity consumers, accounting for more than half of the electricity consumption in Denmark, have installed intelligent electricity meters able to metre the electricity consumption each hour. For further information, please refer to "Bedre Integration af Vind" (www.ens.dk).

Solar energy: technical and commercial issues

Serge Younes
WSP Environment & Energy

1. Introduction

Solar energy can be direct and indirect. Most known energy sources are forms of indirect solar energy. Thus, coal, oil and natural gas derive from biological material that took its energy from the sun, via photosynthesis, millions of years ago. The sun also powers the movements of the wind (which causes waves at sea) and the evaporation of water to form rainfall, which accumulates in rivers and lakes. Therefore, hydroelectric power and wind and wave power are forms of indirect solar energy. Indirect solar energy sources, together with the non-solar environmental energy resource (geothermal), produce energy that dwarfs the combined outputs of all direct applications of radiant solar energy for heating and generating electricity. They will continue to do so for perhaps two more decades.

However, the value of renewable energy resources is not measured only by the kilowatt-hours that are produced. The great economic advantages of many of the uses of solar energy are that, unlike most other forms of renewable and non-renewable power generation, which rely on power distribution, the point of generation of direct radiant solar energy can be in the immediate vicinity of end use.

Energy from the sun can be used directly to heat or light buildings, or to provide domestic hot water to meet basic thermal and hygienic requirements. The sun's radiant energy can also directly:

- provide hot water or steam for industrial processes;
- heat fluids through concentration to temperatures sufficient for a variety of industrial processes (eg, to produce electricity in thermal-electric generators or to run heat engines); and
- produce electricity through the photovoltaic effect.

These applications produce energy only during daylight hours and the more solar irradiation there is, the better they work. Both characteristics constitute serious limitations to the usefulness of solar energy. However, these restrictions can be mitigated as people are usually at work during the daytime, when carefully shaded daylighting can replace the electricity demand and heat output of artificial lighting, and daylighting of buildings works well even in cloudy weather.

Businesses most commonly need industrial process heat during the daytime; the bulk of the demand for electricity is also during daylight hours. In addition, well-insulated water tanks can store solar-heated water for use at any time of the day or night. As a result, the effectiveness of solar energy production is a matter of its ability to meet the needs of users, rather than the time of its collection. This is also true of the coincidence of radiant solar energy with the needs of the electrical grids. These tend to peak in the afternoons – especially on hot, sunny days. Consequently, the 'capacity factor' of solar energy systems – that is, the availability of solar electric energy when it is needed – has little economic meaning. The effective capacity factor of solar electric energy production can sometimes exceed 80%. Other solar applications, such as water and space heating, can work from the heat collected during the day and stored in the water or building over a 24-hour period.

Human behaviour can also influence the effective capacity factor of solar energy systems. Experience gained in Denmark shows that people with photovoltaic systems alter their behaviour in order to maximise the use of photovoltaic electricity at the time of its production.[1] In California, time-of-use metering of net-metered photovoltaic systems affords an economic advantage to the opposite behaviour. There, the sale of photovoltaic electricity back into the grid during peak times is maximised by minimal electricity use in the building during those times, when the return to the producer can be as high as $0.30 per kilowatt-hour; low-cost electricity is then purchased by the customer during non-daylight hours.[2] Most of the world's building stock is residential and thus often unoccupied during the day, when solar energy production is at its peak.

The annual solar resource is surprisingly uniform throughout most populated regions of the world, within a factor of about two.[3] The lower end of this resource availability would be an economic death knell if solar energy were available only in desert climates. However, the economically attractive applications of solar energy are not limited to the sunniest climates, as demonstrated by:

- the extraordinary applications of photovoltaic energy technology in Germany;
- significant solar water heating applications in Germany, Austria and Greece; and
- passive solar and daylighting applications in Scandinavia.

The sun's radiation is sufficient in most locations worldwide to be used as a source of energy.

This chapter focuses on the technical and commercial issues surrounding solar power, also referred to as 'solar electricity', which can be generated via solar photovoltaic and thermal technologies. This chapter contains a short description of these technologies, followed by a history and medium-term forecast of costs and a discussion of the major issues facing project development.

These issues include:

- land requirements;
- the choice of technology, power generation and efficiencies;
- the costs of development and operation; and
- available subsidies.

---

1   International Solar Energy Society White Paper.
2   Findings found during research carried out by WSP on behalf of California Utilities.
3   *Comité International d'Eclairage* (ICE).

2.   Solar resource

The earth follows the rotational forces that also created the solar system. Thus, over the course of 365 days each year and 24 hours each day, every part of the earth's surface receives sunlight. Because the earth is round, the sun hits different locations at different angles. These angles range from 0° (horizontal) at sunrise and sunset or in the winter season at the poles, to 90° (vertical) at noon in summer, and on or near the equator.

The earth travels around the sun in an elliptical orbit. Consequently, the northern hemisphere is closer to the sun during one half of the year and the southern hemisphere is closer to the sun during the other half of the year. The part of the earth that is closer to the sun receives more concentrated solar energy. This time of year is summer. However, regardless of the season, the 23.5° tilt of the earth's axis plays a larger role in determining the amount of sunlight striking the earth at a particular location. The earth's tilting results in longer days in the northern hemisphere during one half the year and longer days in the southern hemisphere during the other half of the year.

Solar radiation incident outside the earth's atmosphere is called extraterrestrial radiation. On average, the extraterrestrial irradiance is 1,367 watts per square metre.[4] This value varies by around 3% as the earth orbits the sun.[5] The earth's closest approach to the sun occurs around January 4; it is furthest from the sun around July 5.

When the sun's rays are vertical, the earth's surface gets the maximum solar energy. The more incident the rays, the longer they have to travel through the earth's atmosphere before reaching the surface (becoming more scattered and diffuse as they travel). The more scattered and diffuse the solar radiation, the less concentrated the solar energy. Because of the earth's round shape, the polar regions never receive incident direct sunlight and during their respective winter months, they receive no sun at all.

However, a few other factors need to be considered when determining the viability of solar energy in any given location:

- geographical location (longitude, latitude, altitude);
- time of day (in solar time, not civil time);
- season, based on the solar cycle and earth tilt;
- local geography; and
- local weather.

The local natural geography can contribute adversely to the amount of energy received. For instance, high mountain ranges obstruct the sun's rays during portions of the day. Large water bodies such as seas and lakes, as well as heavy white snow, have a high impact on diffuse radiation as these highly reflective surfaces tend to scatter sunlight in every direction – this is often referred to scientifically as 'ground albedo'. Artificial landmarks also have an adverse impact: high-rise buildings can block the sun's rays and a building's highly glazed surfaces can scatter sunlight over

---

4   ICE.
5   ICE.

long distances.

As sunlight passes through the earth's atmosphere, some of it is absorbed, scattered and reflected. The following atmospheric components diffuse the sunlight by either absorbing and/or scattering the solar rays:

- air molecules;
- water vapour;
- clouds;
- dust or suspended sand particles in the air; and
- other atmospheric gases and pollutants, such as greenhouse gases.

Sunlight affected by atmospheric components in this way is referred to as 'diffuse solar radiation' or 'diffuse sunlight'. Sunlight that reaches the earth's surface without being diffused is called 'direct beam solar radiation' or 'direct sunlight'. The sum total of all diffuse and direct solar radiation in a given location is called 'global solar radiation'. It is the total amount of sunlight hitting the earth at any specific spot (direct and diffuse combined).

Pollution and other atmospheric conditions (eg, weather patterns) can reduce direct sunlight by as much as 40% on clear, dry days through the mechanisms of absorption and reflection. They can reduce direct beam radiation by 100% on cloudy days. Generally, in clear-sky conditions, global solar radiation can reach a maximum of 75% of the extraterrestrial radiation (approximately 1,000 watts per square metre).

On tilted surfaces, the 'tilted radiation' is thus the sum of the direct, diffuse and ground-reflected solar radiation. The ground-reflected radiation amounts to, on average, about 20% of the global solar radiation.

Based on the above brief explanation of the basics of solar radiation, one can deduce that solar radiation is at its highest near the equator on clear, dry days at midday (solar noon). Global solar radiation is highest when the direct component is highest – perpendicular to the plane of the irradiated surface.

In certain reference books, solar radiation is often referred to as 'solar insolation'. 'Insolation' is a measure of solar radiation energy received on a given surface area in a given time. The name comes from the words 'incident solar radiation'. It is commonly expressed as an average irradiance in watts per square metre or kilowatt-hours per square metre per day. In the case of photovoltaic energy, it is commonly measured as kilowatt-hours per year per kilowatt-peak rating.

Good maps and solar radiation data can be obtained from various sources, including the National Renewable Energy Laboratories for North American Data (www.nrel.gov) and the solar radiation database at www.soda-is.com.

Datasets for long-term solar radiation can also be purchased from local meteorological stations, such as at the airports of major cities. Readers should contact their nearest meteorological service for further information.

3. Photovoltaic power

Photovoltaic technology is the most widely recognised technology in the renewable energy market; it is also developing at a faster rate than any other technology. The result is a rapid improvement of the economic viability of the technology through decreased costs and increased performance. The installation of photovoltaic panels on available roof surfaces is one of the simplest renewable energy solutions.

3.1 Technical summary

'Photovoltaic' is a scientific term used to describe the generation of electricity from natural solar light. The secret to this process is the use of a semiconductor material that can be adapted to release electrons. The most widespread semiconductor material in photovoltaic cells is silicon, an element most commonly found in sand.

Photovoltaic cells convert sunlight into electricity where certain wavelengths of light ionise silicon atoms, separating the positive and negative ions. The flow of negative ions through an external circuit produces a direct current. This is often referred to as the 'photovoltaic junction in a solar cell'.

The direct current can be used by motors in fans and pumps, and for battery charging. Alternatively, it can be converted to alternating current using an inverter. Thus, the principle is that the greater the intensity of the light, the greater the flow of electricity. A photovoltaic system does not, therefore, need bright sunlight in order to operate. It generates electricity even on cloudy days, with its energy output proportionate to the density of the clouds. However, the efficiency of photovoltaic systems is generally inversely proportional to ambient temperature; as such, these systems are generally more efficient on clear, cool days.

The main advantages of solar photovoltaic power are as follows:
- The fuel is free and abundant;
- There are no moving parts to wear out or break down, unless using tracking systems;
- Only minimal maintenance is required to keep the system running;
- Modular systems can be installed anywhere; and
- It produces no noise, harmful emissions or polluting gases.

The most important parts of a photovoltaic system are:
- the cells that form the basic building blocks;
- the modules that bring together large numbers of cells into a unit; and
- the inverters used to convert the electricity generated from direct current to alternating current – a form suitable for everyday use.

Photovoltaic cells are generally made either from:
- crystalline silicon, sliced from ingots or castings; or
- grown ribbons or thin film deposited in layers on a low-cost backing.

Other types of photovoltaic module, including concentrator cells, focus light from a large collector area onto a small area of photovoltaic material using an optical concentrator – a Fresnel lens.

Modules are clusters of photovoltaic cells incorporated into a unit, usually by soldering them together under a sheet of glass. They can be adapted in size to the proposed site and installed quickly. They are also robust, reliable and weatherproof. Module producers usually guarantee their performance for 20 to 25 years.

When a photovoltaic module is described as having a capacity of 125 watts-peak, this refers to the output of the system under standard testing conditions, allowing comparison between different modules. The solar photovoltaic industry typically uses price per watt-peak as its primary unit of measurement. In order to translate watt-peak into kilowatt-hour, an adjustment for the actual location of the solar panel is necessary in order to take into account how much sunlight would be expected in that location over one year. In the United Kingdom and Germany, a 1 kilowatt-peak system would produce between 650 and 850 kilowatt-hours annually, depending on location, whereas in North Africa, South California, southern Spain and the Arabian desert, a similar system would be expected to produce over 2,000 kilowatt-hours per annum.

Inverters are used to convert the direct current power generated by a photovoltaic generator into alternating current which is compatible with the local electricity distribution network. This is essential for grid-connected photovoltaic systems. Inverters are offered in a wide range of power classes, from the smallest residential-scale units to very large industrial units (over 100 kilowatts). Commercially available inverters typically claim efficiencies of between 94% and 97%. Inverters have a shorter useful life than photovoltaic modules; they also generally require major maintenance, or even full replacement, after 15 years.

For off-grid solutions, batteries are often the most sensible method of electricity storage. In addition to batteries, charge controllers are required to protect the battery life. Batteries accumulate excess energy created by a photovoltaic system and store it to be used at night or when there is no other energy input. Ideally, a battery bank should be sized to be able to store power for five days of autonomy in cloudy weather. If the battery bank has a capacity of less than three days, it will cycle deeply on a regular basis and, consequently, have a shorter life. Batteries are rated according to their cycles: they can have shallow cycles of between 10% and 15% of the battery's total capacity or deep cycles of 50% to 80% of the battery's total capacity. Shallow-cycle batteries, such as those for starting a car, are designed to deliver several hundred amperes for a few seconds; then the alternator takes over and the battery is quickly recharged. By contrast, deep-cycle batteries deliver a few amperes for hundreds of hours between charges. These two types of battery are designed for different applications and should not be interchanged. Deep-cycle batteries are capable of many repeated deep cycles and are best suited for photovoltaic power systems. Nevertheless, batteries associated with photovoltaic storage rarely last longer than five years and generally need to be replaced every two to three years.

All in all, the electrical system accounts for up to 80% of the installation's capital cost.

The photovoltaic phenomenon was first discovered by French physicist Alexandre Edmond Becquerel in 1839. Becquerel was experimenting with metal electrodes and electrolytes when he noticed that conductance rises with illumination. While the first photovoltaic cells (selenium based) were created in the late 19th century, it was Albert Einstein who explained the scientific theory of photovoltaics in 1904, for which he was awarded the Nobel Prize in 1921. It was only between the two world wars that experimentation was carried out on a number of materials, which led to the invention of the first monocrystalline silicon and cadmium-telluride cells.

The first commercial application of solar cells occurred in 1955, with the commercialisation by Hoffman Electronics of a 2%-efficient solar cell at a cost of $1,785 per watt-peak.[6] A photovoltaic cell was first installed on a satellite in March 1958 (on the Vanguard 1 satellite). By 1960, the efficiency of solar photovoltaic cells had increased to 14%.[7]

Photovoltaic systems are now considered proven technologies. The average efficiency of the photovoltaic modules depends on the type of cell chosen. It can be up to 23% for the latest high-performance modules, but can drop to 14% for traditional monocrystalline modules. The latest models in sun-tracking concentrated photovoltaic systems can achieve efficiencies in excess of 35%.

Crystalline cells can be either mono or polycrystalline. Mono cells are the most efficient, with conversion efficiencies of 15% to 20%. Poly cells are 13% to 16% efficient, but are less expensive to manufacture than mono cells. Thin-film cells offer the advantage of being cheaper to manufacture than crystalline cells. Thin-film cells include:

- amorphous silicon, which is 5% to 8% efficient and can be mounted on a rigid or flexible substrate, and be fitted on curved or folding surfaces;
- cadmium telluride, which is 6% to 9% efficient; and
- copper indium diselenide, which is 7.5% to 9.5% efficient.

Degradation of cell efficiency occurs in photovoltaic modules, which translates into a decline of yield. High degradation is often observed in thin-film solar cells within the first weeks of use, but degradation then stabilises. This phenomenon in thin-film cells is often considered in the nominal performance information provided by the manufacturers. Crystalline cells experience comparatively low degradation, with some manufacturers observing 1% of annual degradation; other cells degrade by less than 0.5% annually. Studies on photovoltaic installations from the 1990s showed that:

- 20-year-old monocrystalline and polycrystalline cells produced about 90% of their nominal capacity; and
- 30-year-old panels still produced 80% of their nominal capacity.[8]

Manufacturers generally guarantee production of 80% of the nominal capacity over 20 or 25 years. Cell degradation needs to be taken into account when evaluating a photovoltaic facility by using the performance ratio (see section 3.3 below).

---

6   *History of Solar Energy*, International Solar Energy Society. Cost per watt peak is based on actual retail price of Hoffman Electronics' PV module.
7   *History of Solar Energy*, International Solar Energy Society.
8   This value is used as industry standard based on performance of modules in laboratory conditions.

Photovoltaic cells are temperature-sensitive; in the case of crystalline cells, the efficiency drops slightly with increased temperature. Users should make a note of the temperature performance coefficient when selecting photovoltaic modules dependent on the location of the installation and the respective climatic characteristics. The temperature performance coefficient indicates the percentage tby which performance varies if the module is warming or cooling by about 1° Celsius. This value should be as low as possible so that the module still has a high level of performance on warm days. Note that the specified module-rated power is measured under standard test conditions at 25° Celsius.

Photovoltaic modules are generally designed to last 25 years; most manufacturers will provide a 20 to 25-year warranty. While modules have lasted longer than their designed life, with some exceeding 40 years of operation, users should remember that the photovoltaics' efficiency deteriorates with age.

The solar photovoltaic market has reached a record of 5.95 gigawatts-peak in 2008, representing a growth of 110% on the previous year. The market reached a consolidated figure of an additional 6.4 gigawatts in 2009 to reach a total cumulative installed worldwide capacity of approximately 20 gigawatts.[9]

Thus, the sector has sustained an upward trend over the past decade. Europe accounted for 82% of world demand in 2008. Spain's 285% growth pushed Germany into second place in the market ranking, while the United States advanced to number three. Rapid growth in South Korea allowed it to become the fourth largest market, closely followed by Italy and Japan. Further countries are adopting generous subsidies for renewable energy, especially solar photovoltaic. In 2010, following decisions made at the Copenhagen Climate Change Conference, market demand continues to increase exponentially. Photovoltaics in Spain and Germany account for 3% and 2% respectively of the total electricity supply. According to the European Photovoltaic Industry Association, the global photovoltaic market could reach between 8.2 gigawatts (with a moderate growth estimate) and 12.7 gigawatts (with a policy-driven growth estimate) of new installations in 2010.[10]

On the supply side, world solar cell production reached a consolidated figure of 6.85 gigawatts in 2008, up from 3.44 gigawatts a year earlier. Overall capacity use rose to 67% in 2008 from 64% a year earlier. Meanwhile, thin-film production also recorded solid growth, up 123% in 2008 to reach 0.89 gigawatts. Photovoltaic supply has been growing year on year, especially in the past decade, from 50 megawatts-peak in 1990 to 250 megawatts-peak in 2000, breaking the 1 gigawatt barrier in 2004.[11]

In total, the photovoltaic industry generated $37.1 billion in global revenues in 2008. That year it also successfully raised over $12.5 billion in equity and debt, up 11% on the prior year.[12]

3.2   Capital cost

The prices for high power band (ie, over 125 watts) solar modules have dropped from

---

9   Renewable Energy Focus, February 2010
10   European Photovoltaic Industry Association.
11   European Photovoltaic Industry Association.
12   Author's own market research.

$1,750 per watt-peak in 1955 to $150 per watt-peak in 1970, $27 per watt-peak in 1982 and around $4 per watt-peak today.[13] The costs quoted in this chapter are global averages based on 2010 costs from the United States, Asia and the European Union. Prices higher and lower than this are usually dependent upon the size of the order and the technology used. As a rule of thumb, the solar module represents between 50% and 60% of the total installed cost of a solar system. This percentage will vary according to the nature of the application. A complete solar system includes additional components, whether to feed energy into the grid or to be used in standalone off-grid applications. Such additional equipment includes inverters, cabling and support structures.

In 2009 a residential grid-connected solar photovoltaic system cost between $6,000 and $10,000 per kilowatt-peak installed, a large commercial installation (over 100 kilowatts-peak) cost between $5,000 and $8,000 per kilowatt-peak, while industrial installations (over 5 megawatts) cost about $5,000 per kilowatt-peak. A system including batteries for power storage often costs 20% to 30% more than a grid-connected system.[14]

3.3   Cost of generation

The price of solar electricity generation for residential installations (under 2 kilowatts-peak) is around $0.30 per kilowatt-hour in sunny locations; this is typically two to five times the average residential electricity tariffs. The price is $0.30 per kilowatt-hour for commercial facilities (under 50 kilowatts-peak) and under $0.20 per kilowatt-hour for industrial-sized facilities (over 500 kilowatts-peak).[15] These costs are exclusive of any personal or corporate tax and any fiscal or direct financial incentives.

The cost of solar electricity generation is highly dependent on the location and intensity of the solar resource. In northern latitudes and in cloudy weather, the solar generation cost can be as much as double or more that in lower latitudes and in sunny climates, such as in South California or North Africa. The generation cost is on average $0.55 per kilowatt-hour for commercial systems and under $0.65 per kilowatt-hour for industrial systems.[16]

Around 59% of the world's sales of solar energy between 2004 and 2009 were in grid-connected applications. Solar energy prices in these applications are five to 20 times more expensive than the cheapest source of conventional electricity generation, although they may be only three to five times the electricity tariff that utility customers pay. By contrast, photovoltaic installations can be fully cost-competitive on economic grounds in remote (off-grid) industrial and residential locations where diesel generators cost between $0.20 and $0.40 per kilowatt-hour.[17]

The grid-connected market remains the major prize for the solar industry because of the huge scale of the electricity supply market. Indeed, this is where growth has

13   Average price in Q1 2010, based on market research in the United States and the European Union.
14   Author's own market research; this information is correct as at Q1 2010.
15   Author's own market research.
16   Author's own market research.
17   Author's own market research.

been strongest recently. While solar power is a long way from competing with conventional power generation costs at $0.03 or $0.05 per kilowatt-hour,[18] it is much closer to reaching electricity tariffs charged to residential, commercial and industrial consumers. When the photovoltaic system is sited at the consumer's premises, the comparison for the customer is between the tariff rate and the cost of photovoltaic electricity from its system. Precise calculations of solar electricity costs depend on the location and cost of finance available to the owner of the solar installation. However, with the best photovoltaic electricity prices (in the sunniest locations) approaching $0.20 per kilowatt-hour and the highest retail electricity tariffs now exceeding that price, the gap is now closed.[19] Funding programmes, such as feed-in tariffs, that bridge this gap are causing rapid growth in sales of solar photovoltaic, especially across Europe, Japan, Canada and now some US states.

A term often used in the renewable energy, development and investment community is 'performance ratio'. A performance ratio is an evaluation criterion for solar photovoltaic facilities that is independent on the alignment of the power plant, photovoltaic systems and global solar radiation. 'Performance ratio' is defined as the relationship between the actual returns and theoretically potential energy returns of a photovoltaic system. The performance ratio is an appropriate valuation criterion for determining the quality of the plant configuration, because all components and their interaction are considered. A photovoltaic system with high efficiency can achieve a performance ratio of over 70%. The performance ratio is also sometimes called 'quality factor'. In some cases photovoltaic systems can reach a quality factor of 0.85 to 0.95 (ie, a performance ratio of 85% to 95%).

3.4 Grid parity

In order for the solar industry to make a systematic penetration into the electricity segment, installed solar system costs will need to drop from around $6 to $8 per watt-peak to $3 per watt-peak. This would continue the trend shown above of falling solar electricity costs over the past 25 years. Solar Buzz LLC has been carrying out photovoltaic module market research over the past decade.[20] Its analysis shows that the cost of solar modules has been dropping consistently over the past 10 years, from a high in December 2001 of $5.40 per watt-peak in the United States and €5.42 per watt-peak in Europe to $4.20 per watt-peak in the United States and €4.10 per watt-peak in Europe in December 2009. These costs are based on the purchase of single modules (off the shelf) and are exclusive of value-added tax and other taxes.[21]

A push to $2 per watt-peak would bring solar energy costs from the present $0.30 per kilowatt-hour to around $0.10 per kilowatt-hour. This would allow it to compete more strongly with other renewable energy sources and capture a significant share of the electricity market. The solar photovoltaic manufacturing industry is targeting an average retail price of $1.50 per watt-peak by 2020.[22]

---

18   Author's own market research.
19   Author's own market research into renewable energy subsidies, including feed-in tariff mechanisms.
20   www.solarbuzz.com.
21   International Solar Energy Society, White Paper.
22   Author's own market research.

3.5 Sun-tracking devices

As mentioned above, a financially viable solar photovoltaic facility is highly dependent on the site location and the panels' orientation – as much as it is on the capital and operating expenditure. As such, it is preferable to locate solar photovoltaic projects in southern latitudes, but also to ensure that the panels are orientated towards the south and tilted at an optimal angle. It is possible to improve the quantum of received radiation on the panels by mounting the panels on rotating structures that track the sun. There are two major types of sun-tracking mount:

- a single-axis tracker that boasts increased production of 20% to 25% compared to fixed panels; and
- a two-axis tracker that could increase production by 30% to 40% compared to fixed-panel installations.

With an estimated increase in capital cost of 15% to 20%, tracker-mounted photovoltaic installations are quickly becoming the norm for large industrial-scale facilities. The main downside to utility-scale solar tracking power plants is land use. Since the modules are tracking, they must be prevented from shading one another as they follow the sun's path across the sky. Therefore, the panels must be spaced farther apart, thus wasting land and increasing costs for the developer.

The key to deciding whether to use trackers is to consider cost and revenue. The increased revenue generated by a higher energy production can outweigh the cost incurred by the use of larger land areas and the additional investment in trackers. It is expected that by 2012, up to 85% of projects over 1 megawatt-peak will be using tracking systems.

3.6 Operating costs

While the capital expenditures for solar photovoltaic are high (the highest among all mainstream renewable energy sources), the operating expenses for such systems are low. Solar photovoltaic is a passive technology; hence, the operating and running costs are much lower than for most other renewable energy sources.

The modules generally require little maintenance. However, it is often good practice in wet climates to install self-cleaning glazing systems. In desert locations a more aggressive operation and maintenance or cleaning regime needs to be followed, due to the formation of sand cake or excessive deposits of sand during sandstorms. Sand cakes are formed in locations where humidity is high and a solid, sandy surface is created on top of the photovoltaic modules. Partial shading of a photovoltaic module can disable the whole panel; one should therefore ensure that panels are relatively clear of any dirt.

The US National Renewable Energy Laboratory noted in its research in Saudi Arabia in the 1980s and 1990s that glazing delaminating often occurs in sandstorms. Luckily, the latest glazing technology should be able to cope with the effects of sandstorms. Accordingly, the photovoltaic glazing surfaces should not require re-laminating.

Operation and maintenance costs vary and generally range between $0.0002 and $0.001 per kilowatt-hour. The higher reported costs include maintenance costs for generators in remote hybrid photovoltaic systems, as well as capital replacement costs due to environmental factors such as extreme temperatures and vandalism. The experience gained in Europe (mainly Germany) in rooftop installations indicates that operation and maintenance costs are approximately €20 per kilowatt-peak installed per annum. These figures are set to drop by €1 annually to reach €10 per kilowatt-peak by 2020.[23]

The useful life of inverters (10 to 15 years) is shorter than the photovoltaic modules' operational life. It is therefore necessary to allocate funds to replace the inverters. Average inverter costs are between $450 and $500 per kilowatt.[24]

Moving parts, such as tracking devices, should be regularly cleaned and maintained. Most need to be replaced every 15 years. This can be every 10 years or less in sandy-desert locations.

The most significant replacement cost will likely be batteries, where the photovoltaic system includes physical storage.

3.7 Financial incentives[25]

The economic viability of the photovoltaic power system depends on the initial investment, the utility's payback rate and any available incentive mechanism.

Following the Copenhagen Climate Change Conference in 2009, the vast majority of countries have made pledges to achieve carbon reductions and develop renewable energy. In order to meet their targets, different countries have put in place a variety of fiscal incentives to promote renewable energy investment. Incentives mechanisms include:

- tax rebates, including personal, corporate, sales and property taxes;
- capital grants;
- long-term loans; and
- production incentives.

The payment for photovoltaic energy inserted into the grid (compared to the cost of conventional energy taken from the grid) varies from country to country. Some European countries, such as Germany and Spain, apply some of the highest rates.

Many different buy-back schemes are implemented around the world. They include the following:

- Very low buy-back rate schemes apply the same conditions as for other producers. Consequently, the rates are generally low and the ratio (ie, the difference between the payment for photovoltaic energy inserted into the grid and the cost of conventional energy taken from the grid) is well below 1.
- Low schemes are similar to very low buy-back schemes. However, special incentive premiums (of between 10% and 100%) are granted on these general buy-back rates. The resulting total energy payment is still low, with the ratio below 1.
- In parity schemes the price paid for photovoltaic electricity is equal to that

---

23   www.solarbuzz.com.
24   This section should be read in conjunction with the chapter of this book entitled "Renewable energy support mechanisms: an overview".
25   Contact Platforma Solar for more details.

charged by the utility (the ratio is 1).

- High schemes offer attractive prices (the ratio is above 1 – normally between 1 and 2). Restrictions are imposed on the length of payment (high payment during n years/further years at reduced payment).
- Very high schemes have the highest tariffs (the ratio is well above 1 – normally between 5 and 6) and are foreseen strictly for photovoltaics.
- In other schemes, 'green electricity' can be bought by users without a photovoltaic system.

In Europe, production incentives and high buy-back schemes are often guaranteed by the government under the feed-in tariffs system. Since their inception and adoption in Germany in 2000 under the Renewable Energy Act, feed-in tariff mechanisms have proved to be the most successful in promoting solar photovoltaics. They have also been the key contributor to the development of new technologies, increased efficiency and decreased cost of photovoltaic modules.

Feed-in tariffs have been successful because they generally include three key provisions:

- guaranteed grid access;
- long-term contracts for the electricity produced; and
- purchase prices that are methodologically based on the cost of renewable energy generation.

Further, under a feed-in tariff an obligation is imposed on regional or national electric grid utilities to buy renewable electricity from all eligible participants. Feed-in tariffs are also available for a guaranteed 15 to 25 years, depending on the country and technology.

By 2009, feed-in tariff policies had been adopted in 63 jurisdictions around the world, including Australia, Austria, Belgium, Brazil, Canada, China, Cyprus, the Czech Republic, Denmark, Estonia, France, Germany, Greece, Hungary, Iran, Ireland, Israel, Italy, Lithuania, Luxembourg, the Netherlands, Portugal, Singapore, South Africa, South Korea, Spain, Sweden and Switzerland. They have also been adopted in some US states and are gaining momentum in other countries, including China, India and Mongolia.

Feed-in tariff policies are developed typically to target a real, unleveraged rate of return ranging from 5% to 10% – although some developers have achieved returns in excess of 15%.

4.   Solar thermal power

When solar energy is concentrated by reflecting surfaces, the energy density can be dramatically increased. This enables high temperatures to be achieved in fluids in receivers; these high temperatures can then generate electricity in thermal-electric generators. The technology, generically referred to as 'concentrated solar power', comprises three categories: parabolic troughs, power towers and heat engines.

4.1   Technical summary

(a)   Parabolic troughs

Parabolic troughs are long, parabolic-shaped mirrors, usually coated in silver or polished aluminium, with a Dewar tube running their length at the focal point. Sunlight is reflected by the mirrors and concentrated on the Dewar tube. Troughs are generally mounted in rows to heat the fluid that flows in energy-collecting receivers – the Dewar tubes that are maintained along their lines of focus by adjustment of the position of either the mirror or the receiver. The troughs are usually aligned on a north-south axis and rotated to track the sun as it moves across the sky each day. This increases the temperature of the fluid, typically oil, to some 400° Celsius. The heat transfer fluid is then used to heat steam in a standard turbine generator. The process is economical, with thermal efficiency ranging from 60% to 80%. The overall efficiency from collector to grid is about 15%, similar to industrial grid-connected photovoltaic installations.

Typically, about 2.2 hectares are required to install the array of parabolic troughs and the associated steam turbine plant necessary to produce 1 megawatt of electrical capacity.

Commercial plants using parabolic troughs are hybrids (also called 'co-fired'). Fossil fuel – namely, natural gas – is used during night hours. The amount of fossil fuel used is limited to a maximum of 27% of electricity production, allowing the plant to qualify as a renewable energy source. Because these plants are hybrids and include cooling stations, condensers, accumulators and other things besides the actual solar collectors, the power that they generate per square metre of trough area varies enormously. The average space requirement for co-fired trough installations is between 2.5 and 3.2 hectares per megawatt.

Where fossil-fuel sources are not available for co-firing, the usefulness of the steam turbine can be increased by storing heat from the trough array. Single-tank thermocline storage for large-scale solar thermal power plants is among the technologies that are being considered with trough concentrated solar power plants. The thermocline-tank approach uses a mixture of silica sand and quartzite rock to displace a significant portion of the volume in the tank. Then it is filled with the heat-transfer fluid, typically a molten nitrate salt. These systems can extend the operation of the plant by up to six hours, thus extending from a theoretical 12 hours to 18 hours of continuous power production. However, this can be achieved at the expense of power production capacity from a trough array area. For a six-hour storage facility, a 2.2-hectare array would therefore produce only 0.6 to 0.65 megawatts.

The largest operational solar power system, with five fields of 33 megawatts of generation capacity each, is one of the Solar Energy Generating Systems plants located at Kràmer Junction in California. A number of projects planned in North Africa and the Middle East will be over 300 megawatts each.

(b)   Concentrating linear Fresnel reflectors

Concentrating linear Fresnel reflectors, a variation of the parabolic trough technology, use many thin mirror strips instead of parabolic mirrors to concentrate sunlight onto two tubes with working fluid. This technique presents the advantage of using flat mirrors, which are much cheaper than parabolic mirrors. In addition, more reflectors can be placed in the same space, allowing more of the available sunlight to be used. Typical space requirements for concentrating linear Fresnel reflector plants are between 1.8 and 2 hectares per megawatt.

(c)   ***Power towers***

Solar power towers, also known as 'central tower power plants' and 'heliostat power plants', are a type of solar furnace using towers to receive the focused sunlight from mirror fields. They use an array of flat, movable mirrors (called heliostats) to focus the sun's rays upon collector towers.

Early designs used these focused rays to heat water and the resulting steam to power a turbine. However, designs using liquid sodium in place of water have been demonstrated; liquid sodium, which offers high heat capacity, can be used to store the energy before using it to boil water to drive turbines. These designs allow power to be generated when the sun is not shining.

Some of the more recent developments in solar tower technology consist of:

- using flat glass mirrors instead of the more expensive curved glass; and
- introducing molten salt containers to continue producing electricity while the sun is not shining, similarly to parabolic troughs.

The steam produced from the tower receiver is heated to over 500° Celsius to drive steam turbines to generate electricity. The overall efficiency from collector to grid is about 12%.

The first solar tower installations, the 10-megawatt Solar One and Solar Two heliostat demonstration projects in the Mojave Desert, have now been decommissioned. The 15-megawatt Solar Tres Power Tower in Spain builds on these projects. Also in Spain, the 11-megawatt PS10 solar power tower and 20-megawatt PS20 solar power tower have recently been completed.[26]

In South Africa, a 100-megawatt solar power plant is planned with 4,000 to 5,000 heliostat mirrors, each having an area of 140 square metres.

(d)   ***Stirling engines***

Stirling engines direct solar energy with highly focused heliostats onto a piston, which then drives an engine through air expansion. Each Stirling engine is directly mounted onto its own three-axis tracking heliostat. The technical target for the Stirling engines is to be maintenance-free for 50,000 to 100,000 hours of operation.

Dish-Stirling engine/heliostat combinations have been mounted and tested. Developments are moving towards 25 kilowatt-electric modular units, which could be valuable in the future. Until recently, a Dish-Stirling engine/heliostat combination held the world record for efficiency, at about 35%, in converting solar energy to electricity.

More work remains to be done to assure the targeted long life and reliability of the engines and to produce low-cost heliostats.

---

26   Author's own market research.

4.2   Capital cost

The world's largest set of solar-electric generators, 354 megawatts of parabolic trough technology in three fields, continues to operate in Southern California. The first units were installed in the early 1980s and the completed system has been in full operation since 1993. The Harper Lake plant is 160 megawatts and the Kramer Junction plant is 150 megawatts. These projects have demonstrated, among other things, the practicality and reliability of this solar thermal electric technology. Similarly, the 10-megawatt power towers Solar I and II (the second being a rebuild of the first to accommodate liquid sodium heat transfer and storage), also in Southern California, met all research objectives for performance and reliability.

Even though concentrated solar power plants can produce energy at about half the cost of photovoltaic, the technology is being accepted slowly. This is due to various financial and institutional barriers; the primary obstacle is that building a solar plant is like building a fossil-fuel plant, but involves paying for 30 years' worth of fuel at the outset. Consequently, the plant must be fully financed upfront, with an attractive return to investors. The second obstacle is that the physical plant is generally taxed, while fuels for conventional power plants are not. This unfairly penalises solar plants for their 'free' fuel.

These barriers can be removed by:

- providing subsidised, low-cost loans;
- redressing the tax inequities;
- providing energy production incentives; and
- continuing to support research and development that can lead to more efficient reflectors, components and thermal systems.

Concentrated solar power is also extremely sensitive to the solar resource: plants must be built where it is sunniest. The technique is also most economic when built in systems up to 400 megawatts in size. If the barriers are all removed and the best physical conditions can be met, projections suggest that after the installation of a few thousand megawatts, the costs without subsidies could come down to be competitive with fossil fuels. Concentrated solar power plants provide economic certainty for the 30-year warranted life of the plant, free of the highly volatile and unpredictable costs and availability of conventional fuels.

The economics become more attractive and work on a shorter timescale when concentrated solar power provides energy to supplement gas energy in an integrated solar combined-cycle system. The solar energy displaces some of the fuel, as well as some of the combustion emissions; this improves both the fuel economics and the environmental performance of the overall system. Meanwhile, the marginal cost of the solar components adds proportionately less to the overall cost of the gas-fired system. Smaller and more versatile concentrated solar power plant designs are being developed for the 100-kilowatt to 1-megawatt range. The flexibility of application that this technique will provide will make up for the higher kilowatt-hour production costs. Storage techniques are being developed to provide up to 12 hours of energy storage – the economic optimum. This would yield maximum utility for the received solar energy and enhance the economics of concentrated solar power.

Concentrated solar power is a valuable component of the renewable energy portfolios of countries with a sufficient solar resource. It warrants inclusion in governmental policies aimed at stimulating and developing balanced resource portfolios of renewable energy technologies. A worldwide target of 100,000 megawatts of installed concentrated solar power technology for 2025 is an achievable goal, with potentially significant long-term benefits.

World interest in the technology is picking up: significant projects are planned in many countries and valuable funding from the Global Environment Facility (the designated financial mechanism for a number of multilateral environmental agreements, as well as a partnership among 178 countries, international institutions, non-governmental organisations and the private sector to address global environmental issues) is being provided for more. New projects are underway in the United States (Nevada) and Spain, while others are considered in Israel and South Africa. Global Environment Facility funding of $50 million each has been given to Mexico, Egypt, Morocco and India for concentrated solar power projects under development. Iran, Algeria and Jordan are considering integrated solar combined-cycle system projects.

Economic projections suggest that concentrated solar power is also viable in Greece, Italy, Portugal, Australia, Brazil, Liberia, Tunisia and China. Further, these projections show a potential for a cumulative world total of over 100,000 megawatts of concentrated solar power electricity generation in place within 25 years.

4.3   Cost of generation, grid parity, operating costs and financial incentives

The cost of solar thermal power is dropping. Experience in the United States shows that today's generation costs are about $0.15 per kilowatt-hour for solar-generated electricity at sites with high solar radiation, with predicted ongoing costs as low as $0.08 per kilowatt-hour in some circumstances.

There is much pressure for the technology to develop in order to reduce costs. The other factors that will reduce costs are mass production, economies of scale and improved operation. Concentrated solar power is becoming competitive with conventional, fossil-fuelled peak and mid-load power stations. Adding more concentrated solar power systems to the grid can help to keep the costs of electricity stable and avoid drastic price rises as fuel becomes scarce and carbon costs increase.

Most of the cost information available for concentrated solar power relates to parabolic trough technology, which makes up the majority of plants in operation. Estimates indicate that new parabolic troughs using current technology with proven enhancements can produce electrical power for about $0.10 to $0.12 per kilowatt-hour in solar-only operation mode in the conditions available in the southwestern United States. In Spain, the levelled cost of electricity is somewhat higher than this for parabolic trough technology (up to €0.23 per kilowatt-hour), but overall the price is coming down.

Commercial experience from the nine Solar Energy Generating Systems plants, which were built in California between 1986 and 1992 and have operated continuously ever since, shows that generation costs in 2004 dropped by around two-thirds. The first 14-megawatt electric unit supplied power at $0.44 per kilowatt-hour-electric, dropping to just $0.17 per kilowatt-hour-electric for the last 80-megawatt-electric unit. For reference, the cost of electricity from the first 14-megawatt-electric unit was $0.25 kilowatt-hour-electric at 1985 US dollar rates. The cost of concentrated solar power generated electricity is expected to drop even further, thanks to:

- technological improvements;
- scale-up of individual plants' megawatt capacity;
- increased deployment rates;
- competitive pressures;
- thermal storage;
- new heat transfer fluids; and
- improved operation and maintenance.

As with all solar plants, high initial investment is required for new plants. Over the entire lifecycle of the plant, 80% of the cost is from construction and associated debt and only 20% is from operation. Therefore, the confidence of financial institutions in the new technology is critical. Only when funds are available without high-risk surcharges can solar thermal power plant technology become competitive with medium-load fossil-fuel power plants. Once the plant has been paid for, in 25 or 30 years, only operating costs (about $0.03 per kilowatt-hour) will remain and the electricity will be cheaper than any competition. This is comparable only to long-written-off hydropower plants.

In California, there was a 15-year break between the construction of the Solar Energy Generating Systems IX plant in 1992 and the most recent installations – the PS10 and Nevada Solar One grid connection. Consequently, new industry players have had to recalculate costs and risks for concentrated solar power plants for today's market. The data indicates that operating costs have entered a phase of constant optimisation, dropping from $0.08 per kilowatt-hour to just over $0.03 per kilowatt-hour.

The industry has access to a new generation of improved-performance parabolic trough components, which will also lower running costs.

Less is known about the market costs of electricity for the other types of technology because the first examples have been built only recently or are still under construction. However, it is generally thought that solar towers will eventually produce electricity at a cost lower than that of parabolic trough plants.

Various parameters need to be taken into account when calculating the generation costs of solar power. The most important are the capital cost of solar power plants (see above) and the expected electricity production. The second is highly dependent on the solar conditions at a given site, making selection of a good location essential to achieving economic viability. Other important factors include operation and maintenance costs, the lifetime of the turbine and the discount rate (the cost of capital).

The total cost per generated kilowatt-hour of electricity is traditionally calculated by discounting and levelling investment and operation and maintenance costs over the lifetime of a power station, then dividing this by the annual electricity production. The unit cost of generation is thus calculated as an average cost over the lifetime of the power plant, which could be 40 years according to industry estimation. In reality, however, the actual costs will be lower when a power plant starts operating, due to lower operation and maintenance costs, and will increase over the lifespan of the machine.

Taking into account all these factors, the cost of generating electricity from concentrated solar power ranges from approximately €0.15 ($0.19) per kilowatt-hour at high solar irradiation sites up to approximately €0.23 ($0.29) per kilowatt-hour at sites with low average solar resource. With increased plant sizes, better component production capacities and more suppliers and improvements from research and development, costs are expected to fall to between €0.1 and €0.14 ($0.15 and $0.2) per kilowatt-hour by 2020. Beside the estimation of further price drops, the gap with generation costs from conventional fuels is expected to decrease rapidly due to increased prices of conventional fuels at world markets. Competitiveness with mid-load – for example, gas-fired – plants may be achieved in 10 years.

Solar concentrated power offers a number of other cost advantages compared to fossil fuels, which these calculations do not take into account. They include the following:

- Renewable energy sources such as solar offer environmental and social benefits compared to conventional energy sources such as coal, gas, oil and nuclear. These benefits can be translated into costs for society, which should be reflected in the cost calculations for electricity output. Only then can a fair comparison between different means of power production be made. The ExternE project, funded by the European Commission, has estimated:
  - the external cost of gas at around €0.011 to €0.03 ($0.014 to $0.038) per kilowatt-hour; and
  - the external cost of coal at as much as €0.035 to €0.077 ($0.045 to $0.1) per kilowatt-hour.
- The 'price' of carbon within the global climate regime and its regional/national incarnations (eg, the EU Emissions Trading Scheme) will drive up the prices of fossil fuels and, hopefully, reduce that of renewable technologies.
- Since concentrated solar power requires no fuel, it eliminates the risk of fuel price volatility of other generating technologies such as gas, coal and oil. A generating portfolio containing substantial amounts of concentrated solar power will reduce society's risks of future higher energy costs by reducing exposure to fossil fuels' price fluctuations. In an age of limited fuel resources and high fuel price volatility, the benefits of this are immediately obvious.
- The avoided costs for the installation of a conventional power production plant and avoided fossil-fuel costs would further improve the cost analysis for concentrated solar power.

5.   Conclusion

A number of localised factors affect incoming solar irradiation, including local geography and prevailing weather patterns. However, long-term patterns are fairly constant and available long-term measurements from meteorological stations or satellite-derived imagery can provide a strong forecast of radiation for use in the design and financing of solar energy power systems.

Photovoltaic cells are the most commonly used form of solar electricity; the panels are relatively easy to install and simple to maintain. Photovoltaic panels are modular and can be installed on all scales, from a few panels on residential properties (2kWp) to large commercial installations (sub-10MW), all the way to massive industrial installations (100MW or more). The biggest disadvantage of photovoltaic systems is the capital cost. However, continuing research and development into new manufacturing methods and new materials, together with an increase in market demand, has dramatically reduced this cost. Nonetheless, while the technology is now considered mature and is bankable, we are still a few years away from seeing photovoltaic generating cost at grid parity. Today, installations are commercially viable only through state-enforced subsidies which, at a minimum, provide a financial incentive to match grid costs or higher, in order to incentivise investors.

Solar thermal power uses technology for thermal power conversion combined with solar concentrating collectors, whether in the form of an array of mirrors and towers or parabolic dishes. This technology has received less attention from investors in the past, mainly due to the significant capital investment required for large, industrial-sized installations. However, today the technology is considered mature, with the installation of numerous commercial-sized pilot plants in Spain and the United States. One of the advantages of concentrated solar power technology is the ability either to combine the power plant with thermal storage in the form of molten salt, in order to prolong the power generation of the plant at night; or to combine the solar plant with gas co-generation. Solar thermal power is close to achieving grid parity, with generating costs today better than diesel power generation and, in the long term, comparable to the most efficient gas power plants.

An array of new technologies are currently in their infancy, such as organic and exotic metal photovoltaics and other solar thermal-based systems, and will progress through their design development cycle over the next 10 to 15 years. These could potentially be the next evolutionary step in breaking through the technological efficiency barriers of currently available systems.

Solar energy: legal issues

Lucille De Silva
SNR Denton UK LLP

1. Introduction

As discussed in [the previous chapter], the solar energy market exists in areas of the world with high levels of sunlight or insolation (the solar radiation energy received on a given surface area in a given time), where it is growing exponentially. This rapid growth stems from increasing concerns over the environment, climate change and energy security. These fears have led governments to push green policies to the top of their agendas and to seek actively to maximise use of solar energy.

Legal issues specific to the solar industry have emerged alongside the sector's growth. This chapter discusses some of the key legal issues typically found in solar projects, with particular focus on:

- the main types of financial incentive for solar projects and the associated legal risks;
- typical project agreements and their associated legal issues; and
- key regulatory issues and approvals.

In contrast to a number of jurisdictions in Europe (ie, Germany, Italy and Spain), the development of solar energy is in its infancy in the United Kingdom. (The solar sector is poised to develop in the southwest of the country, following an incentive regime established by the UK government in April 2010 for small-scale solar projects.[1]) Consequently, this chapter focuses on international legal trends and issues, rather than the UK market and English law issues.

In addition, although solar energy can be used both to provide heat and to generate electricity, this chapter discusses the legal aspects of solar electricity generation only. Moreover, this chapter does not cover the heat storage issues associated with concentrated photovoltaic technologies.

2. Technologies

Solar generation technologies can be categorised as solar thermal or photovoltaic technologies. Solar electricity can be generated directly using either photovoltaic panels with a photovoltaic array on a roof or a standalone structure.

Photovoltaic technology can be used on a domestic scale, providing electricity for the operator's own use, as well as for sale into the electricity grid (subject to having a supporting industry structure). On a larger scale, solar electricity can be

1   Feed-in Tariffs (Specified Maximum Capacity and Functions) Order 2010/678.

generated by concentrating the sun's rays using lenses or mirrors in concentrated photovoltaic power plants or concentrated solar thermal power plants. Both concentrated photovoltaic energy and concentrated solar thermal energy are generated by focusing light onto a collector, which is then used to heat up fluid. This in turn generates steam to drive a turbine. Concentrated solar thermal plants offer the advantage of being able to store energy as heat. Accordingly, they can avoid the issues stemming from intermittency in power generation which are inherent to other solar technologies.

As further discussed in section 4 below, the type of technology used in a project may affect the legal structure of a project and give rise to different legal issues. The diagram below shows the main types of solar technology within the subsets of photovoltaic and concentrated solar thermal technologies.

Main solar technologies

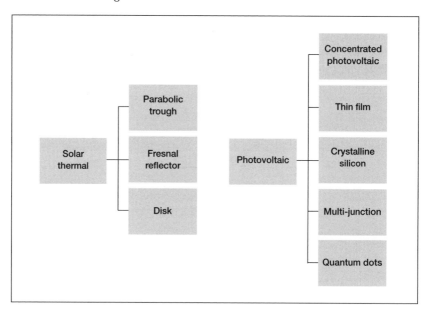

3. Financial incentives and associated legal risks

3.1 The need for financial support

A critical issue for solar projects is the relatively high capital cost of solar photovoltaic and concentrated solar thermal technologies compared to other renewable and conventional power generating technologies. The price differential per megawatt of generation capacity between solar power and other generation technologies is narrowing. However, it is still difficult to get an acceptable and financially viable rate of return in solar projects without governmental (or quasi-governmental) support or incentive.

A number of commonly used financial support and incentive mechanisms seek to encourage the use of solar power and support the development of solar projects. The form of support tends to vary from jurisdiction to jurisdiction and may depend on a number of factors, including:

- the nature of the existing electricity market and the types of existing power generation capacity in that market;
- the need for new capacity, and the availability and price of other fuels for power generation (eg, coal, gas and nuclear power);
- the government's regional policy for renewable power generation and the wish to encourage the use of power generation technologies that generate low greenhouse gas emissions;
- the suitability of the jurisdiction for the deployment of solar power (ie, the levels of insolation and the availability of land, water and interconnection); and
- the relative cost to the government of different power generation technologies.

From a legal perspective, the dependency of a project on financial incentives can give rise to risks that will need to be assessed during any legal due diligence process. Following that, solutions will need to be found to mitigate those risks if the project is to proceed, irrespective of whether the project in question is being debt-financed.

3.2   Initial legal risk assessment

Assessing the governmental support for a solar project is, in all likelihood, the first issue that a legal adviser for a solar project should consider. The key legal issues to consider here are the types of incentive available for solar projects and whether the incentive regime is legally robust.

3.3   Available financial incentives

Financial incentives for solar projects fall into two categories – indirect and direct incentives. Indirect incentives are environmental taxes on other power generation technologies that have an indirect and advantageous effect on the use of solar technologies. Direct incentives are solar tax incentives and credits, investment subsidies, feed-in tariffs and renewable energy certificates which specifically apply to solar technologies.

(a)   *Environmental taxes*

These may be imposed on non-solar electricity production with the effect of incentivising solar projects indirectly. Usually, these taxes allocate a cost to the environmental impact of conventional fuels and power generation technologies, such as coal and gas, used in that jurisdiction. Such taxes render solar energy competitive with conventional fuels and power because solar energy does not incur these environmental impact costs.

(b)   *Solar tax credits*

These are preferential tax schemes which aim to incentivise solar energy. This can be achieved by, for example, a lower level of taxation in the initial stages of a project, during which time the levels of investment and capital expenditure are at their highest. Examples of such credits include a deduction in taxable income or accelerated depreciation, or the lowering of value-added tax payable on solar equipment. Tax credit regimes are commonplace in the United States.[2]

(c)   ***Solar investment subsidies***

These are usually payable directly to the solar energy producer and are based on the rated power production capacity of the plant on completion of the plant's construction. The subsidy decreases the initial capital cost of the plant. Consequently, the electricity produced can be sold into the market at prices that are more likely to reflect the price of competing generation technologies. In practice, investment subsidies for solar energy are not as common as the other types of incentive described in this section. This is because they tend to encourage the construction of over-sized projects which, in the long term, are not sustainable. Subsidy regimes are used in conjunction with tax credits in the US solar market.[3]

(d)   ***Feed-in tariffs***

These tariffs fix and guarantee prices for the sale of solar energy. They are used in various countries (eg, Bulgaria[4], China[5], Czech Republic[6], France[7], Germany[8], Greece[9], Israel[10], South Africa[11] and the United Kingdom[12]) and are proposed in Turkey. Feed-in tariff regimes may vary from jurisdiction to jurisdiction, but are often structured to give the solar energy generator a guaranteed premium price – that is, a tariff for any solar electricity generated for its own use and a separate price for electricity exported and sold via the grid. In this model, the difference between the feed-in tariff and the market price is borne by either the government or the electricity suppliers and subsequently passed on to consumers. Often, feed-in tariffs are reduced over time to encourage early movers into the market.

(e)   ***Tradable certificates***

These certificates (eg, green certificates) are issued, in certain jurisdictions, to producers of electricity from solar energy. The certificates evidence the amount of energy that has been produced from an accredited solar plant. A market for these certificates must exist for them to create value for plant operators. The government usually creates such a market in the relevant jurisdiction, imposing, in parallel, a

---

2   The solar investment tax credit in HR 1424, the Emergency Economic Stabilization Act (2008).
3   *Ibid.*
4   The Renewable and Alternative Energy Sources and Biofuels Act 2007, as amended.
5   Law 180/2005 on the Promotion of Electricity Production from Renewable Energy Sources (2005).
6   REN21, (2009). "Renewables Global Status Report: 2009 Update," Paris: REN21.
7   Article 10 of Act 2000-108 on the Modernisation and Development of the Public Electricity Service, as amended on January 1 2008.
8   Renewables Energy Acts 2004 and 2009.
9   Law 3468/2006 on Production of Electricity from Renewable Energy Sources and High-Efficiency Cogeneration of electricity and heat and miscellaneous provisions.
10   REN21, (2009). "Renewables Global Status Report: 2009 Update," Paris: REN21.
11   *Ibid.*
12   Feed-in Tariffs (Specified Maximum Capacity and Functions) Order 2010/678.

slowly increasing requirement on electricity suppliers to purchase quantities of solar electricity. Thus, the certificates become tradable instruments that can be sold by operators, often separately from the solar electricity itself, to provide an income stream. In turn, this revenue improves the financial viability of the solar project. However, because this scheme can be complex and burdensome, micro-generation solar projects often use it in parallel to a feed-in tariff regime. In contrast, larger projects may use tradable certificates only as an incentive.

3.4   Legal analysis

If a solar project's viability depends on the regulatory regime offering financial support, preferential treatment or less advantageous treatment to other power generation technologies, regulatory risk will be an issue. Even in established and stable regimes which have comprehensive policies for renewable energies, there have been examples of regulatory changes that have affected the viability of either existing solar projects or future market development.

For example, the Spanish government is planning to reduce the solar tariff for photovoltaic and concentrated solar photovoltaic energy.[13] This follows on from the incredible growth of the local solar sector, which has created a financially burdensome obligation on the government. The market's development resulted from the falling costs of solar technology, itself a consequence of an increasingly competitive market which encourages greater than expected numbers of solar project start-ups. Although existing projects have so far not suffered from the change in tariff, the new policy seems to target the whole market in the future.

In Germany, feed-in tariffs were introduced in 2004.[14] They encouraged rapid growth in the solar photovoltaic market. However, in 2010 there have been moves to reduce the level of the tariffs and use the funds thus saved to provide more support to existing nuclear power plants.[15]

In view of the incentive regime relevant for each solar project and taking into account regulatory risk, a legal adviser should consider, at the very least, the questions in the checklist below. The legal adviser should try also to identify ways to mitigate the identified risks, either contractually or commercially.

3.5   Legal checklist

- What are the financial incentive regimes available for solar power?
- Does the financial incentive apply to all solar technologies or particular types of technology only – that is, is it applicable to the project and, if so, who qualifies for it?
- If the financial incentive is technology-specific, do different levels of banding apply to different types of technology?
- What is the legal nature of the available financial incentive and where does it arise?

---

13   Order ITC/3860/2007 regulating electricity tariffs as of January 1 2008.
14   Renewables Energy Act 2004.
15   Renewables Energy Act 2009.

- Who pays or bears the cost of the financial incentive, and is that person robust or creditworthy?
- Is the financial incentive guaranteed?
- For what duration will the financial incentive be payable to the project?
- Could the financial incentive regime change or could the incentive be withdrawn?
- Are there any mechanisms in place that would preserve the financial incentive for an accredited project – for example, grandfathering or stabilisation mechanisms?
- What does the project need to demonstrate to secure the financial incentive? Are any formal reports required?
- Is there a size threshold above (or below) which solar projects will not qualify?
- What is the process and deadline for the application for and registration of the financial incentive and how long does the process take?
- Can the project take advantage of more than one financial incentive regime or are the regimes mutually exclusive?

Once it is clear which incentives apply, the legal adviser should, in consultation with technical and commercial advisers, consider the above questions in relation to the specific incentive(s) that the project sponsors want to take advantage of.

An incentive established by statute will provide a project with more certainty, but such incentive is not necessarily immune to change. However, any future change is expected to be more considered and, therefore, more reasonable and predictable. In certain jurisdictions, change may also require a formal consultation, allowing for lobbying against detrimental changes to a project.

In all circumstances, it is prudent to assess the risk that the incentive may change, either before completion of the construction phase of the project or during the lifetime of the project. If such risk exists and the financial incentive regime is to be altered or reduced for solar projects, the legal adviser should verify whether any legal mechanisms can preserve the incentive(s) available to the project. These mechanisms are often known as grandfathering mechanisms.

Ideally, the applicable incentive will be guaranteed – that is, payable by a robust counterparty or backed by a government guarantee to pay the incentive for the relevant period. Further protection can also be sought for a potential change in law, tax provisions or other stabilisation provisions. These further protections will protect the solar project's revenues contractually, irrespective of a change in law that will detrimentally affect the support of the financial incentive. These provisions can be drafted to secure a right to re-set or re-negotiate the pricing provision of the contract if, in the event of a relevant change in law or taxation process, the contract no longer achieves the same financial outcome. However, these clauses are difficult to negotiate and implement in practice, as they are often contingent on having a governmental or quasi-governmenalt counterparty that is able to bear the risks involved.

In the absence of contractual protection, an investor will look for an incentive that is clearly defined and available for a certain and suitable period of time. There are many examples of incentives available for between 15 and 25 years which allow for the recovery of capital investment and a return on an investment.

In relation to indirect incentives, such as the environmental taxes discussed in section 3.3, as the incentive is not provided directly to the solar project itself, it is unlikely that the sponsor can exert any control over it. However, certain projects may be able to secure a direct contractual promise from the government to protect them from any adverse impact if the indirect incentive is abolished.

3.6   Clean Development Mechanism

A discussion of incentives for solar projects would not be complete without mentioning the Clean Development Mechanism, which is one of the flexible mechanisms under the Kyoto Protocol to the UN Framework Convention on Climate Change.[16] Under the Clean Development Mechanism, industrialised and developing countries can achieve emission-reduction commitments by way of emission-reducing projects such as solar projects. This is done by using emission-reduction credits from these projects, known as certified emission reductions, to offset their own commitments.

To obtain certified emission reductions, a project must comply with the criteria of the Marrakech Accords for the Clean Development Mechanism. These requirements include:

- ratification of the Kyoto Protocol by the country in which the solar project is located;
- approval by the host country (and conformation of the project to national sustainable development criteria);
- demonstration of the solar project's measurable long-term benefits for climate change mitigation;
- demonstration of additionality – that is, a reduction in greenhouse gases in addition to reductions achieved without the project activity; and
- compliance with a detailed process for obtaining Clean Development Mechanism status and the issue of certified emission reductions.

Developers of solar projects need to clear a number of procedural and accreditation hurdles, both nationally and internationally, before they can obtain certified emission reductions. Furthermore, various risks (including regulatory risk) linked to the Clean Development Mechanism regime will apply to solar projects after 2012. However, subject to the resolution of this issue, the generation of certified emission reductions will be a valuable additional income stream for a solar project in a developing country.[17]

---

16   For the full text of protocol see http://unfccc.int/resource/docs/convkp/kpeng.pdf For the status of ratification see http://unfccc.int/files/kyoto_protocol/status_of_ratification/application/pdf/kp_ratification_20091203.pdf.

17   For further information relating to the Clean Development Mechanism, see the discussion of the Copenhagen Sustainable Meetings Protocol in the "Why renewable energy?" chapter.

4. Typical project agreements and issues

4.1 Introduction

The specific project documents required for a solar project will need to be determined on a project-by-project basis. The following types of question need to be asked when considering the documentation:

- Is it, for example, a small-scale, land-based solar photovoltaic project, a roof-mounted solar photovoltaic array (or portfolio of arrays) or a large-scale, desert-based solar thermal power project?
- Will there be multiple sites and offtakers or one main site, offtaker and grid connection?
- Will the project be project-financed?
- Will the construction and operation of the installation be complex or simple?

Key project contracts and arrangements for solar projects

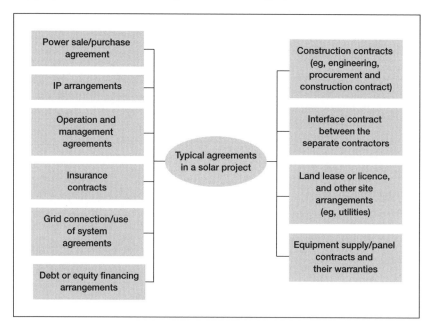

This section focuses on the following arrangements, as they are particularly relevant to solar projects:

- power purchase/sale agreements; and
- construction/engineering, procurement and construction arrangements.

4.2 Power purchase/sale agreements

A power purchase agreement is an agreement for the sale of electricity from a buyer (eg, the solar project investor) to an offtaker (eg, the local supplier or grid company). Such agreements are frequently used in the context of both solar photovoltaic and concentrated solar photovoltaic projects. This is often the case, for example, in large concentrated solar photovoltaic projects, where the entire capacity of the project is sold to an offtaker. This provides a revenue stream which helps to support the scheme's project financing.

Power purchase agreements are also increasingly used in solar photovoltaic projects such as roof-mounted photovoltaic projects. Rather than purchasing the photovoltaic system or equipment itself, the customer or land owner provides the roof space or other land required for the photovoltaic system to the solar project company. The solar project company is then responsible for financing, installing, operating and maintaining the photovoltaic system. It charges the customer a pre-determined rate for the electricity produced. This electricity is sold through a power purchase agreement.

The arrangement has obvious advantages for customers that do not wish to get involved in operating or maintaining the photovoltaic system. In addition, the arrangement may be structured so that the customer has limited or no costs to pay upfront, but then commits to a long-term power purchase agreement. The price and pricing structure under the agreement will obviously be open to negotiation. However, both parties will be looking for a degree of price stability – although in most cases the consumer enjoys the stronger bargaining position.

In turn, the solar project company benefits from an immediate, guaranteed source of income and a good location to generate solar energy. Further, ownership of the photovoltaic system may mean that the project will receive the applicable incentive(s) directly. If this is not the case, the project will need to ensure that either:

- it takes an assignment of the incentive (if it is capable of being assigned); or
- the pricing in the agreement reflects the incentive paid directly to the customer.

For example, in the United Kingdom, the feed-in tariff is made up of two components:

- a generation tariff, which is a fixed payment for every kilowatt-hour generated for a period of 25 years; and
- an export tariff, which is an additional payment for every kilowatt-hour exported to the wider energy market.

For domestic and commercial properties with a solar rooftop installation, the party eligible for the feed-in tariff will be the householder or business occupying the premises. The tariff will be paid by the electricity supplier buying the electricity from the generator. However, the UK feed-in tariff regime also allows for the tariff payments to be assigned to the solar project company, as described above.

In this context, the legal advisers in the solar project should, at least, consider the following:

- If the solar installation is owned by a project company, does that company have sufficient rights of tenure to allow the installation and maintenance of the solar photovoltaic system at the customer's premises?
- What happens if the contractual arrangements between the customer and the

project company terminate early due to a breach by the customer? In these circumstances it may be appropriate to provide for a lump sum payable to reimburse the company for any loss of future/expected income.

• What happens if there is a new owner or occupier of the site?

• Which party will be responsible for all relevant consents and approvals for the solar equipment?

• What happens in case of a change in law that abolishes or reduces the incentive? Can any of this risk be passed onto the consumer and is the consumer sufficiently creditworthy to bear this cost?

4.3   Construction contracts

In solar projects, it is likely that the developer will need to enter into a construction contract for the supply and installation of the solar-specific equipment, such as photovoltaic panels and trackers. However, there will also be other construction or civil work for the plant, and in many cases there may need to be two separate contracts (one contract for the solar specific equipment and one contract for the civil works), which will be entered into by two separate contractors.

Where two engineering, procurement and construction contracts are used, the project company will need to ensure that there is an interface contract between the contractors. The interface contract will set out the rights of one contractor against the other. For example, if the photovoltaic panel contractor suffers delay in the performance of its work as a result of the construction or civil work contractor's failure to keep to the agreed deadline, the panel supplier, in accordance with the interface documentation, can seek compensation from the civil works contractor. Furthermore, the photovoltaic panel contractor will have no right to make a claim against the developer under the engineering, procurement and construction contract.

Key categories of regulatory approval usual in solar projects

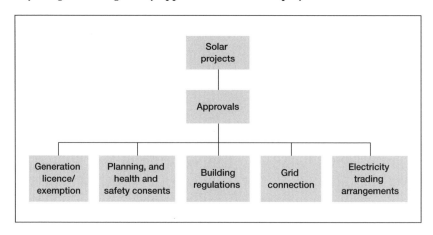

5. Key regulatory issues and approvals

As part of the initial regulatory due diligence, the legal adviser should assess what approvals and other regulatory requirements need to be fulfilled for the proposed solar project to be properly permitted. The relevant regulatory requirements will differ from jurisdiction to jurisdiction. However, there are likely to be common themes for the requirements, which are discussed below.

5.1 Regulatory checklist

(a) *Licences/consents*

- What project-specific licences or consents are needed (eg, for the generation of electricity) and who is responsible for granting them?
- Can any required licences or consents be granted as part of a concession (if any) or will they be granted separately?
- What is the procedure and timescale for getting the necessary licences/consents? Does the government provide one-stop shop facilities for the batching and follow-up of applications on behalf of individual applicants?
- Is it possible to get in-principle approval before grant of the formal licence/consent? For example, could in-principle approval be given on the basis of draft plans and designs?
- What are the fees for any licences or consents?
- What is the duration of the required licences/consents? If they are not coterminous with the solar project, can they be extended or renewed? If so, on what grounds and subject to what procedure?
- Can the required licences/consents be assigned or transferred?
- On what grounds can such licences/consents be terminated? Will a change of control in the project company trigger termination?

(b) *Sector regulation*

- What is the current structure (as regards ownership, operation and regulation) of the market sector in which the solar project is going to operate?
- Is there any legislation that specifically deals with the subject matter of the solar project (eg, electricity or infrastructure laws)? What are the scope and major features of any such law?
- What territory or types of project are covered by such legislation?
- Are there any restrictions on the structure or shareholding of a private sector participant in the solar market?
- Are there any statutory rights or restrictions that apply to companies involved in the solar sector?
- Will a solar project company have any public service obligations, such as duties to supply to particular/nearby customers? Would the project company be able to disconnect in the event of non-payment?
- What activities are regulated (eg, construction standards, technical issues, health and safety, operation, emergencies or financial control)?

(c)   ***Regulator***

- Is there a solar regulator and are there any other regulators with oversight of the project (eg, competition authority or appellate body)?
- What enforcement mechanisms are available to any of these regulators? For instance, can the project licence or concession be revoked? Can fines or other penalties be imposed?

(d)   ***Economic regulation***

- Does the government, a public sector body (eg, regulator) or a third party have the ability to restrict the fees, pricing or tariffs charged by a private sector entity in the solar sector?
- Are there any legal restrictions on the structure of the fees, pricing or tariffs?
- Are there any legal restrictions on the rate or provisions for the increase of payments under the fees, pricing or tariffs?
- Is governmental or regulatory approval of fees, pricing or tariffs required? If so, how is this structured (eg, yearly or for the duration of the licence)? What comfort can the project company and its lenders be given as to the reasonable character of the level of regulation?

5.2   Planning approvals

The planning approval issues relating to solar plants are similar to those that apply to other renewable energy projects. The planning approval process is often time-consuming. Part of the reason for the lengthy nature of this process it that an environmental impact assessment and a subsequent consultation with various stakeholders – such as government departments responsible for environmental protection, and health and safety – may be required.

The difficulty of obtaining planning permission may vary depending on the size of the project (eg, large or small on-site installations). As with other sectors, planning approval issues depend on geography and location, as well as factors such as the visual impact of the project.

5.3   Building regulations

For rooftop installations, in addition to obtaining planning consent, it is essential to determine whether:

- the relevant building regulations allow for the panels to be installed; and
- there are any preventative building regulations that need to be overcome.

Ideally, both the planning and building regulation regimes will address solar rooftop installations. This will reduce the uncertainty associated with determining how the relevant requirements will be applied to a solar installation.

5.4   Connection to the grid

Connection to the electricity transmission and/or distribution systems is a crucial step for any new solar project from which electricity is to be sold into the grid or which may require top-up supplies of electricity from the grid. There are three issues to consider. First, what are the arrangements for the connection? For example, is there a formal application process and, if so, what are the application requirements? Is there any documentation, such as a comfort letter or heads of terms connection agreement, that can be signed with the grid company to give the project comfort regarding the future connection?

Second, what are the timescales for implementing connections? Is there a risk of potential delay for the project arising from the availability of grid connections or the need for reinforcements to the system?

Third, are there any priority arrangements for renewable energy (including solar energy)?

In the United Kingdom, for example, there is a large queue for new projects seeking connection to the transmission network. In 2009 connection dates were being offered for as late as 2023.[18] Until recently, this queue for access to the grid was determined by the date on which the application for connection was made, rather than by how ready the project was for connection. This meant that less advanced projects blocked projects that were more developed and ready for connection. Effectively, developers submitted an application for connection even when there was no realistic prospect of the project in question being connected by the contracted date. To rectify this, the UK government has recently intervened to amend the way that new connections are managed.[19]

Thus, possible delays in connection are a risk that needs to be assessed and catered for in any solar project. In some jurisdictions, priority is given to renewable energy installations to obtain connection to the grid, which will be beneficial to the progress of the solar project.

5.5   Electricity trading arrangements

Where a solar project is being planned in a jurisdiction where there is an electricity trading market in place, the legal adviser to the project should consider how the project will fit into the market arrangements. For example:

- will the solar project need to take steps to comply with any mandatory electricity market arrangements, such as those relating to the sale of electricity from the project into the market;
- would it be beneficial for the solar project to be involved in the electricity market arrangements even if it is not mandatory for it to do so; and
- is there any special treatment of electricity from solar sources, such as priority dispatch?

For larger installations, solar farms in particular, the above questions are critical. In jurisdictions that recognise the intermittent nature of most renewable energy, renewable energy generators have priority dispatch rights. Additionally, for larger on-site installations where some of the surplus electricity is sold into the grid, it is important to

---

18   The government's response to the House of Lords report from the Select Committee on Economic Affairs (2007-2008),"The Economics of Renewable Energy", November 25 2008 (London: The Stationery Office Limited).

19   *Ibid.*

consider whether any special arrangements would make it easier for the installation to participate in the market. Furthermore, micro-generators are often exempt from any electricity market rules. However, this needs to be confirmed in every case.

6.   Conclusion

Two contemporaneous factors have made the development of solar energy projects more complex. First, there is an increased multinational and multi-jurisdictional element to these projects. For example, due to the saturation of and reduced financial incentives in the solar markets of Germany and Spain, developers in these respective markets have been looking to use their expertise outside their national boundaries. Therefore, it is not uncommon to find projects involving parties from a variety of jurisdictions, each with a different understanding of how a solar project is structured.

Second, awareness of climate change, environmental concerns and the finite character of supplies of fossil fuels has increased. This awareness and the search for alternative energy supplies have meant that governments in areas where there has traditionally been limited solar energy development have been investing time in drafting new legislation to structure and incentivise the development of solar energy.

In essence, competing ideas about solar project development and breaking new ground with new legislation have fashioned a number of new issues for solar project developers and investors alike. The international element and variety of government-backed regimes mean that solar project developers and investors need, more than ever, to have a clear understanding of:

- the financial incentives on offer;
- the project documents required; and
- the consents and approvals needed both to protect the financial viability of a solar project and to cover all aspects of the apportionment of risk for the solar project in question.

This chapter aims to provide an insight into the issues and themes listed above. Through generic advice and specific checklists, it should give solar project developers and investors food for thought. However, readers are warned that there will be differences in regulation between jurisdictions and each project will need to be addressed on a case-by-case basis.

Small-scale hydropower in Europe: legal issues

María José Descalzo
Juan I González Ruiz
Uría Menéndez

1.   General overview

Increasing concerns about sustainability, security and diversification of energy supply, as well as pressure to reduce the dependence on oil and natural gas, mean that renewable energy is promoted not only in Europe, but also in the rest of the world.

Many consider hydropower as the world's largest source of affordable renewable energy. It represents about 20% of the world's electricity supply[1] and around 19% of Europe's electricity supply.[2]

Hydropower facilities have been developed in the past century all over the world. Large plants are usually developed by state-owned companies or, at least, under governmental schemes that guarantee the remuneration of the investment in the facilities.

This type of facility is well developed in industrialised countries, while it encounters financial difficulties in developing countries. Consequently, the promotion of hydropower energy nowadays usually focuses on small-scale facilities, and the upgrading and rehabilitation of old plants.

Despite the lack of consensus on the definition of 'small hydropower' (eg, in Canada it includes facilities up 25 megawatts (MW),[3] in the United States up to 30MW[4] and in Spain up to 10MW[5]), the largest countries of the European Union consider that small hydropower facilities have an installed capacity of 10MW or less.[6] The European Union boasts 13 gigawatts (GW) of installed small hydropower facilities, with 23GW of potential installed capacity.[7]

Large hydropower facilities are competitive in themselves, but small facilities (as well as other renewable technologies) still need economic support to compete in the market.

This chapter considers the legal issues surrounding the development of small-scale hydropower projects in Europe, with a case study of the Spanish hydro market. Many of the matters considered in this chapter may apply in other parts of the world.

---

1   Data from the World Bank (www.worldbank.org).
2   Data from the European Small Hydropower Association (www.esha.be).
3   European Small Hydropower Association, *Renewable Energy World* n°253.
4   *Id.*
5   Spanish Royal Decree 661/2007.
6   Data from the European Small Hydropower Association (www.esha.be).
7   *Id.*

2. EU regulatory framework

2.1 Renewable energy framework

The first issue to consider is the EU regulatory regime for the promotion of renewable energy. In 1997 the European Commission published a white paper on renewable energy setting forth the strategy of the EU member states. Further, the European Union (and therefore its member states) became a signatory to the Kyoto Protocol[8] on May 31 2002; by doing so the European Union set the objective of reducing its greenhouse gas emissions to 8% below its 1990 levels by 2012.

Prior to adhering to the Kyoto Protocol, the European Union had adopted the Renewable Energy Directive (2001/77/EC) to encourage the development of electricity produced from renewable energy sources (non-fossil fuel sources, including hydropower, wind, solar, geothermal, wave, tidal, biomass, landfill gas, sewage treatment plant gas and biogas).

Under the Renewable Energy Directive:

- member states are required to set national indicative targets for the consumption of electricity produced from renewable sources consistent with the European Union's target of generating 12% of its energy and 22% of its electricity from renewable energy sources by 2010;
- member states may introduce support schemes;
- a system of guarantee of origin was introduced to enable generators to demonstrate the renewable origin of electricity;
- member states had to introduce measures to simplify administrative procedures; and
- clear, transparent and objective access to the grid had be implemented for such facilities.

In addition, Directive 2003/87/EC (amending Directive 96/61/EC) established a scheme for trading greenhouse gas emission allowances within the European Union. The aim of the EU Emissions Trading Scheme is to help EU member states comply with their commitments under the Kyoto Protocol. As further explained below, hydropower facilities contribute to the reduction in carbon dioxide emissions.[9]

The European Union reaffirmed its commitment to the promotion of energy from renewable sources on January 10 2007 with the presentation of a long-term Renewable Energy Roadmap (COM (2006) 848 final). The roadmap, which proposes a mandatory target of generating 20% of energy from renewable sources by 2020, explains the necessity of generating energy from renewable sources and establishes a pathway for

---

8   The Kyoto Protocol is an international agreement adopted on December 11 1997 and linked to the United Nations Framework Convention on Climate Change, which was adopted in New York on May 9 1992. Under the protocol, 37 industrialised countries and the European Union legally committed to reduce gas emissions by an average of 5.2% below 1990 levels by 2012. For further information on this, please read section 3.5(b) in the chapter of this book entitled "Renewable energy support mechanisms: an overview".

9   According to the data published by the European Small Hydropower Association on its website, 1 gigawatt-hour of hydropower supplies electricity to about 220 European households. This avoids the emission of about 700 tonnes of carbon dioxide.

bringing renewable energy into mainstream EU energy policies and markets. The roadmap also proposes a new legislative framework for the promotion and use of renewable energy in the European Union, to provide participants in the renewable energy sector with long-term stability to make rational investment decisions.

In March 2007, following the presentation of the Renewable Energy Roadmap, the EU Council:

- endorsed the binding target to achieve 20% of overall EU energy consumption from renewable energy sources by 2020;
- made a firm, independent commitment to achieve at least a 20% reduction in greenhouse gas emissions compared to 1990 levels by 2020; and
- stressed the need to increase energy efficiency in the European Union so as to reduce its energy consumption by 20% compared to current projections for 2020.

To that end, the European Union adopted a new directive on the promotion of the use of energy from renewable sources (Directive 2009/28/EC), in line with the Renewable Energy Directive. The new directive establishes certain targets to be reached by 2020. This directive:

- establishes a common framework for the promotion of energy from renewable sources;
- sets mandatory national targets for the overall share of energy from renewable sources in gross final consumption of energy and for the share of energy from renewable sources in transport;
- provides different support schemes;[10]
- establishes national action plans;[11]
- imposes the reduction of administrative barriers on member states;[12] and
- introduces the possibility of statistical transfers, joint projects and joint support schemes between member states, as well as with third countries.

---

10   Pursuant to Article 2 of Directive EC/2009/28:

"support scheme, means any instrument, scheme or mechanism applied by a Member State or a group of Member States, that promotes the use of energy from renewable sources by reducing the cost of that energy, increasing the price at which it can be sold, or increasing, by means of a renewable energy obligation or otherwise, the volume of such energy purchased. This includes, but is not restricted to, investment aid, tax exemptions or reductions, tax refunds, renewable energy obligation support schemes including those using green certificates, and direct price support schemes including feed-in tariffs and premium payments".

11   Member states need to adopt a National Action Plan that sets targets for the proportion of energy from renewable sources in transport, electricity, heating and cooling in 2020 and adequate measures that must be taken to reach these targets. The European Union approved a format of report to be submitted by March 31 2010. Member states had to include in such a report the measures provided to simplify the administrative procedures in order to set up a single authority responsible for granting all the administrative permits and licences needed.

12   In respect of the administrative procedures, Directive 2009/28/EC aims to ensure that national rules on authorisation procedures are proportionate and necessary, and promote the coordination with the different authorities (local, regional and national). The rules should also comprise:
- comprehensive information on the processing of applications;
- streamlined administrative procedures;
- objective, transparent and non-discriminatory rules;
- transparent costs; and
- less burdensome procedure for small installations.

2.2 Support schemes

The renewable energy industry benefits from government subsidies and incentives in the European Union. For the purposes of the new Renewable Energy Directive, all hydropower facilities (except the pumping storage[13]) are considered renewable energy.[14]

As stated in the chapter of this book entitled "Renewable energy support mechanisms: an overview", there are different types of incentive and subsidy to benefit the producers of electricity from renewable energy sources.

According to the European Union, the different mechanisms can be divided into investment support (eg, capital grants, tax exemptions or reductions on the purchase of goods) and operating systems (eg, price subsidies, green certificates, tender schemes and tax exemptions or reductions on the production of electricity).[15] The latter category can be further divided into quantity-based market instruments and price-based market instruments.

(a) Quantity-based market instruments

Quota obligation: Governments impose an obligation on consumers and suppliers to acquire a certain percentage of their electricity from renewable energy. This obligation is usually facilitated by tradable green certificates. To satisfy this obligation, the suppliers must acquire environmental certificates in amounts equal to their quotas. Suppliers that fail to do so are usually subject to a penalty.

In this type of system, renewable energy producers receive revenue from two sources:

- purchases of electricity sold to the market; and
- purchases of tradable green certificates sold to suppliers that need to satisfy their quotas (ie, Belgium, Latvia, Poland, Romania, Sweden and the United Kingdom).

Public auctions: A tender is announced for the provision of a certain amount of electricity from a certain technology source. The process of bidding should ensure that the cheapest offer is accepted.[16]

(b) Price-based market instruments

Feed-in tariffs and premiums are the preferred system in the European Union (18 member states use this system, including Denmark, Germany and Spain). Feed-in tariffs and premiums are granted to generators for the electricity that they sell to the system. Each government sets the preferential, technology-specific feed-in tariffs and premiums that it pays to producers. Typically, there are two types of feed-in tariff:

- Regulated tariffs – generators have the right to receive a regulated fixed tariff

---

13   "Conventional pumped hydro uses two water reservoirs, separated vertically. During off peak hours, water is pumped from the lower reservoir to the upper reservoir. When required, the water flow is reversed to generate electricity" – Electricty Storage Association.

14   Article 5(3) of Directive 2009/28/EC.

15   Commission Staff Working Document, "The support of electricity from renewable energy sources" (23.1.2008 SEC(2008) 57).

16   *Id.*

that may vary depending on the installed capacity of the facility or the date of start of operation.

- Premiums – the renewables generator has the right to receive the market price plus a premium. However, some member states (eg, Spain) have introduced caps and floors so that the total remuneration received by such facilities will be limited to the reference market price or the price freely negotiated, plus the reference premium, subject to the applicable caps and floors.

The cost of this system is normally covered by end consumers.

The feed-in tariffs are guaranteed for a limited period only, varying – depending on the member state – from 10 to 25 years. In the recent years tariffs have depended on technology too, with some members states introducing a decreasing tariff system for certain technologies (eg, Germany and Spain for solar photovoltaic facilities) to promote investment in such technologies and therefore cost reductions.

Fiscal incentives, such as tax exemptions or reductions, are also used. Under such schemes, producers of renewable electricity are exempted from certain taxes (eg, carbon taxes).

In addition, most EU member states have reinforced the position of renewable energy generated in their electricity systems by recognising a dispatching priority for renewable energy and priority of access to the grid (eg, Denmark, Germany and Spain).

2.3 Water Framework Directive

As hydropower projects require the use of water resources, it is necessary to consider the EU Water Framework Directive (2000/60/EC). The main purpose of the directive is to establish a framework for the protection of inland surface waters, transitional waters, coastal waters and groundwater. The directive:

(a) prevents further deterioration and protects and enhances the status of aquatic ecosystems and, with regard to their water needs, terrestrial ecosystems and wetlands directly depending on the aquatic ecosystems;

(b) promotes sustainable water use based on a long-term protection of available water resources;

(c) aims at enhanced protection and improvement of the aquatic environment, inter alia, through specific measures for the progressive reduction of discharges, emissions and losses of priority substances and the cessation or phasing-out of discharges, emissions and losses of the priority hazardous substances;

(d) ensures the progressive reduction of pollution of groundwater and prevents its further pollution, and

(e) contributes to mitigating the effects of floods and droughts
  and thereby contributes to:
  – the provision of the sufficient supply of good quality surface water and groundwater as needed for sustainable, balanced and equitable water use,
  – a significant reduction in pollution of groundwater,
  – the protection of territorial and marine waters, and
  – achieving the objectives of relevant international agreements, including those

which aim to prevent and eliminate pollution of the marine environment, by Community action to cease or phase out discharges, emissions and losses of priority hazardous substances, with the ultimate aim of achieving concentrations in the marine environment near background values for naturally occurring substances and close to zero for man-made synthetic substances.

To reach these targets, the directive imposes restrictive conditions on the use of water (eg, it forbids any decrease in the ecological quality of the water). The directive allows economic uses of water, as long as negative impacts are fully mitigated and uses are included in advance in the relevant river basin plan.[17] (A river basin is a natural geographical and hydrological unit for water; under the directive, a river basin management plan needs to be set up for each river basin district.)

A general requirement for ecological protection and a general minimum chemical standard were introduced to cover all surface waters. These are the 'good ecological status' and 'good chemical status' elements defined in the directive. Specific protection zones are also defined.

For more information on the relevance of the directive to hydropower projects, see section 3.6 below.

3. Key issues for small hydropower projects

Hydropower facilities offer significant advantages (eg, they are affordable and highly efficient, do not emit greenhouse gases during operation and run on mature, reliable technology). However, they also present disadvantages that may substantially hinder their development.

3.1 Geographical constraints

Not all sites are suitable for implementation. The development of a hydropower facility requires water in certain conditions (ie, rainfall and precipitation, a minimum flow of water) and certain geographical characteristics (ie, a geographical set-up that allows the construction of a dam; river vallies; often in remote areas). The construction of a hydropower facility, which may entail the resettlement of people and the displacement of land, needs to balance all of the different uses of the river water (eg, human consumption, irrigation, fishing). Accordingly, the sponsors of a hydropower project must adopt specific measures to coordinate the different interests and mitigate the negative effects of their project on the remaining uses.

For this reason, in recent years the main efforts of investors have focused on the upgrade and refurbishment of existing hydropower plants. The potential future of this technology is typically linked to the reutilisation of existing infrastructures. The possibility of upgrading an existing facility is especially relevant to large hydropower plants, as most of them were built more than 50 years ago. According to data published by the European Small Hydropower Association, upgrades may represent

---

17   For more information on the impact of the EU Water Directive on hydropower facilities, please refer to Mike Landy and James Craig, "Small-scale hydropower: how to reconcile electricity and environmental protection goals?" (ETC/ACC Technical Paper 2009/13- December 2009) (The European Topic Centre on Air and Climate Change (ETC/ACC).

5% of the potential of the existing European hydroelectric production - that is, 5% of the electricity produced could come from refurbished facilities.[18]

3.2   Relevant initial investment

Hydropower facilities are capital intensive because:
- they require substantial civil works;
- there often raise geographical and land issues;
- access to the site is usually difficult, as is connection to the grid.

These aspects, among others, may increase costs. By contrast, operating and maintenance costs are low and the useful life of the facilities is long (ie, normally at least 30 years for the equipment and indefinite for the civil works). By way of example, in Spain, large hydropower facilities were built during the 1960s and are still in operation. In 2005 the Spanish Ministry of Industry, Tourism and Commerce calculated the generation cost per kilowatt-hour of a hydropower facility. The cost for facilities with an installed capacity of 10MW or less was between €0.045 and €0.061. The cost for facilities with an installed capacity of more than 10MW was between €0.041 and €0.056.[19]

3.3   High construction risks

These follow from:
- the nature of the works;
- the specific access conditions and nature of the sites; and
- the potentially long construction periods, compared to other generation projects.

Because of these risks, investors will typically opt for an engineering, procurement and construction contract or a turnkey contract under which the contractor assumes responsibility for construction and delivery of the project, receiving in return a fixed price. The construction risk is therefore theoretically passed onto the contractor. However, while much depends on the negotiating skills of the sponsors, the complexity of constructing a hydropower project makes it unlikely that all aspects, costs and risks may be foreseen and thus transferred to the contractor.

Further, a long and complex construction phase is likely to have the following consequences:
- The engineering, procurement and construction contractor may be reluctant to accept a fixed price in advance without the inclusion of a price adjustment mechanism. In addition, a fixed price is likely to be determined on the basis of assumptions so that if such assumptions are modified, the fixed price can be adapted accordingly.

---

18   European Small Hydropower Association, "Current status of small hydropower development in the EU-27" (September 2008).
19   Spanish Ministry of Industry, Tourism and Commerce, *Plan de Energías Renovables en España 2005-2010* (www.mityc.es).

- A long construction period may increase the possibility of changes in regulations, which may result in an increase in construction and operation costs. Thus, the engineering, procurement and construction contractor must take into account such circumstance when determining the fixed price or in order to include price adjustment mechanisms.

3.4 Hydrological circumstances

Hydropower facilities depend on water resources that are rarely managed by the parties involved. Promoters of hydropower facilities will have access only (if at all) to historic and statistical data, making it difficult to assess future programmes. In addition, dry years may cause production deficits, which may in turn have a material impact on the financing of the project.

3.5 Complex administrative procedures

As outlined in more detail in section 4 below, a hydropower facility usually requires a variety of authorisations. The administrative procedure to obtain such authorisations is often complex and time consuming, especially if the facilities are to be located at a new site (as opposed to repowering or expanding an existing hydropower site). Data published by the European Small Hydropower Association shows that administrative procedures take on average from 12 months in Austria to 12 years in Portugal.[20] The fact that hydropower entails the use of public resources (in particular, land and water) adds a layer of complexity that must be taken into account when planning a project. The procedure becomes even more complex if two or more jurisdictions are involved: each may need to issue specific authorisations to use the water resources (as well as other permits that may be required). It is also likely that each jurisdiction will follow different procedures.

In Spain, a negative report (eg, in respect of the environmental impact) by an administrative authority involved in the procedure will likely put the entire project on hold. Thus, at the time of writing several projects were still awaiting authorisation.

3.6 Environmental issues

Large dams have the potential to inundate lands and affect ecosystems by interrupting the flow of water, reducing the quality of the water, creating barriers for fish and increasing noise, among other things. When developing a hydropower facility, it is important to consider environmental measures to mitigate its negative impact.

As discussed in section 2.3 above, the EU Water Framework Directive imposes strict restrictions on the use of water. A project will not be allowed unless relevant and significant corrective and predictive measures are implemented (which typically represent a considerable increase in project costs).

According to the European Small Hydropower Association, the Water Framework

---

20   European Small Hydropower Association, "Administrative barriers for small hydropower development in Europe" (June 2007).

Directive and the Renewable Energy Directive have contradictory impacts on the further development of small hydropower facilities.[21] On the one hand, the Renewable Energy Directive intends to promote renewable facilities. On the other hand, the Water Framework Directive aims to preserve a "good ecological status" that effectively forbids any decrease in ecological quality; any changes should aim at increasing the ecological quality. The construction of hydropower facilities requires the implementation of relevant mitigation measures since such construction affects the river basin.

However, in recent years new technologies (eg, fish-friendly turbines) have been developed which allow the implementation of new measures that may mitigate or even eliminate the negative impact of hydropower facilities.

3.7 Financing constraints[22]

The construction and financing of hydropower facilities (especially large ones) have been traditionally carried out by the public sector or with the support of the public sector. However, the liberalisation of the market, in particular within the European Union, means that the public sector is now unlikely to undertake the development of new facilities in the majority of member states, other than through financial and tariff support mechanisms. As the World Bank pointed out in its guidelines "Financing of Hydropower Projects" (July 2000), the key issues in the financing of private hydropower projects are bankability and affordability. However, there are multiple costs (eg, the construction phase costs noted in section 3.3 above) that may make it difficult to finance these projects on a private basis.

The majority of the projects that the World Bank examined were financed through project finance structures in which a special purpose company had been incorporated and in which the equity was provided by the sponsors.[23]

In practice, many of the private projects in the recent past have been financed through limited-recourse loans – that is, project finance. The term 'project finance' is used to refer to "a non-recourse or limited recourse financing structure in which debt, equity and credit enhancement are combined for the construction and operation, or the refinancing of a particular facility in a capital-intensive industry."[24]

This includes electricity generation, in which the lenders rely on the cash flow of the project (typically a power purchase agreement with the utility company that guarantees the acquisition of the electricity output generated at a set price) for debt repayment.[25] Project finance presents important advantages; for instance, a pure project finance structure does not require the sponsor to guarantee repayment of the debt. However, this structure also presents disadvantages, such as a complex risk allocation, higher interest and fees, and the ongoing supervision of the lenders.

---

21   See, for example, European Small Hydropower Association, "Contradictions between the Water Framework Directive and the RES-electricity Directive" (August 3 2004).
22   This subsection should be read in conjunction with the chapter of this book entitled "Issues for financiers".
23   World Bank, "Financing of Hydropower Projects", Section 5, p46.
24   Scott L Hoffman, *The Law and Business of International Project Finance*, Third Edition.
25   *Id*.

To facilitate the assessment by financial institutions of the social and environmental issues that may arise in this type of project, some leading international financial institutions have adopted a set of principles to ensure that projects are developed in a socially responsible manner and reflect environmental management practices. These principles, known as the Equator Principles, are intended to be used as a general framework by financial institutions in their project financing activities.[26] Under the principles, the bank entities provide only loans that conform to such principles.

Together with project finance, the following debt sources are available for generation projects (not all the sources listed will be available for every project):

- government funding, through grants, loans and guarantees;
- equity from sponsors or third parties (eg, the engineering, procurement and construction contractor) that may support the project until completion;
- corporate loans from banks (local and international) and institutional lenders. Loans are secured against assets and personal guarantees of the sponsor;
- support, usually in the form of loans, insurance or guarantees, from bilateral agencies. Typically, such agencies are designed to promote trade or other interests in a country. However, as hydropower projects rarely contain an export element, they rarely use bilateral agencies;[27]
- credit support for project finance in developing countries, from multilateral development banks such as the World Bank Group, the Inter-american Development Bank and the Asian Development Bank to regional development banks and other international agencies. Pursuant to its articles of association, the World Bank grants loans to finance specific projects provided that the borrower is unable to obtain a loan from any other sources on reasonable terms regarding the conditions to repay the loan. The World Bank also provides guarantees, although a counter-guarantee from the host country is required in this case.[28]
- World bank-approved financing, given to some hydropower projects in Africa and Asia, along with rehabilitation projects in Eastern Europe. Bank approvals for hydropower have raised over $800 million in 2008.[29] It seems that in response to demand to promote such projects, the financing for energy infrastructure by the World Bank has been strong, reaching $8.2 billion in 2009 (more than $2.147 billion for Europe and Central Europe); and[30]
- bond issues. Although they are similar to loans in many respects, differences abound.[31] Besides, in practice capital markets are accessible only to established and financially strong companies.

---

26   www.equator-principles.com.
27   World Bank, "Financing of Hydropower Projects", p66.
28   Article III(4)(i) of the World Bank's articles of association.
29   World Bank.
30   World Bank.
31   This subsection should be read in conjunction with the chapter of this book entitled "Issues for financiers".

3.8 Project finance risk allocation[32]

As project finance is non-recourse to the sponsor, the projects risks need to be allocated to the different players in the project. Four different risk stages can be distinguished:

- development risks (eg, failure to obtain permits or licences) that correspond to the project sponsors and, if any, the development lenders;
- construction risks (see section 3.3 above);
- start-up risks (eg, non-compliance with the performance tests), which are usually assumed by the engineering, procurement and construction contractor; and
- operating risks that may appear during the operation of the project and that probably will be borne by the operator under the operation and maintenance contract or the equipment supplier if the equipment (eg, turbine) guarantee has not expired.

Besides development and technical issues, the project may also be affected by currency-related risks (eg, unavailability of foreign exchange, inability to transfer foreign exchange abroad, currency devaluation risks).[33]

3.9 Regulatory issues

Besides the barriers raised by the lengthy administrative procedures,[34] local governments' positions and the modification of the applicable regulations raise particular problems in this type of project. Thus, according to the European Small Hydropower Association, recent measures taken by some member states have halted the development of hydropower facilities (eg, the Polish authorities approved a moratorium on small hydropower facilities in November 2009).[35]

The legal risk (ie, the risk of changes in law), which is often a trigger for claims relating to expropriation, is a vast issue which this chapter merely touches upon. From a project finance perspective, the risk entails that the change in law:

- affects the ability of the project to service debt or be profitable; and
- may affect the remuneration of the electricity output, the possibility to sell the energy produced to the grid or the establishment of a tax, among other things.

Protection exists against such risk. For instance, Spanish lenders usually require a specific guarantee (ie, certain recourse against the sponsor) to protect the base case of the project from any potential change in law whenever they suspect that a change

---

32 This subsection should be read in conjunction with the chapter of this book entitled "Risk allocation in renewables projects".

33 For more information on this subject, please refer to Scott L Hoffman, *The Law and Business of International Project Finance*, Third Edition.

34 For more information on the administrative barriers within the European Union, please refer to the European Small Hydropower Association report, "Administrative barriers for small hydropower development in Europe" (June 2007).

35 Mike Landy and James Craig, "Small-scale hydropower: how to reconcile electricity and environmental protection goals?" (ETC/ACC Technical Paper 2009/13- December 2009) (The European Topic Centre on Air and Climate Change (ETC/ACC).

is likely which may affect the hydropower project's cash flow. In developing countries, the sponsors and their lenders may require both specific grandfathering provisions in the concession arrangements and security from the host government.

4.   Required authorisations

As noted in section 2.1 above, one of the aims of Directive 2009/28/EC is to simplify the administrative procedures applicable to the construction and operation of renewable energy facilities. However, in the authors' view, this aim has not yet been reached in a number of member states where many steps must still be taken at a federal, state and local level.

Although the authorisations required and administrative procedures to follow vary between jurisdictions, this section sets out features common to most jurisdictions. One must consider the specific requirements of the relevant jurisdiction(s) in order to determine what authorisations are needed from which authorising bodies and the administrative process (including time and costs) involved.

4.1   Water rights

To have the right to use the water needed to operate a hydropower facility, an authorisation or concession is necessary. For example, in Spain the private use of the hydraulic public domain not specifically authorised under the 2001 Law on Water or any other piece of regulation is subject to obtaining a concession granted by the competent water authority. Typically, any modification of the initial concession granted would require the approval of the water authority. However, even where the initial concession was granted following a competitive process, a modification to that concession that is not considered substantial does not require a new competitive process.

Key issues relating to the concessions of water rights include:
- the duration of the concession (lenders usually look for long concessions);
- the subordination of the granting of such water rights to other water uses (including the use of water for drinking or agricultural purposes), which – if upstream of the hydropower plant – may increase the hydrological risk referred to in section 3.4 above; and
- the number of jurisdictions through which the relevant water resource (eg, river) flows. If more than one jurisdiction is involved, this will require consideration as to whether any existing bilateral or multilateral treaties concerning such water resources exist and, if so, whether such treaties require amendment. As noted in section 3.5 above, it may also require that the developer comply with the administrative procedures of more than one jurisdiction.

4.2   Authorisations to design, construct, own, operate and maintain

Hydropower facilities typically require an array of authorisations relating to their design, construction, ownership, operation and maintenance. In addition to the other authorisations mentioned in this section, key authorisations include:
- authorisations to construct the facility, which usually apply to all

hydropower projects (unlike thermal power projects, which in certain jurisdictions are exempt from seeking authorisations if they are below a certain output capacity); and

- authorisations to carry out the activity of generating electricity. The authorisation is often in the form of a licence or permit (ie, a legal instrument that authorises the generation activity), but it imposes obligations on the licensed party only (ie, it imposes no obligations on the issuing authority or any other government body).

In addition, municipalities and local authorities may request further licences to occupy the land, undertake construction or operate the business (eg, works or activity licences).

4.3 Authorisations to qualify for economic benefits

In some jurisdictions, including Spain, only hydropower plants whose capacity does not exceed a certain level (eg, 50MW) are entitled to receive the economic support given to renewable energy facilities. Any such support is often reserved to smaller-scale facilities. Thus, in Spain higher feed-in tariffs apply to facilities with installed capacity not exceeding 10MW.

Many jurisdictions, including Spain, require that to qualify for economic benefits, the developer either:

- prove (usually by application to the relevant authority) that it falls within a specific class of facilities that qualify for such benefits; or
- if permitted, make an application as a special type of facility.

4.4 Rights of connection and access

The process for obtaining authorisation to build and operate small hydropower facilities needs to start with confirmation by the local distribution or transmission company that access and connection to the transmission and/or distribution grids are feasible. Access and connection to the grid in the majority of EU member states are regulated (not negotiated). In Spain, access can be refused only if the distribution or transportation company can prove that it is not technically feasible. Effective and non-discriminatory access by third parties to the electricity networks is considered by the European Union an essential condition for the existence of a genuine internal electricity market.

4.5 Planning approval

Hydropower facilities typically require planning approval. If such approval is needed, the competent authority may, among other conditions, set additional conditions to protect the land.

4.6 Environmental consents

Hydropower facilities will invariably require environmental consents. The main step involved in obtaining such consents is to conduct an environmental impact assessment.

In broad terms, the purpose of the environmental impact assessment is to identify, describe and assess on a case-by-case basis the direct and indirect effects of a particular project on:

- human beings, as well as the flora and fauna;
- the soil, water, air, climate and landscape;
- material assets and cultural heritage; and
- the interplay between these factors.

The environmental impact assessment is a process that ends with an environmental impact declaration, by means of which the competent environmental authority determines:

- whether the project is viable from an environmental point of view; and
- the conditions and other limitations required, if any, in order to protect the environment and natural resources.

As the impact of hydropower plants on the environment is significant, the environmental impact declaration needs to deal with additional permits to construct the facilities.

To analyse the feasibility of the project from an environmental standpoint, the lenders usually use the Equator Principles as standard. In that regard, if after a preliminary analysis the project presents a high level of risk, the Equator Principles will require a management environmental plan to monitor the project development.

4.7 Waste production and management restrictions

The installation of industries or of any other activity that produces hazardous waste must comply with certain obligations regarding segregation of waste, labelling and delivery to authorised managers, among other things. As for any other activity, the promoter of a hydropower facility must take into account the waste regulations.

4.8 Protected natural areas

If special areas of protection, such as roads, cattle routes and railways, are affected by the project or if the project may negatively affect a natural area or ecosystem, additional permits may be required and further conditions may be imposed.

4.9 Land rights

Land rights are also necessary to develop hydropower facilities. To occupy the land, sponsors will need to reach agreement with the owner of the site, which may be a public administration or a private owner. In many countries, a statutory compulsory purchase regime exists to allow the generators to acquire the relevant land. For example, in Spain, generation facilities such as hydropower facilities are legally declared to be 'installations of public utility'. This status entitles sponsors to:

- the mandatory expropriation of the owners of the land required for the project;
- the temporary occupation of the land;
- the rights necessary to build and operate the facilities; and

- the access easements rights and other rights in connection with the property that are held to be necessary to operate the facilities.

Expropriation follows a public utility declaration by the project sponsors. Upon expropriation, a fair, one-off compensation payment must be paid to the affected landowners. In addition, the relevant sponsors may reach agreement with the landowners as to the terms of any easements required, thus avoiding the need to follow the entire expropriation proceedings. Such private agreements can contemplate the payment of royalties or, more typically, a one-off compensation payment.

5.   Key contracts
During both the development and the operating phase, a complete set of contracts must be executed by the sponsors to develop, construct, operate and maintain the hydropower facility.

5.1   Land use agreements
The sponsor must ensure that it has secured all the land rights needed to construct and operate the hydropower project. These must include not only the particular location of the hydropower facility and the interconnection facilities that connect the hydropower to the transportation or distribution grid, but also the right to access the site and cross or occupy (even temporarily) sites owned by third parties (whether public or private owners). In this regard, together with the land lease or land sale and purchase agreements (or other equivalent contract that ensures the use of the land needed for the lifetime of the project), the sponsor must have in force the relevant use rights agreements (eg, cross-site agreements and easements).

5.2   Supply and installation agreements
The sponsor must contract the supply and installation of the main elements of the hydropower facility (eg, turbine) from a technically and financially robust supplier that can provide appropriate warranties to satisfy the lenders. Typically, to comply with the timescale of the project, it is usual for long lead items such as the turbine to be purchased early; this usually requires the sponsor to pay a deposit to the supplier. If the facility is to be project financed, the lenders' funds will normally be unavailable until all permits to begin construction are in place; this first payment will typically be made by the sponsor or through a bridge loan or corporate loan by a bank.

5.3   Engineering, procurement and construction contract
As mentioned in section 3 above, sponsors of this type of project typically opt for an engineering, procurement and construction contract (especially if the project is to be project financed), under which the contractor assumes responsibility for construction and delivery of the project in exchange for a fixed price. Under such a contract, the contractor is responsible for constructing and putting into operation the hydropower facility. Once the facility is deemed finalised (after the relevant trials

and performance tests have been completed to the satisfaction of the sponsors and the lenders' technical advisers), the parties typically execute a provisional completion acceptance certificate. The assumption of risk by the sponsors for the equipment, components and elements of the hydropower facility takes place upon execution of such provisional completion certificate. Therefore, the engineering, procurement and construction contractor will be responsible for the care and conservation of the hydropower facility until provisional completion. An additional guarantee period for mechanical issues is also usually established under the contract until execution of the definitive completion certificate.

5.4   Operation and maintenance contract

Under this contract, the operator is responsible for operating and maintaining (both preventive and corrective operation and maintenance) the hydropower facility. The contract may also contain, in the first years of operation, certain performance or availability guarantees.

5.5   Financing contracts

Please refer to section 3 above and the chapter of this book entitled "Issues for financiers".

5.6   Connection and access to the grid

If the access is either regulated or negotiated, a third-party access contract with the owner of the distribution or transportation network must be executed.

5.7   Power purchase agreements

On hydropower projects, the parties usually execute a power purchase agreement that sets out the obligation of a party to purchase the electricity produced and the obligation of the sponsor to sell it. The cash flow of the project is therefore contained in the power purchase agreement.[36]

5.8   Others

In addition to the contracts mentioned above, the characteristics of a project may warrant the execution of other contracts or documents, including incorporated or unincorporated joint venture agreements, corporate guarantees, mortgages and construction agreements with the owners of the distribution network to reinforce the grid.

In jurisdictions that do not have an authorisation regime (eg, licences) to carry out the functions of generation, distribution and supply, the sponsors will also require a form of concession agreement from the government which grants the sponsor the necessary rights to undertake the hydropower project.

---

36   In Spain, a power purchase obligation is imposed on the distribution company (provided that the purchase is technically feasible). However, following the implementation of Directive 2009/72/EC and the repeal of Directive 2003/54/EC, the cash flow is no longer derived from such contract, but from a complex settlement procedure in which the Spanish energy regulator (the National Energy Commission) acts as an intermediary or trustee between the generators and the distribution companies.

6. Case study: Spanish regulations on hydropower

6.1 Characteristics

Since the early 1980s, a special regulatory regime has applied to cogeneration facilities, small hydropower stations and other facilities using renewable energy sources. The Spanish 1997 Electricity Law draws a distinction between:

- 'ordinary generation facilities', which must bid their electricity output into a mandatory pool or otherwise sell their electricity through physical bilateral contracts; and
- 'special regime generation facilities', which include cogeneration, small hydropower facilities, biomass and all other types of renewable energy facility with an installed capacity of up to 50MW.

Ordinary generators (eg, large hydropower facilities) accrue no special premiums or special tariffs. By contrast, special regime producers benefit from the availability of feed-in tariffs and from the choice of selling their output in the pool, while accruing special incentives and premiums set out by the government (although not all special regime facilities are entitled to such choice). Electricity generation remains an unregulated activity (ie, it is not subject to the same constraints, supervision and remuneration schemes as transmission and distribution) – although construction and operation permits and licences need to be obtained from different administrations. There is no generation licence or charter requested of the sponsor, but rather a licence referred to the facility: generation facilities are licensed and the person responsible for its operation will be registered alongside the particular generation facilities.

The main features attached to special regime generation facilities, as opposed to ordinary generation facilities, can be summarised as follows:

- Special generation facilities are entitled to priority access and connection to the distribution and transmission grids, and to a parallel connection so that their facilities may also use electricity taken from the grid.
- Such facilities are entitled to supply their entire output to the grid, provided that the grid can technically do so.
- The electricity supplied by special generation facilities is to be paid at the prices and under the terms and conditions expressly set forth in Royal Decree 661/2007 on renewable energy production, as long as the facilities retain their status and are in operation.
- Special generation facilities may also sell all or part of their output through direct lines to consumers (commonly the case with cogeneration facilities).

The development regulation provides a beneficial remuneration scheme for renewable energy facilities, which explains the development of renewable projects in Spain. In contrast to other countries that use green certificates or mandatory purchase obligations imposed on supplies, the remuneration scheme for special regime generators in Spain has traditionally been based on feed-in tariffs.

The remuneration framework applicable to the special regime facilities depends on the sale option chosen by the generator. Generators subject to the special regime may sell the electricity generated at any of their facilities by using either of the following options:

- delivering their output to the electricity system through the transmission or distribution grid and receiving in return a regulated single tariff for all scheduling periods (hydropower facilities with installed capacity between 10MW and 50MW are not eligible for this option and may make their offers only through the pool or to third-party purchasers on a physical bilateral basis); or

- selling electricity in the pool at the sale price set by the organised market or the price freely negotiated by the owner or representative of the facilities, supplemented, as the case may be, by a premium in euro cents per kilowatt-hour (€ cents/kWh). In this case, caps and floors are provided in respect of certain technologies. This means that the total remuneration received by such facilities will be limited to the reference market price or the price freely negotiated, plus a reference premium, subject to the applicable caps and floors.

Owners of special regime facilities are entitled to choose between either sale option, but once they have chosen they need to remain within the relevant remuneration system for at least a year.

The tariffs are determined on the basis of the category, group and subgroup to which the facility belongs, as well as to the installed power and, as the case may be, according to the date of commissioning of the facility. For hydropower facilities the tariffs on the previous page will apply (valid as at January 1 2010).

Capacity	Duration	Regulated tariff € cents/kWh	Reference premiums € cents/kWh	Cap € cents/kWh	Floor € cents/kWh
Up to 10MW	First 25 years	8.2519	2.6495	9.0137	6.8978
	Subsequently	7.4268	1.4223		
Between 10MW and 50MW[37]	First 25 years	N/A	2.2263	8.4635	6.4746
	Subsequently	N/A	1.4223		

---

37   These hydropower facilities are not entitled to elect the tariff option.

6.2   Installed capacity

According to data published by the Spanish system operator (Red Eléctrica de España SA),[38] in 2008 the total installed capacity of large hydropower facilities, (above 50MW) was 16,657MW (a figure that has not varied for the last five years). The aggregate installed capacity of small facilities was 1,979MW (up from 1,638 MW in 2004).

7.   Conclusion

Hydropower technology presents indubitable advantages, in particular by helping to achieve the Kyoto Protocol targets (and those set out by the Copenhagen Climate Summit), diversifying the geographical origin of electricity and reducing the European Union's dependence on oil and natural gas.

However, the future growth of new facilities in the EU member states is probably limited due to the economic, physical, social and environmental risks and constraints that the development of new hydropower facilities presents in Europe and, in particular, in northern and western Europe.

The development of large hydropower facilities may be a thing of the past. In Europe, only the development of small hydropower facilities would appear to have some (limited) future.

The development of hydropower facilities outside Europe (eg, in some countries in South America, Africa and Asia) seems more likely and attractive. But this will have to be implemented through a more environmentally conscious attitude, from both sponsors (and local governments) and financial lenders. The Equator Principles, which enable lenders to analyse projects from a harmonised standpoint and define the environmental risks and mitigation measures that must be taken into account, and the World Bank's stand on hydropower facilities bear witness to this more sensible attitude.

---

38   www.ree.es.

Energy efficiency: the other side of the coin

Gillian Davies
Mariya Kuchko
Janet Laing
Alejandro Saenz-Core
Mott Macdonald Limited

---

1. Introduction

While renewable energy is receiving widespread attention in the fight against climate change, energy efficiency is playing a fundamental role in reducing greenhouse gas emissions and enhancing energy security. Nearly half of the reductions in greenhouse gas emissions targeted by the European Union for 2020 are expected to come from energy efficiency measures.

Energy efficiency plays a prominent role because major reductions are, in theory, abundant and affordable. By contrast, the deployment of renewable energy on a scale required to decarbonise society remains a colossal technical and financial challenge. Beside requiring the construction of new power plants, renewable energy requires the reconfiguration of national electrical grid systems. Such investment is hindered by energy prices that generally favour conventional fossil-fuel power and reflect the uncertainty in long-term renewable energy tariffs and government policy.

Consequently, energy efficiency becomes a crucial tool in reducing greenhouse gas emissions, alongside carbon capture and storage and nuclear power as long-term prospects.

This chapter focuses on both technical measures and policy incentives. It considers international policies with regard to energy efficiency, with an emphasis on three key areas:

- power generation and energy-intensive industries;
- buildings and electrical appliances; and
- transport.

As these challenges have become ever more pressing, governments and markets are seeking innovative solutions in new policies, regulation and technologies. This chapter illustrates good practices in different sectors, from industry to residential, from power generation to transport.

2. The potential of energy efficiency

Energy efficiency is an essential mitigation tool against climate change. Its importance to the global greenhouse gas mitigation agenda is reinforced by the fact that the opportunities to reduce global greenhouse gas emissions from energy conservation and efficiency measures outweigh renewable energy, carbon capture and storage, and forest sinks.

Figure 1 illustrates the potential for global emissions reduction using different mitigation measures as calculated by four models (IMAGE[1], MESSAGE[2], AIM[3] and IPAC[4]). Some models do not consider mitigation through forest sink enhancement (AIM and IPAC) or CCS (AIM), and the share of low-carbon energy options in total energy supply is also determined by inclusion of these options in the baseline. CCS includes carbon capture and storage from biomass. Forest sinks include reducing emissions from deforestation

The Intergovernmental Panel on Climate Change finds that while greenhouse gas stabilisation levels should be achievable by "deployment of a portfolio of technologies that are currently available and those that are expected to be commercialised in the coming decades", energy conservation and efficiency needs to play a major role within this portfolio.

Fortunately, significant gains in energy efficiency need not, in theory, be driven by government regulation or fiscal incentives. In the United States alone, McKinsey estimates that profitable (net profit value positive) energy efficiency opportunities could lead to a 23% reduction in energy consumption against a 'business-as-usual' projection by 2020.[5]

Recognising the potential opportunities that energy efficiency presents, the International Energy Agency[6] put energy efficiency on the agenda of high-profile events such as the G8 meetings in 2006, 2007 and 2008. The agency recommends that governments implement specific energy efficiency policy measures across seven priority areas: cross-sectoral activity, buildings, appliances, lighting, transport, industry and power utilities.

If the recommended energy efficiency measures were implemented globally and immediately, the International Energy Agency calculates that by 2030 they could lead to around 96 exajoules or 2,300 megatonnes equivalent of oil in annual energy savings per year. This would reduce greenhouse gas emissions by 8.2 gigatonnes of carbon dioxide a year, which is equivalent to twice the European Union's current annual emissions.

---

1   Integrated Model to Assess The Global Environment – Model Type: Market Equilibrium; Solution Concept: Recursive Dynamic; Time Horizon: Beyond 2050; Modelling team and reference: Netherlands Env Assessment Agency Van Vuuren et al, Energy Journal, 2006a. (IMAGE 2.2) Van Vuuren et al, Climatic Change, 2007 (IMAGE 2.3) .

2   MESSAGE-MACRO (Model for Energy Supply Strategy Alternatives and Their General Environmental Impact) – Model Type: Hybrid: Systems Engineering & Market Equilibrium; Solution Concept: Recursive Dynamic; Time Horizon: Beyond 2050; Modelling team and reference: International Institute for Applied Systems Analysis, Austria Rao and Riahi, 2006.

3   AIM (Asian-Pacific Integrated Model) – Model Type: Multi-Sector General Equilibrium; Solution Concept: Recursive Dynamic; Time Horizon: Beyond 2050; Modelling team and reference: NIES/Kyoto Univ., Japan Fujino et al, 2006.

4   IPAC (Integrated Projection Assessments for China) – Model Type: Multi-Sector General Equilibrium; Solution Concept: Recursive Dynamic; Time Horizon: Beyond 2050; Modelling team and reference: Energy Research Institute, China Jiang et al, 2006.

5   McKinsey, "Unlocking energy efficiency in the US economy" (2009).

6   International Energy Agency, "Implementing Energy Efficiency Policies: are IEA member countries on track?" (2009).

Figure 1: Sources of reductions in emissions

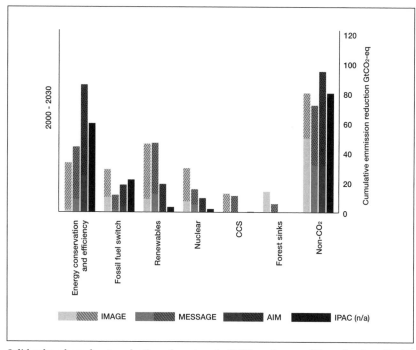

Solid colour bars denote reductions for a target of 650 parts per million of carbon dioxide equivalent.

Cross-hatched coloured bars denote the additional reductions to achieve between 490 parts per million and 540 parts per million of carbon dioxide equivalent.

Source: Intergovernmental Panel on Climate Change[7]

3. Energy efficiency policy

While there are, in theory, abundant opportunities to implement profitable energy efficiency measures, energy costs are not high enough to drive large-scale improvements without government intervention.

The current international climate framework, the Kyoto Protocol, does not spell out energy efficiency requirements.[8] However, individual countries may implement energy efficiency policies to meet their Kyoto obligations. The Kyoto framework can also help member countries to meet their targets via the Clean Development Mechanism (CDM). The CDM provides an opportunity for emissions reduction

---

7   Intergovernmental Panel on Climate Change, "Fourth Assessment Report: Climate Change 2007 (AR4): Mitigation of Climate Change".

8   The Kyoto Protocol is an international agreement adopted in Kyoto on December 1997 which is linked to the United Nations Framework Convention on Climate Change adopted in New York on May 9 1992. Under the protocol, 37 industrialised countries and the European Union legally committed to reduce gas emissions by an average of 5.2% below 1990 levels by 2012. For further information on the protocol, please read section 3.5(b) of the chapter of this book titled "Renewable energy support mechanisms: an overview".

projects to be developed in countries not listed in Annex I to the protocol (ie, non-industrialised countries) to generate certified emission reduction carbon credits, which can be traded on international carbon markets.

In March 2007 the European Commission endorsed an integrated approach to climate and energy policy that aims to combat climate change and increase the European Union's energy security while strengthening its competitiveness.

To kick-start this process, the EU heads of state and government set a series of demanding climate and energy targets to be met by 2020. These targets are:

- to reduce EU greenhouse gas emissions by at least 20% below 1990 levels;
- to bring the proportion of renewable energy in overall EU energy consumption to 20%; and
- to reduce primary energy use by 20% compared with projected levels, to be achieved by improving energy efficiency.

Collectively known as the '20:20:20 targets', they are to be reached through a series of policy instruments, including:

- the EU Emissions Trading Scheme – a cap-and-trade programme that affects the European Union's largest emitting businesses, such as power plants, refineries, cement kilns, and the pulp and paper and steel industries;
- an Effort Sharing Decision (Decision 406/2009/EC) that sets binding member state emissions targets for sectors outside the Emissions Trading Scheme, such as transportation, agriculture and waste;
- Directive 2009/28/EC on the promotion of the use of energy from renewable sources, including individual national targets for renewable energy specified in its Annex I;
- Directive 2006/32/EC on energy end-use efficiency and energy services, and national energy efficiency action plans in accordance with Article 14(2) of the directive, where member states show how they intend to reach the 9% indicative energy savings target by 2016;
- the recast of Directive 2002/91/EC on the energy performance of buildings (adopted on May 18 2010);
- binding agreements and policies on renewable energy at a country level; and
- a framework for carbon capture and storage.

It is clear that Europe is now committed to transforming itself into a highly energy efficient, low-carbon economy.

Some EU member states are using tradable white certificates to promote energy savings from energy efficiency improvements using market-based mechanisms. These schemes can allow an aggregate target for energy saving from a particular target group to be achieved more cost effectively. Tradable white certificates are a relatively new form of regulatory instrument in operation only in the United Kingdom, Italy, Denmark and France. This means that there is little practical experience with the design and implementation of such schemes, and little theoretical analysis of their behaviour and effects.

As regulations continue to emerge from the European Union, energy efficiency will remain a priority for the following reasons:

- Environmental – burning fossil fuels is the main source of greenhouse gas emissions, contributing to climate change and other environmental issues.
- Sustainability – the aim is to reduce dependence on non-sustainable energy sources and make business and products more competitive.
- Security – the European Union covers over 50% of its energy needs through imports and this percentage is expected to rise up to 70% within the next 20 to 30 years.

Energy efficiency measures related to the 20:20:20 targets are trickling down to member state policy. For example, the United Kingdom has introduced the Carbon Reduction Commitment Energy Efficiency Scheme to regulate energy users that fall outside the EU Emissions Trading Scheme (eg, government organisations, manufacturers, property portfolios, retailers, the health, food and drink sectors, and leisure businesses, including hotels). Beginning in 2010, organisations that consume over 6,000 megawatt-hours of electricity annually but are not covered by the EU Emissions Trading Scheme will need to buy allowances to cover the greenhouse gas emissions associated with all electricity and primary fuel use.[9] Funds from the allowance purchases will be recycled to participants based on emissions reductions achieved over the course of a year.

Renewable energy cannot contribute to emission reductions where the generation is eligible under the Renewables Obligation (the UK renewable energy incentive scheme). Accordingly, the mechanism will primarily encourage organisations to take up energy efficiency opportunities for which there may not otherwise be an economic justification.

4. Effect of the regulatory regimes

The CDM regime under Kyoto appears to favour supply-side energy efficiency projects, which account for 11% of total certified emission reduction generation up to 2012 – a total volume of 314 megatonnes of carbon dioxide during this period (Figure 2).

The market share of demand-side energy efficiency projects is much lower at 1% of total certified emissions reductions. This is due to the small-scale nature of demand-side projects that need to be replicated across a large number of sites, making project transaction costs too expensive. The United Nations Framework Convention on Climate Change is considering a number of measures, such as programmatic CDM/Joint Implementation or programme of activities, to address the high overheads associated with multiple applications of small-scale mitigation measures.

Carbon projects can also be established under the voluntary carbon market for purchasers of carbon credits that do not need to fulfil compliance obligations such as the EU Emissions Trading Scheme or Kyoto. For voluntary emission reduction and programmatic CDM, much of the voluntary carbon market is made up of demand-

---

9   Department of Energy and Climate Change, "The Carbon Reduction Commitment: User Guide" (2009).

side energy efficiency projects, but prices can be higher and monitoring regimes less stringent.

Figure 2: CDM pipeline

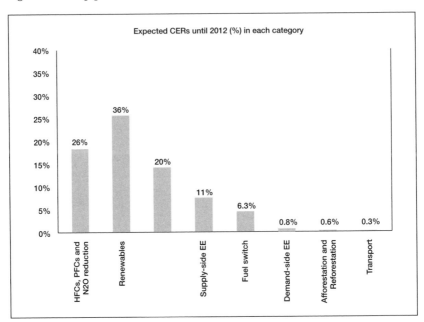

Expected CERs until 2012 (%) in each category

Source: UNEP/Risoe CDM Pipeline – August 1 2010

Under the European framework, the impact of an energy efficiency policy can be dramatic. According to recent research conducted by IHS Cambridge Energy Research Associates, if the European Union meets all the policy targets and implements energy efficiency measures in combination with renewable targets, it will not need to use the cap-and-trade scheme, as total emissions will fall substantially below the EU emissions cap and business-as-usual scenarios.

However, there are some doubts as to whether all EU renewables and energy efficiency targets can be met in full by 2020. Applying a more realistic scenario, IHS reports that projected 2020 EU emissions will be higher than the allocated cap (or Green Realistic Agenda), with the deficit under the cap covered by the EU Emissions Trading Scheme.

While energy efficiency offers the opportunity for deep emission cuts, governments face the challenge of delivering a large number of small changes across the whole society.

This delivery can broadly be categorised into:

- supply-side measures, which involve the production of increasingly efficient or alternative lower-energy technologies; and
- demand-side measures, where individuals and organisations make changes that reduce their energy consumption.

On the supply side, the development of energy efficient technologies requires investment and targeted innovation. Mandatory regulation, such as minimum energy performance standards for appliances and equipment, can help to push technological innovation.

Demand-side policies need to encourage the take-up of solutions on offer. This is where policies that constrain emissions, such as the Kyoto Protocol, the EU Emissions Trading Scheme and the UK Carbon Reduction Commitment Energy Efficiency Scheme, come into play.

For individuals and small businesses, subsidy schemes can encourage:

- the uptake of energy-efficient appliances;
- the installation of better insulation; and
- the installation of more efficient lighting systems.

Widely available energy support services, in the form of information dissemination or professional support provided by energy service companies, can be helpful. Policy makers can also require or incentivise utility companies to deliver energy efficiency solutions.

Increased focus on energy efficiency from policy makers and the public is beginning to gain traction. That said, the International Energy Agency states that on the whole, member countries are not on track to take full advantage of the opportunities to reduce greenhouse gas emissions through energy efficiency.

The International Energy Agency has highlighted three policy elements to drive energy efficiency across a national economy successfully:

- potential for cost-effective energy savings;
- improvements in technology; and
- consumer access to resources, products and skilled assistance to help them make informed decisions.[10]

Additional policy examples are provided in the remainder of this chapter.

5.   Energy efficiency economics

The McKinsey abatement cost curve demonstrates that energy efficiency measures provide the most economical opportunities for reducing carbon emissions. Insulation improvements, fuel-efficient vehicles, improved lighting and air conditioning systems, and reduced standby losses avoid negative costs per tonne of carbon dioxide, thanks to the significant energy savings that they achieve compared to the capital expenditure required.

By contrast, implementing other greenhouse gas mitigation measures (eg, renewable and nuclear energy, forestation and fuel-switching) adds a cost for each tonne of carbon dioxide reduced. The total cost of meeting the EU 20:20:20 targets for mitigating electricity generation emissions alone is approximate €48 billion annually, with 60% of that sum invested in renewables.

Barriers to implementation of many energy efficiency measures have less to do

---

10   International Energy Agency, "Energy Efficiency Initiative" (1998).

with the ultimate cost and more to do with:

- ease of installation;
- the priorities of the energy user; and
- short-term economics, such as initial affordability and payback period.

Another challenge to reducing greenhouse emissions is the so-called 'rebound effect', which is when consumers increase their energy use as efficiencies and costs decrease.

An important exception is the cost of retrofitting industrial motor systems. Although this is one of the most expensive mitigation measures, it is also one of the most important, due to the sheer volume of energy consumed. This reinforces the need for policy instruments that offer an added incentive for energy efficiency measures that are not readily adopted on the basis of economics. The public-private partnership Carbon Trust in the United Kingdom, for example, helps to tip the economic equation for business by offering interest-free loans for capital equipment that reduces carbon emissions.

The economic benefits of becoming more efficient often extend beyond direct energy costs. Energy efficiency measures often make processes more efficient, reducing the cost of production, waste disposal, administration and maintenance. Less tangible benefits, such as employee comfort and reduced risk, also come into play.

The economics of energy efficiency measures become more attractive as the price of carbon begins to penetrate business and a business's position in the market hinges on the ability to demonstrate emission reductions.

Under the UK Carbon Reduction Commitment, organisations will be ranked according to their ability to reduce greenhouse gas emissions through energy efficiency measures, leading onto reputational consideration for businesses facing consumer or supply-chain pressures.

Emerging carbon market implications also shift the energy efficiency economics for many businesses. The national allocation plans for 27 EU member states for Phase 2 of the EU Emissions Trading Scheme (which runs from 2008 to 2012) suggests that the price and cost effects for improvements in carbon and energy efficiency in the energy and industry sectors will be stronger than in Phase 1 (2005 to 2007). This is mainly because the European Commission has substantially reduced the number of allowances to be allocated by member states. To the extent that companies from these sectors (notably, power producers) pass on to consumers the extra costs for carbon, higher prices for allowances translate into stronger incentives for demand-side energy efficiency.

6. Power generation and distribution

Energy efficiency measures are some of the most economical of all greenhouse gas mitigation options. However, to achieve a significant reduction on a national or global level, it is important to target the energy-intensive sectors of industry and power generation, especially when confronted with the challenge of diffusing energy efficiency techniques such as lighting and insulation across society.

According to the International Energy Agency, significant potential remains across all sectors. For example:

- in industry, the global application of best available technologies and best practices in the five most energy-intensive sectors could lead to energy savings of 15 exajoules per year, while also reducing carbon dioxide emissions by 1.3 gigatonnes per year;
- in public power generation, the global adoption of current best-practice levels of efficiency for fossil fuels would result in annual fuel savings of between 28 exajoules (roughly the total primary energy supply of Canada, Italy and the United Kingdom combined) and 35 exajoules (the equivalent of the total primary energy supply of Japan and Germany combined). Corresponding carbon dioxide savings would be about 2.3 gigatonnes to 2.9 gigatonnes.[11]

In 2006 the production of electricity was responsible for 33% of global fossil-fuel use and 38% (10.7 gigatonnes of carbon dioxide) of fuel combustion-related carbon dioxide emissions.[12] Improving the efficiency of electricity production offers a significant opportunity to reduce dependence on fossil fuels, which also helps to combat climate change and improve energy security.

Figure 3: Electricity generation by energy source

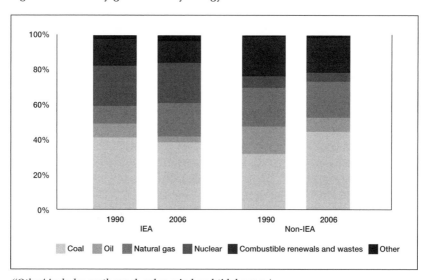

('Other' includes geothermal, solar, wind and tidal energy)
Source: International Energy Agency/Organisation for Economic Cooperation and Development[13]

---

11   International Energy Agency, "Towards more energy efficiency future 2009" energy efficiency future 2009, © OECD/IEA.
12   International Energy Agency, "Annual Energy Outlook 2008".
13   International Energy Agency/Organisation for Economic Cooperation and Development, *Energy Balances of OECD Countries*, Paris, 2009.

For the past decade there has been a strong move toward adopting more sustainable practices. In addition, attention has focused on increasing renewable technology to tackle power plant emissions (Figure 3). However, switching fuels, by increasing the share of natural gas in the mix of fossil fuels, has created some immediate benefits in International Energy Agency member countries by:

- improving fossil-fuel efficiency from 37% in 1990 to 40% in 2006; and
- significantly reducing greenhouse gas emissions.

By contract, the continued reliance on a larger share of coal (the least efficient fossil fuel) resulted in much lower average efficiency (33%) in 2006 in non-International Energy Agency countries (mainly because of India and China).

Burning natural gas rather than coal not only is more efficient, but also emits about 40% less carbon dioxide than coal per unit of energy generated. Most of the reductions in carbon dioxide emissions achieved in the United Kingdom during the 1990s were due to fuel switching, in the so-called 'dash for gas' to replace coal in power generation.

Despite the environmental attraction of gas-fired power plants, most new plants continue to rely on coal because that fuel is the least expensive (in case of no/minimum flue gas treatment techniques) and most secure option for many countries. Considering that the technologies for carbon capture and storage are still nascent, the efficiency of operating coal-fired power plants becomes an important factor in near-term greenhouse gas mitigation.

The Fourth Assessment Report of the Intergovernmental Panel on Climate Change (Table 1) gives an overview of potential carbon dioxide savings by improving plant efficiency and fuel switching. For example, a 27% reduction in emissions (in grams of carbon dioxide per kilowatt-hour) is possible by replacing a 35% efficient coal-fired steam turbine with a 48% efficient plant using advanced steam, pulverised coal technology. Replacing a natural gas single-cycle turbine with a combined cycle of similar output capacity would help to reduce carbon dioxide emissions per unit of output by around 36%.

While power plants have become steadily more efficient, further improvements could be achieved by introducing new coal-fired power plants that operate at more efficient temperatures. Coal-fired power plants operating within supercritical parameters usually have efficiencies in the range of 45%, compared with 40% for subcritical systems.

The most advanced form of fossil-fuelled power plant is the combined-cycle gas turbine. Such turbines are more than 50% efficient, compared to the commonly used older steam turbine, which achieves efficiency rates of only about 30% and dumps roughly two-thirds of the energy input of fuel into the atmosphere via cooling towers.

Combined-cycle gas turbines are more climate friendly than older, coal-fired plants. This is not only because they are more efficient, but also because natural gas generates less carbon emissions per unit of energy than coal.

In some countries, the waste heat from power stations is widely used in district schemes to heat buildings. In 2000 some 72% of Denmark's electricity was produced in such combined heat and power plants. By using such plants, energy plant efficiencies can overall reach 85%.[14]

Table 1: Reduction in carbon dioxide emission coefficient by fuel substitution and energy conversion efficiency in electricity generation

Existing generation technology			Mitigation substitution option			Emissions reduction per unit of output
Energy source	Efficiency (%)	Emission coefficient (gCO2/kWh)	Switching option	Efficiency (%)	Emission coefficient (gCO2/kWh)	(gCO2/kWh)
Coal, steam turbine	35	973	Pulverised coal, advanced steam	48	710	-263
Coal, steam turbine	35	973	Natural gas, combined cycle	50	404	-569
Fuel oil, steam turbine	35	796	Natural gas, combined cycle	50	404	-392
Diesel oil, generator set	33	808	Natural gas, combined cycle	50	404	-404
Natural gas, single cycle	32	631	Natural gas, combined cycle	50	404	-227

Source: Intergovernmental Panel on Climate Change[15]

Clearly, there is room for further improvements in the supply-side efficiency of electricity systems – in particular, by:

- further increasing the efficiency of generating plant; and
- ensuring that whatever waste heat remains is piped to where it can be used.

There are also opportunities to find new ways to reduce the heat loss associated with steam production. However, policies to reduce air pollutants sometimes require the installation of pollution abatement equipment that consumes additional energy at the power plant, thereby offsetting some of the efficiency gains achieved through new generation technologies.

After fuels have been converted to electricity, further losses occur in the transmission and distribution systems used to convey electricity to customers. Reducing such losses is an important aspect of improving efficiency in the electricity

---

14   For more information regarding combined heat and power, please read the chapters of this book titled "Combined generation: technical, commercial and legal aspects" and "Combined generation: commercial aspects".

15   Intergovernmental Panel on Climate Change, "Fourth Assessment Report: Climate Change 2007 (AR4): Mitigation of Climate Change" (table 4.9 at p.295).

generation sector. In 2006 these losses represented 10% of final electricity consumption worldwide. The situation has improved in International Energy Agency member countries, where overall losses fell from 9% in 1990 to 7% in 2006.[16]

Further greenhouse gas reductions related to power supply can be achieved though demand-side management measures. For example, smart metering and smart grids wide application is under development. This is likely to become best practice internationally, especially as renewable energy and electric cars place new demands on power grids.

Nowadays, power utilities try to help customers to improve their energy efficiency. It is a radical movement from the traditional utility business model: making more money by selling more power. And it is not the only challenge to the old ways of doing things.

Here is how a utility can make more by selling less: instead of spending €2 billion on a new 1,000 megawatt (MW) power plant, say, the utility decides to spend €2 billion or less insulating homes, paying customers to install more efficient equipment and making the grid smarter. Taking those steps would slash power consumption by more than 1,000MW, thus eliminating the need to build the power plant and cutting greenhouse gas emissions at the same time.

7.   Energy-intensive industries

Industry is responsible directly and indirectly (via consumed power) for about one-third of global greenhouse gas emissions, of which over 80% is from energy use. Total energy-related emissions (9.9 gigatonnes of carbon dioxide in 2004)[17] have been growing by 65% since 1971 (an average annual growth of 2%). This is despite intentions of the industry continuously to implement energy efficiency measures over the past decades.

Some 30% of industrial emissions come from the iron and steel industry, 27% from non-metallic minerals (mainly, cement and lime production) and 16% from chemicals and petrochemicals production. In the near future, energy efficiency will potentially be the most important and cost-effective means of mitigating greenhouse gas emissions from industry.

Over the past few decades, the world's most energy-intensive industries have migrated to developing countries. In 2006, developing countries accounted for:

- 74% of global cement manufacture;
- 63% of global nitrogen fertiliser production;
- about 50% of global primary aluminium production; and
- 48% of global steel production.[18]

This signals a shift in energy-intensive greenhouse gas emissions to developing countries.

From a technology standpoint, the use of more efficient motor systems and steam supply systems is broadly applicable across all industries, while others are process-specific. The International Energy Agency estimates that the use of best

---

16   International Energy Agency, "Annual Energy Outlook 2008".
17   International Energy Agency, "Tracking Industrial Energy Efficiency and CO2 Emissions" (2007).
18   International Energy Agency, "Annual Energy Outlook 2008".

available technologies worldwide would result in savings of about 19% to 32% on current carbon dioxide emissions in this sector.

Industries have the option of mitigating their greenhouse emissions in several ways:

- by reducing their direct emissions;
- by reducing their energy consumption;
- by reducing the emissions associated with the whole lifecycles of the materials that they handle via improvements in their supply chains, distribution systems or waste streams;
- by implementing carbon reduction projects outside the boundaries of their business or purchasing carbon credits on the market; and
- by developing new, lower-greenhouse emission products or services.

According to the International Energy Agency, "[t]he energy intensity (amount of energy used per unit of product) of most industrial processes is at least 50% higher than the theoretical minimum".[19] This provides a significant opportunity for reducing energy use and its associated carbon dioxide emissions within a business or its supply and distribution chains. A wide range of technologies has the potential to reduce industrial greenhouse emissions, of which energy efficiency is one of the most important – especially in the short to mid term. Other opportunities include fuel switching, material efficiency, renewable energy and reduction of greenhouse gas emissions other than carbon dioxide.

All enterprises consume energy in some way. Even if energy costs represent a relatively low proportion of total costs, an effective energy management system can help to reduce energy costs substantially. Measures aiming at effective energy utilisation could help:

- to reduce direct production costs and indirect administrative costs; and
- to improve enterprises' profitability.

In the past, energy efficiency solutions have been found largely through technological instruments. However, now integrated solutions including energy policy, management of energy and supply chains, IT solutions and process optimisations are considered. This is reflected in several directives and standards such as EN 16001:2009[20] and PAS 2050.[21]

---

19   International Energy Agency, "Annual Energy Outlook 2005: With projections to 2025".
20   EN 16001 represents the latest best practice in energy management building upon existing national standards and initiatives. The standard applies to activities that are under the control of an organisation; it specifies the requirements for an energy management system to enable an organisation to:
      • develop and implement a policy;
      • identify significant areas of energy consumption; and
      • target energy reductions.
      It is a useful document for all types and sizes of organisation and accommodates diverse geographical, cultural and social conditions.
21   PAS 2050 is a publicly available specification for assessing product lifecycle greenhouse gas emissions. PAS 2050 has been developed by BSI British Standards and co-sponsored by the Carbon Trust and the UK Department for Environment, Food and Rural Affairs. PAS 2050 is an independent standard, developed with significant input from international stakeholders and specialists across academia, business, governmental and non-governmental organisations through two formal consultations and multiple technical working groups.

The International Organisation for Standardisation (ISO) is strengthening its commitment to developing standards that offer practical tools for achieving sustainability through new initiatives on energy efficiency and renewable energy sources. The new ISO 50001 standard on energy efficiency is under development and will be published by the end of 2010.[22] It is expected that corporations, supply chain partnerships, utilities, energy service companies and others will use ISO 50001 as a tool to:

- reduce energy intensity and carbon emissions in their own facilities (as well as those belonging to their customers or suppliers); and
- benchmark their achievements.

7.1 Energy-intensive industrial sectors

(a) Iron and steel

Iron and steel is one of the largest industrial consumers of energy and, as a result, the largest industrial emitter of carbon dioxide. The four largest national producers of iron and steel (China, the European Union, Japan and the United States) accounted for 67% of the sector's carbon dioxide emissions in 2005.[23]

Steel is produced through at least a dozen processing steps with different technological configurations, depending on product mixes, available raw materials and energy supply. The main ways to improve the industry's energy efficiency and reduce its greenhouse gas emissions are by:

- optimising the smelting process;
- improving blast furnace efficiency;
- applying direct casting techniques;
- increasing the use of recycled steel direct casting;
- substituting fuel and feedstock (eg, coal instead of coke); and
- using the direct reduced iron process instead of the conventional blast furnace and basic oxygen processes.

In general, the blast furnace and basic oxygen process is about three times more energy intensive than the direct reduced iron process. This is because the ore preparation and the coke and iron making steps are unnecessary in the direct reduced iron process. Accordingly, switching from one process to the other could bring significant energy savings.[24]

The most widely used steel process is often rendered less efficient by local constraints. For instance, some sites use less efficient blast furnace and basic oxygen processes because of a lack of power to fire electrical arc furnaces and access to recycled scrap material.

---

22   ISO 50001 will establish an international framework for industrial plants or entire companies to manage all aspects of energy, including procurement and use. The standard will provide organisations and companies with technical and management strategies to increase energy efficiency, reduce costs and improve environmental performance. Based on broad applicability across national economic sectors, the standard could influence up to 60% of the world's energy demand.
23   International Energy Agency 2007, "Tracking Industrial Energy Efficiency and CO2 Emissions".
24   "BREF: Ferrous metal processing industry".

Over recent decades iron production has been steady while scrap/electric arc furnace process production has been growing in the United States and the European Union. Direct reduced iron process is widespread in the Middle East, South America, Mexico and India.

(b)   *Non-metallic minerals*

The non-metallic minerals industry includes the production of cement, bricks, ceramics, glass and other building materials. It is one of the largest industrial consumers of energy and is the second largest emitter of carbon dioxide. The four largest national producers of non-metallic minerals (China, the European Union, India and the United States) accounted for 75% of the sector's carbon dioxide emissions in 2005.[25]

Within this sector, cement production accounts for 83% of total energy use and 94% of emissions. Around 60% of emissions result from the production of carbon dioxide as a chemical byproduct of limestone calcination during the clinker-making process, as opposed to combustion for energy.[26] The most energy-consuming process is the production of clinker from limestone and chalk by heating limestone. Clinker makes up 95% of Portland cement, the most widely used cement type.[27] Energy costs make up between 20% and 40% of the cement prime costs.

The main ways to improve the sector's energy efficiency and reduce its greenhouse gas emissions consist of:

- switching from wet to dry kilns;
- adding a number of pre-heaters to existing rotary kilns;
- using pre-calcination technology;
- substituting some or all fuel and feedstock;
- using the roller process and high-efficiency classifiers for grinding; and
- blending cement with non-clinker feedstocks (eg, volcanic ash, granulated blast furnace slag from iron production, fly ash from power generation).

It can also help to reduce energy intensity from 3 gigajoules per tonne of clinker (dry kiln, six stages pre-heater and pre-calcining) to 6.8 gigajoules per tonne of clinker (wet kilns).

(c)   *Chemicals and petrochemicals*

The chemicals and petrochemicals sector uses a number of distillation, evaporation, direct heating, refrigeration, electrolytic and biochemical processes to separate and convert materials into final products. It is one of the largest industrial consumers of energy and is the third largest emitter of carbon dioxide. The four largest national producers (China, the European Union, India and Russia) accounted for 62% of the sector's carbon dioxide emissions in 2005.[28]

25   International Energy Agency 2007, "Tracking Industrial Energy Efficiency and CO2 Emissions".
26   "BREF: Large volume inorganic chemicals – Ammonia, acids and fertilisers industries, large volume organic chemical industry".
27   "BREF: Cement, lime and magnesium oxide manufacturing industries".
28   "BREF: Large volume inorganic chemicals – Ammonia, acids and fertilisers industries, large volume organic chemical industry".

The chemical and petrochemical industry consumed 34 exajoules in 2004, which represented 30% of all industries' energy use worldwide. This share has increased sharply from 15.5% in 1971 at an average annual growth rate of 2.2%.[29] Over half of this consumption is for feedstock use, which cannot be reduced through energy efficiency measures. The industry's intensive emissions of carbon dioxide are low because significant amounts of carbon are stored in the products produced. As the energy and feedstock savings potential for the industry are limited, other measures are needed to reduce its energy consumption.

The most energy-consuming processes within the sector are the production of methanol, ammonia and high-value chemicals from steam-cracking of naphtha, ethane and other feedstock. Separations, chemical synthesis and process heating are the main energy consumers in the chemical industry. Technological advances that could reduce energy consumption include:

- improved membranes for separation;
- more selective catalysts for synthesis; and
- greater process integration to reduce heating requirements.[30]

(d)   *Pulp and paper*

The pulp and paper sector is the fourth largest industrial consumer of energy and emitter of carbon dioxide. The four largest national producers (China, the European Union, Japan and the United States) accounted for 80% of the sector's carbon dioxide emissions in 2005.[31]

In addition to conventional energy sources, the pulp and paper industry generates energy as a byproduct through waste energy recovery. It covers about 50% of its energy needs by combusting biomass residues. The efficiency of heat consumption in the industry as a whole improved significantly between 1990 and 2003.

The opportunities to mitigate carbon dioxide emissions in the pulp and paper industry consist of:

- implementing energy efficiency improvements;
- developing cogeneration;
- increasing the use of (self-generated) biomass fuel; and
- increasing the recycling of recovered paper.

Energy efficiency could improve by 14% by using the best available technologies. However, the prospects for future energy efficiency gains are limited due to the relatively low-energy intensity of the remaining pulp and paper processes. Power is used mainly for mechanical pulping and paper drying. This is an area where further efficiency gains are available. Chemical pulp mills produce large amounts of black liquor that is used to generate electricity at relatively low efficiencies. New technologies that promise higher conversion efficiency could provide important energy benefits, particularly for electricity and possibly biofuels. The need for large

---

29   International Energy Agency, "Industrial Emissions Report" (2008).
30   "BREF: Pulp and paper industry".
31   International Energy Agency 2007 "Tracking Industrial Energy Efficiency and CO2 Emissions"

amounts of steam makes combined heat and power an attractive technology.

7.2   Industry-wide technologies

A number of pan-industry systems offer substantial opportunities for energy optimisation. They include:

- boilers;
- power and heat/cold production;
- compressed air systems;
- electrical drives;
- lighting;
- ventilation and air conditioning systems; and
- energy recovery.

Most significantly, approximately 65% of the electricity consumed by industry is used by motor systems. The efficiency of motor-driven systems can be increased by:

- reducing losses in motor windings;
- using better magnetic steel;
- improving the aerodynamics of the motor; and
- improving manufacturing tolerances.

However, maximising efficiency requires that:

- all components be sized properly;
- the efficiency of end-use devices (eg, pumps and fans) be improved;
- electrical and mechanical transmission losses be reduced; and
- proper operation and maintenance procedures be used.

Variable speed drive motors offer significant potential for energy savings. Implementing high-efficiency motor-driven systems or improving existing ones in the European Union (not including Bulgaria and Romania) could reduce energy consumption by 30%.[32]

According to the International Energy Agency, steam generation represents about 15% of global industrial energy use.[33] The efficiency of current steam boilers can be as high as 85%, thanks to general maintenance, improved insulation, combustion controls and leak repair, improved steam traps and condensate recovery.

The energy recovery techniques of waste heat and low-calorie fuel are old and well known. However, large energy savings can still be achieved at a variety of industrial sites – especially in developing countries and within the Commonwealth of Independent States.

Waste heat can be re-used in other on-site processes or used to preheat incoming water and combustion air. New, more efficient heat exchangers or more robust (eg, low-corrosion) heat exchangers are being developed continually, thus improving

---

32   De Keulenaer, H, Belmans, R, Blaustein, E, Chapman, D, De Almeida, A and De Wachter B, *Energy efficient motor driven systems* (2004).

33   International Energy Agency, "Annual Energy Outlook 2008".

heat recovery profitability. Typically, cost-effective energy savings of 5% to 40% can be achieved when these systems are implemented.

Compressed air systems are widely applied in the industry. Meanwhile, the efficiency of the compressor unit is about 10%. Thus, the production of 1 kilowatt of compressed air requires 10 kilowatts of energy. Therefore, it is crucial, among other things, to:

- optimise the load and size of the compressors;
- tighten the compressed air pipeline; and
- optimise the compressed air pressure level requirements.

Lighting accounts for about 10% of the energy costs at industrial sites, but can be optimised with natural lighting, automatic control systems, and energy-efficient bulbs, ballasts and luminaries. An example of an efficient lighting system is one that uses:

- separate (sometimes automatic) controls for different lighting zones;
- task or ambient lighting (relatively low background light levels) where appropriate; and
- greater lighting when and where needed.

8. Buildings and electrical appliances

Residential, business and public buildings consume large volumes of energy through lighting, space and water heating, air conditioning and other electrical equipment. The largest contributors to high energy demand vary by building type and country. In the domestic sector, in countries that are members of the Organisation for Economic Cooperation and Development (OECD), for example, use of gas and oil products for space heating accounted for 54% of total energy consumption per household in 2004.[34] In developing countries, by contrast, traditional biomass use for heating and cooking accounted for 56% of total energy consumption.[35] In the United Kingdom, buildings are responsible for nearly 50% of the nation's energy consumption.[36]

A 2009 report[37] by E3G and the Climate Group found that improving standards for new builds and modernising existing building stock through building codes could save 1.3 gigatonnes of carbon dioxide equivalent by 2020. Table 2 on the next page shows the policies, measures and instruments that the Intergovernmental Panel on Climate Change has found to be effective in reducing the overall energy consumption of buildings and appliances.

8.1 Buildings

According to the Intergovernmental Panel on Climate Change's Fourth Assessment Report, the buildings sector has the largest potential for low-cost carbon dioxide mitigation in the short to medium term (based on a bottom-up study of technology options applied across a number of sectors).

The energy use of buildings is mainly defined by how the various energy-using

---

34   Buildings and Climate Change: Status, Challenges and opportunities. UNEP, 2007
35   International Energy Agency, "Energy Technology Perspectives: Scenarios & Strategies to 2050" (2008).
36   Buildings and Climate Change: Status, Challenges and opportunities. UNEP, 2007
37   E3G/The Climate Group, "Breaking the Climate Deadlock: Technology for a Low Carbon Future" (2009).

Table 2: Effective building and appliance-related energy efficiency instruments

Policies, measures and instruments shown to be environmentally effective	Key constraints or opportunities
Appliance standards and labelling	Periodic revision of standards needed
Building codes and certification	Attractive for new buildings Enforcement can be difficult
Demand-side management programmes	Need for regulations so that utilities may profit
Public sector leadership programmes, including procurement	Government purchasing can expand demand for energy-efficient products
Incentives for energy service companies	Success factor: Access to third party financing

Source: Intergovernmental Panel on Climate Change[38]

devices (eg, pumps, motors, fans, heaters, chillers) are put together as a complete building infrastructure system, rather than the efficiencies of the individual devices. The savings opportunities at the system level are generally much higher than those that can be achieved at the device level; these system-level savings can often be achieved as a net investment-cost savings.

The system approach requires a so-called 'integrated design process', where the building performance is optimised through an interactive process that involves the whole design team from the beginning of the project. However, the approach to building remains traditional, with little or no consideration for energy consumption and with the project passing from the design team to the engineers, whose primary concern is to make the building habitable through mechanical systems. The design of mechanical systems is also a step-by-step process whose components are, in some cases, specified before all information needed in order to design an efficient low-carbon solution is available.

This does not mean that the traditional design process involves neither integration nor teamwork. Rather, it means that the integration is not primarily concerned with minimising total energy consumption through the modification of either design solutions or concepts.

The following steps should be included in a basic integrated design process:

- considering the building orientation, form and thermal mass;
- designing a high-performance building envelope to maximise passive

---

38   Intergovernmental Panel on Climate Change, "Fourth Assessment Report (AR4) IPCC 2007: Mitigation of Climate Change".

heating, cooling, ventilation and day-lighting income;
- installing energy efficient systems to meet remaining energy loads;
- ensuring that individual energy-using devices are as efficient as possible and properly sized; and
- ensuring that systems and devices are properly commissioned and will be operated as designed.

By focusing on the building shape and a high-performance envelope:
- heating and cooling loads are minimised;
- day-lighting opportunities are maximised; and
- mechanical systems loads can be significantly reduced.

Such solution generates cost savings that can offset the additional investments into a high-performance envelope and cost of installing premium (high-efficiency) equipment throughout the building.

Following these steps usually generates energy savings of between 35% and 50% for a new commercial building, compared to standard practice. The use of more advanced or less conventional approaches has often achieved savings of between 50% and 80%.[39]

The main sources of energy consumption in residential and commercial dwellings are heating and cooling loads. Consequently, common approaches exist to reduce energy consumption. They include:
- using high-performance thermal envelopes in combination with passive heating. These envelopes can reduce heat losses to the point where a large part of the remaining heat loss can be offset by internal heat gain (from people, lighting, appliances) and passive solar heat gain, with the heating system required only for the residual. For example, the European Passive House Standard[40] requires a heating energy use of no more than between 10 kilowatt-hours per square metre per year and 20 kilowatt-hours per square metre per year (depending on country, budget and climate). This is typically achieved by reducing the heat loss to about 45 kilowatt-hours per square metre per year, with one-third of the heat loss covered by internal heat gains and one-third by passive solar heat gains;
- reducing the cooling load by up to 50% by:
  - orienting buildings to minimise the wall area facing east or west (the directions most difficult to shade from the sun);
  - clustering buildings to provide some degree of self shading (as in many traditional communities in hot climates);
  - providing fixed or adjustable shading;
  - using highly reflective building materials;
  - improving insulation;

39   LD Danny Harvey, "Reducing energy use in the buildings sector: measures, costs, and examples", *Energy Efficiency* (2009) 2:139–163.
40   www.passivhaustagung.de/.

- using windows that transmit a relatively small portion (25%) of the total (visible and invisible) incident solar radiation, while a larger percentage of the visible radiation is permitted to enter for day-lighting purposes;
- using thermal mass to minimise day-time interior temperature peaks;
- using night-time ventilation to remove day-time heat; and
- minimising internal heat gains by using efficient lighting and appliances; and
- using passive and low-energy cooling techniques, consisting of:
  - passive ventilation;
  - evaporative cooling techniques;
  - desiccant dehumidification (using a specific material (desiccant) to remove moisture from the air and regenerating it using heat); and
  - earth-pipe cooling, when ventilation air can be pre-cooled by drawing outside air through a buried air duct.

Deutsche Bank's Greentowers project is an example of efficient energy use in office building.[41] The project consists of the biggest refurbishment of a building in Europe and will result in one of the most eco-friendly high-rises in the world.

The core elements of the modernisation of the Deutsche Bank buildings are:

- a lighting control system;
- efficient use of daylight;
- solar technology to cover 50% of hot water consumption;
- breathing façades and upgraded glazing to reduce heat loss;
- chilled ceilings in combination with thermal mass storage;
- district heating (combined heat and power);
- heating and cooling with a combined cooling and heating system;
- rainwater collection and internal water recycling; and
- material recycling (98%).

The expected outcomes of the refurbishment are:

- a 67% reduction in heating energy;
- a 55% reduction in power consumption;
- a 55% reduction in carbon dioxide emissions; and
- a 15% reduction in freshwater consumption through the use of rainwater and recycled water (43%).

As standards of living improve in developing (and often warm) countries, careful building design that includes, for instance, passive cooling systems and natural ventilation can play a significant role in energy efficiency in upcoming decades.

Unfortunately, in many developed countries the efficiency of new infrastructure will be outweighed by existing infrastructure, which is often old – especially in Europe.

Besides improving the energy efficiency of existing buildings, simple

---

41   www.greentowers.de.

improvements to the building envelope (eg, wall and roof insulation and more efficient windows) and smart metering systems can lead to significant and extremely cost-effective savings for existing dwellings. For commercial buildings, significant and low-cost energy improvements can be gained by properly:

- inspecting the building's mechanical systems; and
- operating the building according to its design specifications.

The reduction in electricity consumption in buildings requires both technological improvements and effective awareness programmes to influence behaviour. For the latter, programmes communication from smart meters can help to improve consumer understanding of energy consumption patterns.

In the European Union, the Energy Performance of Buildings Directive (2002/91/EC) introduced various measures to be implemented across the member states. Article 7 of the directive requires that:

- energy performance certificates be made available when buildings are constructed, sold or rented out; and
- display energy certificates be displayed in public buildings with a total useful floor area over 1,000 square metres.

The certificates show an energy efficiency rating of between A and G (G being the poorest-performing buildings) and recommendations for improvement. Certificates make it easy to compare buildings. The fact that the cost of utilities such as heating will be lower per unit area for higher-efficiency buildings works as an incentive.

8.2   Appliances

Increasing the energy efficiency of white goods and other appliances (eg, televisions and computer equipment) could reduce emissions by 0.3 gigatonnes globally.[42] This is not a large figure in global terms. However, appliance efficiency becomes more relevant when one considers:

- the implications of the emergence of a middle class in developing nations;
- technical advancements such as high-resolution television; and
- carbon reduction strategies of energy-intensive countries such as the United States.

Appliances also play a role in generating peak-level electrical demand, with potential implications for emerging renewable energy, electrical vehicle and grid systems infrastructure requirements.

The EU Energy Performance Directive introduced minimum standards and energy performance certificates for appliances similar to those for buildings. In Japan's reduction scheme, standards are set for 21 items of equipment ranging fromtelevisions to electric toilet seats.[43] The scheme also sets out the most stringent

---

42   International Energy Agency, "Implementing Energy Efficiency Policies: are IEA member countries on track?" (2009).

43   Murakami *et al*, "Overview of energy consumption and GHG mitigation technologies in the building sector of Japan", *Energy Efficiency* (2009) 2:179-194.

unit standards for air conditioning equipment. The UK government has announced a £50 million boiler scrappage scheme that aims to replace up to 125,000 old and inefficient boilers with better performing systems.[44] Australia was the first country to impose a ban on incandescent light bulbs. The policy, introduced in early 2007, requires a complete phase out by the end of 2010. Incandescent light bulbs are replaced by compact fluorescent bulbs, which use approximately 20% less electricity thanks to reduced waste heat generation.

Some complexities need to be considered when estimating the net impact of energy efficiency solutions for the consumer market. For example, a rebound effect can occur where the economic savings achieved by more efficient technologies can be associated, directly or indirectly, with a resultant increase in the overall number of energy-using devices due to the lower service costs per appliance.[45] Thus, the replacement of cathode ray tube television screens with more efficient LCD (liquid crystal display) screens in developed countries has not resulted in any net energy savings in recent years because it has been accompanied by an overall increase in:

- the number of televisions per household;
- the size of individual screens; and
- the number of viewing hours.[46]

In commercial buildings, information and communications technology equipment is a key and rapidly growing consumer of electricity. Globally, the sector's emissions were 0.53 gigatonnes of carbon dioxide equivalent in 2002; they are projected to rise to 1.43 gigatonnes of carbon dioxide equivalent in 2020.[47] However, information and communications technology also provides a significant opportunity for achieving increased efficiency in other high energy demand sectors such as power transmission, transportation systems and manufacturing processes. Climate Group estimates that information and communications technology could generate emission reductions of 7.8 gigatonnes of carbon dioxide equivalent in 2020 – around five times the sector's projected 2020 footprint. One way of reducing carbon dioxide emissions is to use smart grids: it is estimated that their use could reduce emissions by 2.03 gigatonnes of carbon dioxide equivalent.[48]

Consumer appliances play an important role in the increase of residential energy consumption and greenhouse gas emissions. Many strategies could be implemented to pursue energy conservation and the reduction of greenhouse gas emissions from these appliances – for instance, by promoting:

- non-chlorofluorocarbon refrigerants;
- high-efficiency refrigerators using a variable speed compressor and a vacuum-heat insulating material; and
- compact fluorescent light bulbs.

---

44   HM Government, "Pre-budget Report" (2009).
45   4CMR (2006), "The Marco-Economic Rebound Effect and the UK Economy".
46   International Energy Agency, "Energy Technology Perspectives: Scenarios & Strategies to 2050" (2008).
47   The Climate Group, "SMART 2020: Enabling the low carbon economy in the information age" (2008).
48   Climate Group "SMART2020: Enabling the low carbon economy in the information age" (2008)

9. Transport

9.1 Status and trends

At the time of writing, the transport sector was responsible for over 25% of the world's carbon dioxide emissions.

Since 1990, the transport sector's carbon dioxide emissions worldwide have increased by 36%, despite increasing oil prices and growing concern about emissions. In 2005 transport accounted for 23% of the world's energy-related carbon dioxide emissions, up from 21% in 1990.

The main factors of these emissions growth rates are changes in the volume of travel and the efficiency of the modes of transport used.

The International Energy Agency tracked the changes to the sector's emissions from 1990 to 2004. The agency recorded significant changes in the volume of travel in OECD countries:

- Travel in light-duty vehicles increased by about 20% annually to around 14,000 kilometres per person per year;
- Transportation by truck increased by 36% annually;
- Air travel grew by over 5% annually; and
- The shipping sector grew dramatically because of the growth of the international trade sector.

More recently, the economic recession of 2008/2009 has probably affected these growth rates. However, there are no indications that the upward trend in passenger miles will reverse. Indeed, in recent studies the International Energy Agency has forecast high growth rates for all modes of transportation for decades to come.

In developing countries, particularly China, India, Brazil, Mexico and South Africa, growth rates in every transportation mode and related carbon dioxide emissions have risen sharply. This is a result of the high growth rates of these countries' gross domestic products experienced in recent years.

The transport sector accounts for over 50% of the world's oil demand and nearly 25% of the energy-related carbon dioxide emissions.

From feedstock and fuel distribution to vehicles, greenhouse gas emissions from the transport sector accounts for over 27% of the world's emissions.

The recorded trends between 1990 and 2004 in energy intensity by mode per passenger and freight have shown slow improvements in average efficiency, apart from passenger air travel. Energy intensity among transportation modes varies, with some modes of transport – such as mass transit modes (buses and rail) and bulk freight modes (shipping and rail) – offering low energy intensities, and others (aviation and automobile transportation) offering high energy intensities.

According to the baseline scenario evaluation made by the International Energy Agency in its "Energy Technologies Perspective 2008", global transport energy use and emissions will increase by over 50% by 2030 and more than double by 2050.

The fastest growth is expected to come from air travel, road freight and light-duty vehicle (car, small van and sport utility vehicle) travel. Regionally, growth will be led by the developing world, especially China and India. This will be a consequence of increased vehicle ownership caused by rising incomes. Renewable fuels are not expected to make a significant impact on the market. Fuel in transport is likely to continue to come predominantly from fossil fuels. While conventional oil production is expected to peak and begin to decline, the shortfall is likely to be made up by non-conventional oil (eg, sand tar) and other fossil resources (eg, gas-to-liquids and coal-to-liquids) – especially in China, South Africa and India. These fuels are likely to be more carbon intensive than oil, which makes their use even less sustainable than the current practice. It is urgent to shift to a more sustainable, low-carbon transport system.

9.2   Curbing the trends

The sector presents enormous challenges for achieving deep cuts in fuel use and greenhouse gas emissions. In the absence of significant technological change, the only way to achieve significant reductions in greenhouse gas emissions over the next 50 years will be heavily dependent on achieving high improvements in efficiency rates and lower growth rates in travel, especially for the most energy-intensive modes. Passengers and freight shifts and very low-carbon fuels such as biofuels, hydrogen and electricity will also contribute to reducing the carbon intensity of the transportation sector.

The International Energy Agency's Energy Technologies Perspective proposes the following key recommendations and comments for curbing the current transportation trends:

- Technologies such as fuel cells and vehicle on-board energy storage, via batteries, ultra-capacitors and hydrogen storage should be developed. These technologies are not yet mature and it may be many years before they can deliver carbon dioxide reductions at a reasonable cost.
- The fuel economy of light-duty vehicles should be improved. With strong policies, available technologies have the potential to reduce the energy use per kilometre of new light-duty vehicles by up to 30% in the next 15 to 20 years.
- Increased investments in advanced public transit systems, such as bus rapid transit, can help cities to ensure that their residents have low-cost, high-quality mobility options.
- Substantial efficiency improvements should be made in freight movement using medium-duty urban-use trucks, including hybridisation, and improving routing and logistic systems. Reductions in energy use by these means may be expected to be as high as 40% by 2030.
- Rail accounts for a small share of transport energy use and greenhouse gas emissions (about 3%). However, it holds the potential for significant growth in the future, particularly if passengers and freight are shifted from other high-intensive carbon modes of transportation.
- International shipping accounts for 80% of marine energy use and about 3% of global greenhouse gas emissions. The sector may grow significantly if global trade continues to expand. Through energy-saving devices and more efficient operation, the sector may reduce energy intensity by up to 30% by 2050.
- Rising fuel costs give airlines and aircraft manufacturers an incentive to improve aircraft energy efficiency. New aircraft models incorporate many

cost-effective efficiency technologies; an energy intensity reduction of up to 40% may be expected by 2050.

- Biofuels could dramatically reduce emissions from the transportation sector.

It is increasingly understood that first-generation biofuels produced primarily from food crops are limited in their ability to achieve targets for oil-product substitution, climate change mitigation and economic growth, considering that they may have a negative impact on biodiversity and the biomass feedstock may not always be produced sustainably.

Accordingly, their sustainable production is under review, as is the possibility of creating undue competition for land and water used for food and fibre production. A possible exception that appears to meet many of the acceptable criteria is ethanol produced from sugar cane.

Second and third-generation biofuels (from non-food biomass) that are under development have the potential to offer significant carbon reduction compared to current cane ethanol or maize-ethanol production.

Other points to note include the following:

- Hydrogen fuels and fuel-cell vehicles demand a high investment in research and development to become commercially widespread. It is unlikely that they will emerge before 2030.
- Vehicle electrification is re-emerging as a potentially viable long-term option. Plug-in hybrids offer a potential near-term option as a means of transition to full electric vehicles. However, development in battery technology and cost reduction is necessary before electric vehicles can become industry standard.
- If electric vehicles become more widespread, more power generation will be needed and it will be necessary to use low-carbon generation technologies to displace high-greenhouse gas electric generation.

In summary, the potential solutions for curbing the trends of greenhouse gas emissions in the transportation sector focus on:

- improving the efficiency of engines;
- developing cleverer vehicle use;
- using lower-carbon content fuels; and
- shifting modes of transport from high-carbon intensity to low-carbon intensity modes.

The key methods for reducing the carbon intensity of transportation through energy efficiency measures are described further in the following sections.

9.3 Automobiles

(a) Fuel efficiency
Efficiency can be expressed in terms of consumption per unit distance:

- per vehicle;
- per passenger; or

- per unit mass of cargo transported.

Energy efficiency improvement devices would be installed in standard engines in order to improve the energy use of the fuel, such as waste heat recovery devices in the exhaust system, turbo or air compressed feed to the burners and regenerative brakes, among other devices.

(b)   *Fuel economy – maximising behaviours*
The energy in fuel consumed in driving is lost in many ways, including engine inefficiency, aerodynamic drag, rolling friction and kinetic energy lost to braking (and, to a lesser extent, regenerative braking). Driver behaviour can influence all of these. One way of changing behaviour would be to discourage people from making city journeys that require more fuel per distance travelled than highway driving.

(c)   *Fuel economy in automobiles*
Outside of the European Union, cars typically average a consumption of 5.6 litres to 15 litres per 100 kilometres. Due to concerns over carbon dioxide emissions, new EU regulations are being introduced to reduce the average emissions of cars sold from 2012 to 4.5 litres per 100 kilometres for diesel-fuelled cars and 5 litres per 100 kilometres for a petrol-fuelled cars (which corresponds to 120 grams of carbon dioxide per kilometre).
The EU strategy to reduce carbon dioxide emissions from passenger cars and improve fuel economy is based on three pillars:
- the commitments from the automobile industry to make fuel-economy improvements;
- the labelling of new cars; and
- the promotion of fuel efficiency through fiscal measures.

Since carbon dioxide emissions are linked to fuel consumption, a car that emits less carbon dioxide will consume less fuel and, thus, enjoy lower running costs.

(d)   *Car manufacturers' promises*
The European, Japanese and Korean automobile manufacturers' associations, which supply some 98% of the cars on the EU market, are compromised in implementing key voluntary commitments. All parties are working towards the same quantified average emission objectives for new passenger cars. Targets are to be achieved mainly through technological developments and market changes, including energy efficiency applications.

(e)   *Consumers can choose*
In addition to car manufacturers achieving technical advances, consumers still need to be convinced to buy more efficient cars if the benefits are to be enjoyed. This means providing better information on fuel economy at the point of sale. This is the aim of the EU car labelling scheme. Thus, Directive 1999/94/EC requires that:
- a label on fuel consumption and carbon dioxide emissions, as well as a poster, be displayed at the point of sale of each new passenger car;

- a guide presenting this data be freely available for all new vehicles offered for sale; and
- fuel consumption and carbon dioxide emissions data be included in printed material used to market, advertise and promote new cars.

9.4 Shipping

The international shipping industry accounts for around 3% of global greenhouse gas emissions.

The industry needs to reduce its carbon footprint, as carrying on at the current rate means that ship emissions would grow by between 150% and 250% by 2050 compared with 2007 levels.

International shipping has not been included in the Kyoto Protocol to limit global greenhouse gas emissions. However, the protocol gives the International Marine Organisation, the UN agency responsible for shipping, the authority to develop rules to deal with shipping's greenhouse gas emissions from industrialised countries.

The International Maritime Organisation has been working on various measures since July 2009. The organisation's Marine Environment Protection Committee has agreed on a package of interim and voluntary technical and operational measures to reduce greenhouse gas emissions from the global shipping industry. The agreed measures include:

- interim guidelines on the method of calculation and voluntary verification of an energy efficiency design index for new ships. This is intended to stimulate innovation and the technical development of more energy-efficient ships;
- guidance on the development of a ship energy efficiency management plan for new and existing ships. The aim here is to develop best practices for the fuel-efficient operation of ships; and
- guidelines for the voluntary use of a ship energy efficiency operational indicator for new and existing ships, which enables operators to measure the fuel-efficiency of a ship.

These energy efficiency measures would mean on average that shipping will be 15% to 20% more efficient in 2020.

At its next meeting in October 2010, the committee will define the methodology and criteria for the feasibility studies and impact assessments to be done. The committee is likely to give a clear indication of which market-based measures it wishes to evaluate further and identify the elements.

It is uncertain whether climate action should:

- continue to be taken under the auspices of the International Maritime Organisation; or
- be taken through other international forums, such as the UN Framework Convention on Climate Change, or even on a regional basis, such as at the EU level.

In October 2009 the European Union agreed that the global reduction targets for emissions from the international maritime sector should be 20% below 2005 levels by 2020.

The European Commission has started preparatory work in case the International Maritime Organisation fails to come up with an adequate solution in time. The commission is considering several options for the reduction of maritime transport emissions, including inclusion in the EU Emissions Trading Scheme.

Contrary to expectations, shipping was not part of the overall agreement at the December 2009 summit in Copenhagen and thus no cap on the sector has yet been set. The expectation was that the UN Framework Convention on Climate Change and the International Maritime Organisation would work together. To do so, the International Maritime Organisation would need the UN Framework Convention on Climate Change to agree to some principle timelines and a strengthening of the whole process or joint collaborative action, with the International Maritime Organisation dealing with further details.

For the time being, however, the International Maritime Organisation will continue to determine the regime to reduce emissions from the shipping industry.

9.5   Aviation

The international aviation industry accounts for around 2% of global greenhouse gas emissions.[49]

The growth of the global economy and the popularity of budget airlines have resulted in a remarkable expansion of air transport. Passenger traffic has risen by nearly 9% per annum over the past 45 years – equivalent to 2.4 times the average increase in gross domestic product – and by 14% in 2004 alone.[50]

Air travel grew by over 5% annually between 1990 and 2004 in OECD countries.[51] Further, by 2020 the world's aircraft fleet is expected to double in size.[52] The flipside of this expansion is an unwelcome increase in greenhouse gas emissions and their impact on climate change. Thus, while the European Union's total greenhouse gas emissions fell by 3% between 1990 and 2002, carbon dioxide emissions from international aviation increased by almost 70% during this period.[53]

Airlines and aircraft manufacturers are incentivised to improve aircraft energy efficiency fuel costs. New aircraft models incorporate many cost-effective efficiency technologies; future aircraft are expected to improve further on this count. The industry is expected to reduce its energy intensity by up to 40% by 2050.[54]

In addition, biofuels could become part of the jet-fuel blend as early as 2012 and be ready for large-scale use by 2015.

One of the drivers is that the aviation sector will be included in the EU Emissions Trading Scheme from 2012. The first test flight using (first generation) biofuels was

49   International Air Transport Association, "A global approach to reducing aviation emissions", November, 2009
50   Chapter 5: Transport and its infrastructure of IPCC 4th assessment report.
51   Commission for Integrated Transport.
52   International Air Transport Association.
53   http://ec.europa.eu/environment/etap/inaction/sectors/Transport/245_en.html.
54   David McCollum, Gregory Coul, David Greene "Greenhouse Gas Emissions from Aviation and Marine Transportation: Mitigation Potential and Policies", December 2009.

carried out by Virgin Airlines in February 2008. Aviation is now focusing on a second-generation, sustainable biofuel.

Additionally, aircraft emissions of nitrogen oxides (NOx) form ozone when emitted at cruise altitudes. These oxides trigger the formation of condensation trails and cirrus clouds, which also contribute to global warming. In 1999 the Intergovernmental Panel on Climate Change estimated the total impact of aviation to be two to four times greater than from its carbon dioxide emissions alone, even without counting the highly uncertain, but potentially large, cirrus cloud effects.

To put the problem into perspective, a return flight for every two passengers from London to New York produces about as much carbon dioxide as an average European domestic car does in a whole year.

(a)   EU Emissions Trading Scheme

While the Kyoto Protocol is not set to tackle greenhouse gas emissions from international aviation, the European Union has examined ways to include the sector in its overall climate change strategy. On July 8 2008 the European Parliament approved a proposal for a directive amending Directive 2003/87/EC to include aviation activities in the EU Emissions Trading Scheme.

The aim is to reduce carbon dioxide emissions from the aviation industry in the European Union. The directive's key aspect is that all aircraft operators whose flights arrive at or depart from EU airports will be included in the EU Emissions Trading Scheme from January 1 2012.

As airfares increasingly reflect external costs, consumers will also increasingly take account of the overall costs of flying. In turn, this will become an incentive for aircraft manufacturers and carriers to invest in climate-friendly technology and for aircraft operators to change how they work.

This approach will be accompanied by cost-efficient policies to encourage technological innovation, including:

- energy efficiency applications; and
- improvements in energy intensity in new aviation engines and the use of low-carbon/clean fuels.

This means that every tonne of carbon dioxide emitted from flights arriving at and departing from airports within the European Union (over 2,700 aircraft operators in 2009) be verified by a third party and covered by an EU carbon allowance. If the verified emissions exceed the operator's allocated amount of free allowances, the operator will have to buy allowances at market prices – €15 per tonne of carbon dioxide at the end of 2009.

(b)   New obligation for operators

While the compliance period will not start until 2012, aircraft operators must comply with the following requirements in order to be eligible for a free allocation:

- Operators needed to provide a monitoring plan to the competent European authority before September 1 2009 describing:
  - how they monitor and report their carbon dioxide emissions; and

- how they will measure their tonne-kilometres data.
- Operators must report the verified amount of aviation activity in tonne-kilometres for 2010 before April 1 2011.

The new obligation to monitor and report verified annual emissions of carbon dioxide from January 1 2010 is creating a driver to incorporate new efficiencies in technical and operational issues in the aviation industry.

10. Conclusion

Industry activity, power generation and transport are responsible for most greenhouse gas emissions worldwide. There are many mechanisms, tools and schemes to tackle and mitigate their impact on the environment and thus ensure a better future. Consumer behaviours are already shifting, while climate change awareness increases. The European Union promotes and widely supports the development of renewable energies, energy efficiency and greenhouse gas reduction measures within its member states. In a few years it will be clear whether:

- all these policies and regulations can help to achieve the European Union's ambitious 20:20:20 targets; and
- greenhouse emissions can be decoupled from economic growth.

Effective energy management is a priority because of the significant potential to save energy and reduce greenhouse gas emissions worldwide. Abundant opportunities to reduce greenhouse emissions are available using existing technology and, thus, without substantial capital.

The challenge of deploying energy efficiency is largely associated with the challenge of diffusing best practice through society. Even when energy represents a significant cost for an industry, opportunities for improvement may be missed because of organisational barriers, or a lack of focus or priorities. An energy management system creates a foundation for improvement and provides guidance for managing energy throughout an organisation. The implementation of international standards such as BN 16001, PAS 2050 and ISO 50001 allows organisations to structure their energy use and optimise energy costs. Innovative policies such as the United Kingdom's Carbon Reduction Committee are intended to:

- place energy and carbon on the agenda of businesses;
- improve in-house knowledge; and
- provide an incentive to reduce energy consumption.

Significant savings in energy use in new buildings of all types and in all climate zones are possible using existing technologies. The main obstacles to achieving high energy savings in new buildings is the lack of knowledge at the design stage. Extensive training programmes in the integrated design process and in techniques for reducing hearting, cooling, ventilation and lighting loads in buildings, as well as training all the trades involved in building construction, are urgently required, even with a significant strengthening of building codes.

The transport sector presents huge challenges for achieving deeps cuts in fuel use and related reductions in greenhouse gas emissions. Therefore, beside industries, power generation and individuals, transport has become a target for greenhouse gas reductions. Step by step, the European Union is introducing binding regulations to reduce the environmental impact of transport. Meanwhile, vehicle manufacturers, driven by consumers, are also working towards a more fuel-efficient and environmentally friendly future.

Energy efficiency plays a critical role in addressing business-as-usual practices and making near-term gains in reducing greenhouse gas emissions. However, gains derived from energy efficiency ultimately run the risk of being outpaced by longer-term modernisation and deforestation. A combined approach to greenhouse gas mitigation and adaptation is required. Changing the way that society develops and power is produced is essential for long-term success.

Combined cooling, heat and power generation: technical, commercial and legal issues

Thorsten Mäger
Dirk Uwer
Hengeler Mueller

1. Combined generation and demand for green power

Energy issues represent some of this century's most critical challenges and opportunities. Serious problems such as climate change, erosion of energy security and increasing dependence of economies on oil call for a change in the core strategies of industries. Policy makers show a heightened interest in renewable energies in their search for innovative and cost-efficient solutions to these problems.[1] Although energy resources are unevenly distributed across the globe, all countries – regardless of the stage of development of their economy – are concerned by the need to improve energy efficiency.[2] This is because efficient energy production not only creates the conditions to reduce greenhouse gas emissions, but also enables a broader energy supply without increasing fuel consumption. Furthermore, the savings resulting from more efficient energy generation also account for a more stable energy market. This is achieved by reducing the dependence on oil and gas that has become a burden for industrialised economies. Companies are also increasingly sensitive to these challenges. Thus, environmental trust has become a fundamental asset for reaching business targets, underpinning companies' reputation and gaining competitiveness.[3]

Energy efficiency and the reduction of greenhouse gas emissions are key drivers of the European Union's energy policy.[4] The union has ambitious objectives. By 2020 it wants to cut greenhouse gas emissions by 20% on 1990 levels and to bring the share of energy consumption in the union stemming from renewable energy sources to 20%.

This can be achieved only if the energy industry also adheres to these goals and invests in innovative and efficient energy generation technologies.

---

1   Article 2(a) of the EU Renewable Energies Directive (2009/28/EC) defines 'energy from renewable sources' as "energy from renewable non-fossil sources, namely wind, solar, aerothermal, geothermal, hydrothermal and ocean energy, hydropower, biomass, landfill gas, sewage treatment plant gas and biogases".
2   World Energy Council, "Energy efficiency policies around the world: review and evaluation" (2008), p14.
3   As the European Commission underlines: "environmental investments are normally referred to as win-win opportunities – good for business and good for environment." Green paper: "Promoting a European Framework for Corporate Social Responsibility" (COM (2001) 366 final), p10.
4   The European Commission emphasises that: "the renewable sector is the one energy sector which stands out in terms of ability to reduce greenhouse emissions and pollution, exploit local and decentralised energy sources, and stimulate world-class high-tech industries".
     Communication from the Commission to the Council and the European Parliament, "Renewable energy road map – renewable energies in the 21st century: building a more sustainable future" (COM (2006) 848 final), p3.

These technologies include combined cooling, heat and power (CCHP) generation. CCHP offers a high potential for using renewable sources of energy.[5]

Over the past decade, EU policy has aimed at liberalising and unbundling the electricity and gas markets. It has also aimed at designing new processes to accommodate high levels of energy generation by using renewable energies and improving efficient technologies. Recently, the energy generation field has developed new systems based on multiple generation processes. These include a large variety of techniques and technologies, such as co-generation (eg, combined heat and power (CHP) generation and CCHP generation).

These systems are based on the principle of simultaneous generation of multiple forms of useful energy (mechanical and thermal) in a single system. They use various integrated components, such as:

- a heat engine;
- a generator;
- heat recovery;
- cooling equipment (if cooling is also required); and
- an electrical connection.[6]

CCHP generation seems to combine the key ingredients needed by modern energy industries – namely, competitiveness, technical performance, environmental sustainability and end-use efficiency. This explains the enthusiasm of the European Association for the Promotion of Co-generation (COGEN Europe) when it refers to co-generation as "the single biggest solution to the Kyoto targets".[7] This interest is easily explained by taking a closer look at the advantages offered by combined generation. The Co-generation Observatory and Dissemination Europe and the International Energy Agency list the following:

- increased efficiency of energy conversion and use;
- lower carbon emissions;
- large cost-savings for the energy consumer;
- opportunity to move towards more decentralised forms of electricity generation;
- reduction of the dependence on imported fossil fuel;
- improvement of local and general security of supply;
- opportunity to foster competition in generation; and
- a way to increase employment.[8]

---

5   *Id*, p8 and Communication from the Commission to the European Parliament and the Council, "Europe can save more energy by combined heat and power generation" (COM (2008) 771 final), pp3-4.

6   'Co-generation' is usually defined as the simultaneous production of electricity and heat, both of which are used. The same single generation process can have three outputs instead of two, but the basic technology and the fuel needs do not differ significantly. The main difference is that tri-generation processes require, in addition, the use of either a compressor or an absorption chiller. In other words, tri-generation (ie, combined cooling, heat and power (CCHP) generation) can also be seen as a co-generation system. See UN Environment Programme, "Energy technology fact sheet: cogeneration" (www.cogeneurope.eu/wp-content/uploads//2009/02/fact_sheet_chp.pdf). Article 3(a) of the Co-generation Directive (2004/8/EC) also defines 'co-generation' as the "simultaneous generation in one process of thermal energy and electrical and/or mechanical energy".

7   COGEN Europe, "What is cogeneration?" (www.cogeneurope.eu/category/about-cogen/what-is-cogeneration/).

CCHP generation enables the production of more than one output (ie, cold, heat and electricity). The input can vary and includes:

- natural gas;
- biomass (wood waste or animal combustible biomass); and
- biogas (combustible gases produced from the biological degradation of organic wastes).[9]

Even when relying on fossil fuel, CCHP results in carbon savings and enhanced energy efficiency.

In July 2009 co-generation made up 11% of the total electricity production within the European Union.[10] Some EU member states have acknowledged the advantages of a more extensive use of combined generation, while others seem reluctant.[11] The total installed capacity ranges from 110 megawatts (MW) in Ireland to 20.84 gigawatts (GW) in Germany. Outside the European Union, Russia and China also enjoy high levels of installed combined generation capacities (65.1GW and 28.15GW respectively). Other countries, such as India and Brazil, boast lower levels of installed capacity, but can be regarded as future markets.[12]

In this context, this chapter describes the key technical, commercial and legal features of CCHP generation. This study first considers the technical and commercial aspects of combined generation. To address what is generally the principal concern of any would-be investor, this chapter looks, in particular, at the balance between technical reliability and efficiency of the technology on the one hand, and profitability on the other. Further, the chapter considers the legal aspects of CCHP generation, describing some of the main risks and contingencies to be addressed. It also suggests ways both to mitigate these risks and to underpin the economic and environmental advantages of CCHP generation.

2. Technical aspects

The trade-off between technological sustainability and efficiency and commercial feasibility is perhaps one of the main challenges faced by CCHP generation. From a technical point of view, the system can be described as a generation process that allows for:

- the production of energy on-site (or near the site); and
- the recovery of the waste heat from power generation for cooling or heating, by using absorption chillers, compressors or heat recovery equipment for producing steam or hot water.[13]

---

8 Co-generation Observatory and Dissemination Europe, "What is cogeneration?" (www.code-project.eu/category/info-tools/about-cogeneration/); International Energy Agency, "Combined heat and power. Evaluating the benefits of greater global investment" (OCDE/IEA, 2008), p7.
9 Co-generation Observatory and Dissemination Europe, *op cit*; International Energy Agency, *op cit*, p7. For further explanations, see also COGEN Europe, "A guide to cogeneration" (March 2001), p4.
10 COGEN Europe, "Briefing note" (November 16 2009), p1.
11 The share of co-generation in the total national power generation is higher in Finland and Denmark than in Spain and the United Kingdom. See International Energy Agency, "Combined heat and power. Evaluating the benefits of greater global investment" (OCDE/IEA, 2008), p8.
12 International Energy Agency, *op cit*, p16.

The process consists of the simultaneous generation of different forms of energy, which can have different electrical capacities – starting below 1 kilowatt (KW).[14] Most of the generated energy is consumed on-site, while the remainder is injected into the electricity grid, to be sold to third parties via transmission and distribution systems.

Thus, CCHP technology generates electricity, heat and cold from a single source (ie, natural gas, biomass, biogas, coal, waste heat or oil). In some cases, the technology can also rely on a fuel-neutral integrated energy system that can be adjusted to the energy needs of the end user. It requires a turbine or an engine to create mechanical energy to be transferred to an electricity generator.[15] The heat is then transferred to the heat recovery boiler, or to a cooling compressor or absorption chiller.

Combined generation integrates the output of energy production and does not discard the heat resulting from the generation process. The heat recovered can be used for thermal purposes or can be chilled, so that the cold can also be used without the need for any other source of production or energy demand. This explains why multiple-generation systems are more energy efficient than conventional technologies.[16]

From a technical perspective, energy efficiency can be understood as the ratio between the power generated and the heating value of the fuel consumed. In conventional electricity generation, approximately one-third of the energy potential contained in the fuel consumed is converted into electricity, with the remainder lost as waste heat.[17] Hence, only one-third of the primary energy consumed is delivered to the end user in the form of electricity, taking into account the losses that occur during conventional generation and the transmission and distribution of electricity.[18]

In contrast, the average overall efficiency of CHP is 75% – an impressive figure when compared with the average overall efficiency rate of 49% when producing heat

---

13   As described by the US Department of Energy: "integrated energy systems combine on-site power or distributed generation technologies with thermally activated technologies to provide cooling, heating, humidity control, energy storage and/or other process functions using thermal energy normally wasted in the production of electricity/power".
US Department of Energy, "CHP technologies" (www.eere.energy.gov/de/chp/chp_ technologies/).

14   Articles 3(l)(m) and (n) of the Co-generation Directive draw a distinction between co-generation units, micro-cogeneration units (with a maximum capacity of 50 kilowatts) and small-scale co-generation units (with an installed capacity below 1 megawatt). National legal frameworks make a similar distinction between the sizes of co-generation units (eg, the German 2002 Combined Heat and Power Act, as amended).

15   Usually, only reciprocating engines and steam turbines allow for the use of different types of fuel. Gas turbines, micro-turbines and sterling turbines frequently use natural gas as the only fuel. See World Alliance for Decentralised Energy, "Building integrated cooling, heat and power for cost-effective carbon mitigation" (December 2005) (www.localpower.org/documents/ report_de_in_buildings.pdf), pp2-3.

16   According to the US Department of Energy: "in conventional conversion of fuel to electricity, over two-thirds of the energy input is discarded as heat to the environment. By recycling this waste heat, CHP systems achieve efficiencies of 60% to 80% — a dramatic improvement over the average 33% efficiency of conventional fossil-fuelled power plants. Higher efficiencies reduce air emissions of nitrogen oxides, sulphur dioxide, mercury, particulate matter, and carbon dioxide."
US Department of Energy, "CHP technologies" (www.eere.energy.gov/de/chp/chp_ technologies/).

17   UN Environment Programme, "Energy technology fact sheet: cogeneration" (www.cogeneurope.eu/wp-content/uploads//2009/02/fact_sheet_chp.pdf). From an economic standpoint, energy efficiency includes technological, behavioural and economic changes that may cause a reduction of the amount of energy used to produce one unit of economic activity. See World Energy Council, "Energy efficiency policies around the world: review and evaluation" (2008), p9.

18   See also International Energy Agency, "Combined heat and power. Evaluating the benefits of greater global investment" (OCDE/IEA, 2008).

and electricity separately. The figures for tri-generation (rather than co-generation) do not differ significantly.

The different generation systems use similar types of equipment:

- an engine or a turbine;
- a generator;
- a boiler; and
- an absorption chiller, if the system is also used to generate cold.

The differences between technologies come from the prime mover (ie, engine or turbine). Different movers require different cycles, use different fuels and achieve different efficiency levels.[19] The most commonly used movers are gas turbines, reciprocating engines and steam turbines. Reciprocating engines use spark ignition engines or compression ignition engines and, in some cases, can have a dual-fuel configuration. They can reach system efficiencies of between 65% and 80%.[20] However, these engines:

- carry high maintenance costs;
- can be used only for lower temperature co-generation; and
- produce high levels of low-frequency noise.

Gas turbines are the most extensively used mover for large-scale generation. They can reach overall system efficiencies (electricity and useful thermal energy) of between 70% and 80%. They are considered one of the cleanest means of generating electricity.[21]

Both reciprocating engines and gas turbines rely on mature technology and are fuelled by oil or gas. The following chart describes their functioning.

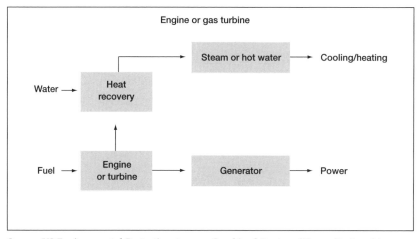

Source: US Environmental Protection Agency, Combined Heat and Power Partnership

---

19   From a commercial perspective, the equipments and associated costs are also different.
20   Energy and Environmental Analysis, "Technology characterization: reciprocal engines" (December 2008) (www.epa.gov/chp/documents/catalog_chptech_reciprocating_engines.pdf), p1; Resource Dynamics Corporation, "Cooling, heating and power for industry: a market assessment" (August 2003) (www.eere.energy.gov/de/pdfs/chp_industry_market_assessment_ 0803.pdf), pp2.1-2.7.

Steam turbines generate power as a by-product of heat. Their efficiency levels are usually lower than those of gas turbines.[22] The high investment in the equipment is balanced by a possible reduction of long-term costs as steam turbines can operate on a wide variety of fuels, from clean natural gas to solid waste, including all types of coal and biomass. Furthermore, steam turbines can provide different heat grade levels and have proved to be reliable. Steam is extracted from the turbine at a low pressure and can then be either used directly for heating purposes or converted into other forms of thermal energy, notably chilled water. The following chart illustrates the generation system of a steam turbine.

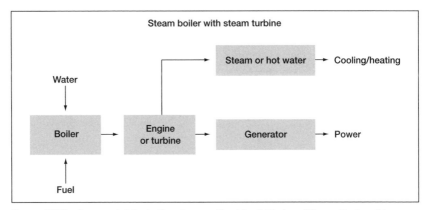

Source: US Environmental Protection Agency, Combined Heat and Power Partnership

The combination of a gas turbine with a steam turbine is called 'combined-cycle generation' – a recent development in energy generation. Here, the steam generated in the gas turbine is used to drive a steam turbine to generate further electricity that can be sold as excess energy and improve the plant's overall efficiency.

More recent technologies, such as micro-turbines and fuel cells, have been developed. Micro-turbines can be highly efficient and generate low carbon emissions.[23] These turbines carry high investment costs and their application is limited to low temperature co-generation sites. However, the advantages mentioned above (high efficiency and low carbon emissions) and the market applications of these micro-turbines explain the recent surge of financial incentives to further their development. Fuel cells convert hydrogen and oxygen into electricity without combustion. This technology generates high levels of energy efficiency, together with low emissions and low noise. These more recent systems may be seen as the future of combined generation. However, they are not mature enough for the reliability levels envisaged for them to be confirmed.

21   For more technical information, see Resource Dynamics Corporation, *op cit*, pp2.7-2.11, and Energy and Environmental Analysis, "Technology characterization: gas turbines" (December 2008) (www.epa.gov/CHP/documents/catalog_chptech_gas_turbines.pdf).

22   For more information, see Energy and Environmental Analysis, *supra*.

23   For technical and commercially available solutions, see ENER-G, "Micro cogeneration – a compact and efficient solution" (http://www.energ.co.uk/micro_cogeneration).

The choice of a co-generation technology depends on whether the type of application envisaged is industrial, commercial or residential. For instance, food processing and pulp and paper industries typically host steam turbines to achieve economies of scale by using diversified sources of fuel. Other features of the technology used are similarly important. They include noise, carbon emissions, reliability and maintenance requirements and, of course, the costs associated with both the equipment and its long-term operation and maintenance. Operators tend to favour gas turbines, steam turbines and combined-generation turbines.[24]

3. Commercial aspects

Different potential co-generation hosts face different potential drawbacks. Small to medium-sized companies can rarely afford the strategic electricity generation management required by co-generation plants. Industries outside the energy sector (or related fields) are often reluctant to enter into the energy generation business on their own.

Investing in co-generation implies initial costs and long-term costs. The upfront costs include:

- engineering design;
- building;
- tests;
- commissioning; and
- associated services (eg, site modifications and electricity services).

These costs are usually high and the payback period may exceed five years.[25]
Long-term costs are associated with:
- fuel supply and storage; and
- the operation and maintenance of the equipment, including consumables and back-up electricity.

However, some upsides of combined generation investment should also be underlined. First, the electricity not consumed on-site can be marketed. The additional revenue generated may balance the high initial capital outlays. Second, the long-term costs can be reduced by using alternative fuels, such as biomass or biogas. This option not only side-steps the high and instable costs related to the procurement of natural gas, but also allows the internalisation of costs through the EU Emissions Trading Scheme. Third, financing options from third parties are available. They include private equity, debt financing and project financing, all of which avoid the costs of upfront investment and maintenance of the unit. Lastly, contractual structures may contribute to mitigating the financial costs. Some solutions, including design, build and own contracts, may be found to circumvent initial large capital outlays (see Section 4.4 below). Tax credits and other incentives

---

24   For a technical comparison between the different technological options, see COGEN Europe, "A guide to cogeneration" (March 2001), p26-27.
25   COGEN Europe, "Briefing note" (November 16 2009), p5.

may also foster the development of combined generation.

Thus, when analysing the commercial aspects and cost-effectiveness of combined generation, one can disregard neither the emerging trends among generation systems nor the existing and forthcoming market opportunities which may very well support such investments.

A preliminary distinction should be drawn between three types of application for co-generation:

- industrial generation;
- commercial and institutional generation; and
- decentralised energy generation.

Industrial applications usually consist of large-scaled energy generation projects installed near industrial plants for food processing, pulp and paper, chemicals, metals and oil refining.

Within the commercial and institutional fields, healthcare units and hospitals are the most common applications.[26] However, other opportunities have been recently identified in hotels, buildings, supermarkets, airports, waste water treatment plants, bio-refineries (ie, ethanol production) and utility power plants.[27] Essentially, all commercial and institutional facilities with a consistent demand for both thermal energy and power are potential hosts for combined generation plants. A study conducted by the US Department of Energy in 2003 revealed that the market penetration of combined generation was only one-third.[28]

As mentioned above, the last type of application is decentralised energy, cooling and heat production, and district heating.[29] In 2005 some 1.8 billion tonnes of carbon were generated by producing separately the heating, cooling and powering of commercial and residential buildings. This corresponds to one-third of that year's total carbon emissions.[30] The trend can be reversed by developing larger, efficient co-generation projects: building-based CCHP systems are now widely recognised as efficient carbon-saving systems that also avoid the costs related with the

---

26   Energy and Environment Analysis, "Market potential for advanced thermally activated BCHP in five national account sectors" (May 2003), p6.
27   International Energy Agency, "Combined heat and power. Evaluating the benefits of greater global investment" (OCDE/IEA, 2008), p14.
28   For a US market analysis, see Resource Dynamics Corporation, "Cooling, heating and power for industry: a market assessment" (August 2003) (www.eere.energy.gov/de/pdfs/chp_industry_ market_assessment_ 0803.pdf), pp3-7.
29   Article 2(g) of the Renewable Energies Directive defines 'district heating' or 'district cooling' as: "the distribution of thermal energy in the form of steam, hot water or chilled liquids, from a central source of production through a network of multiple buildings or sites, for the use of space or process heating or cooling".
30   World Alliance for Decentralised Energy, "Building integrated cooling, heat and power for cost-effective carbon mitigation" (December 2005) (www.localpower.org/documents/report_de_in_ buildings.pdf), p1.
31   Helsingin Energia, "Join district heating" (www.helen.fi/services/liity_kl.html). This system was developed on the basis of a liberalised market. Although in some cases small co-generation investments may receive government subsidies, the International Energy Agency considers that Finland's success is driven by the efficiency of the network and opportunity given to investors to recoup their investments by selling electricity to the grid in a cost-effective manner. See International Energy Agency, "Combined heat and power. Evaluating the benefits of greater global investment" (OCDE/IEA, 2008), p16.
32   Energi Styrelsen, "Oil and gas production in Denmark 08" (www.ens.netboghandel.dk/PUBL.asp? page=publ&objno=16332159).

transmission and distribution of energy. Finland, a country with large heating needs, has successfully implemented a decentralised use of co-generation. District heating covers about 93% of Helsinki's needs.[31] In Denmark, the figures are also striking. According to the Danish Energy Authority, 80% of the district heating is co-produced with electricity.[32] The Copenhagen district heating system comprises four co-generation plants and supplies an area of around 50 million square metres. This makes it one of the largest heating systems in the world.[33]

Tri-generation can be applied to all cases of co-generation. In addition, it may create new opportunities to meet district cooling requirements, as well as cooling demand in industrial and individual buildings. District cooling is suitable for urban areas where it is necessary to cool buildings and other facilities during certain periods of the year. Usually, district cooling is implemented together with district heating, with cooling generated in the hottest months and heating in the coolest periods.[34] Industrial cooling is associated with certain sectors, such as food processing and brewing, and is also a common application for CCHP generation. Finally, cooling in individual buildings is particularly successful in hotels and airports. The Shanghai Pudong International Airport is a well-known case. This airport operates a natural gas-fuelled co-generation plant to generate electricity, heating and cooling for the airport's terminals at peak hours, with a total energy efficiency of 74%.[35]

In the forthcoming years, combined generation is likely to play a more important role in energy generation. In 2007 the potential for industrial co-generation in India was identified as being over 7.5GW. Some EU governments have set a target for the share of combined generation in their overall power generation as a way to comply with the European Union's objectives to reduce carbon emissions. This is the case in Germany, where the Combined Heat and Power Act sets the target for the share of co-generation in power generation by 2020 at 25%.[36]

4.   Legal aspects

4.1   Setting a global policy
Various legal considerations intrinsic to CCHP generation weigh heavily in investment decisions. These considerations relate to:
- licence, permit and other installation requirements;
- the tax benefits and other incentives that may support the capital expenditures necessary to install and operate a combined generation unit; and
- the contractual structure of the project.

---

33   Danish Board of District Heating, "The development in Denmark" (www.dbdh.dk/artikel.asp?id=463&mid=24).

34   COGEN Europe, "A guide to cogeneration" (March 2001), p30.

35   International Energy Agency, "Combined heat and power. Evaluating the benefits of greater global investment" (OCDE/IEA, 2008), p14.

36   Section 1 of the Combined Heat and Power Act, as amended, and COGEN Europe, press release (pr.euractiv.com/press-release/eu-member-states-reports-reveal-cogeneration-economic-option-search-measures-close-ene) (October 13 2009).

Before tackling these issues, this section provides a brief overview of the legal framework and global policies that have been implemented to foster combined generation.

The cornerstone of the EU legislative framework for combined generation is provided by the EU Co-generation Directive (2004/8/EC). This directive aims to increase energy efficiency and improve the security of supply by setting out rules applicable to:

- the criteria for energy-efficiency levels (Article 4); and
- the guarantee of origin of the electricity generated by an efficient source (Article 5).

Other relevant aspects concern the electricity grid system and the tariffs applicable to back-up and top-up electricity (Article 8), as well as the authorisation procedures (Article 9).[37] Another key piece of legislation is the EU Renewable Energies Directive (2009/28/EC), which the EU member states must implement by December 5 2010. This directive provides new rules regarding the guarantees of origin of the electricity, heating and cooling produced from renewable energies (Article 15). New information and training rules are also included (Article 14).[38]

4.2   Licences and connection to the grid

In most EU countries the installation of a co-generation facility requires an environmental licence, a site licence and a generation licence. The environmental licence is typically granted by the governmental department responsible for environmental issues. The licence usually needs to be obtained before starting the development of the project. The site permit concerns the construction of the plant and is usually issued by the local authority charged with zoning approvals. The environmental licence and site permit may also be part of a single authorisation granted by one authority only. Lastly, in liberalised markets the generation of electricity is not subject to particular restrictions, but the plant operator must usually hold a generation licence for the generation facility.[39] The duration and complexity of the licensing process depend on the legal rules in force, and the local policies and institutions.

Ensuring a connection to the grid is also a crucial aspect of every combined generation project. Even in liberalised markets, operators of co-generation plants incur costs for connection to the grid in addition to their initial investment. According to COGEN Europe, connection costs vary at between 5% and 20% of the

---

37   As recognised by the European Commission:

the Member States and the Commission are closely working to promote cogeneration. To this end, after the adoption of the CHP Directive in 2004, important steps have been taken to ease its full implementation throughout the European Union (e.g. determination of reference values and elaboration of detailed guidelines for the calculation of the electricity from CHP).

European Commission, Combined Heat and Power Generation, MEMO/08/695, November 13 2008. See also Commission Decision 2008/952, November 19 2008, establishing detailed guidelines for the implementation and application of the Co-generation Directive.

38   The EU Energy Services Directive (2006/32/EC) may also have an impact on combined generation.

39   In certain jurisdictions, smaller-scale CHP plants may not require a generation licence. For example, in the United Kingdom a generation licence is not required if the operator falls within the (general) exemption regime of the Electricity (Class Exemptions from the Requirement for a Licence) Order 2001 or obtains a specific exemption (pursuant to Section 4 of the Electricity Act 1989).

overall installation costs.[40] Furthermore, a grid connection involves delicate technical requirements to ensure that the plant's stability is not desynchronised when affected by load changes or system breakdowns.[41] In addition, the regulatory frameworks governing the connection of CCHP-generating facilities vary – also with regard to their complexity – between countries. In some countries, connecting utility plants to the grid requires a large amount of paperwork and compliance with governmental regulations. It can thus be a lengthy and burdensome process. In other countries (in particular, those in which third-party access is regulated), the process is swifter, but the standardisation of technical and contractual requirements is still necessary.[42] Promoting equal and fair access to co-generation will not only enable the sale of excess energy, but also enhance reliability by using the grid as a supplier of the back-up and top-up electricity needed.[43]

For all market players, but particularly for small, independent companies, a straightforward licensing process is crucial to enable new multiple generation investments. Ensuring economically fair and technically consistent licensing procedures and grid access is paramount to the development of combined generation. This is recognised by the Co-generation Directive (Articles 8 and 9).

Usually, co-generation regulations provide that the CHP producer enjoys priority access to the electricity that it generates for its own use. In liberalised markets, one possible way to guarantee the sale of any energy surplus is for the CHP producer to enter into a power purchase agreement with an electricity supplier. The limitation of risk within the agreement will be crucial to the plant operator.[44] Unfortunately, CCHP production is subjected to the price instability of both electricity and gas. This is exacerbated by the fact that the price of natural gas in most EU gas supply contracts is linked to the price of oil.

4.3 Programmes and other incentives

As discussed above, CCHP operators will increase their chances of commercial success by conducting a thorough risk assessment at the time of drawing up their development plans. Operators may also benefit from favourable governmental policies.

Governments can create incentives at three different levels: through subsidies, loans and tax credits. Policies and measures will depend on each government's stance on co-generation. However, a minimum level of fair remuneration should be

---

40   In Germany, the operator of a combined heat and power (CHP) or CCHP plant has to bear the costs of connection to the grid. However, this does not extend to those assets necessary for grid connection, for which title passes to the grid operator (Section 4(1)(a) of the Combined Heat and Power Act, in connection with Section 8(3) of the Ordinance on Grid Connection of Power Plants).
41   COGEN Europe, "A guide to cogeneration" (March 2001), p31.
42   In Germany, the network operator of the grid to which the CHP or CCHP plant is most closely situated is obligated to give preferential treatment to that plant's connection to the grid and the offtake of the (excess) electricity produced.
43   The Co-generation Directive defines 'back-up electricity' as "the electricity supplied through the electricity grid whenever the cogeneration process is disrupted, including maintenance periods, or out or order". It defines 'top-up electricity' as "the electricity supplied through the electricity grid in cases where the electricity demand is greater than the electrical output of the cogeneration process".
44   See also Tim Foster, "On-site power generation: how to sell excess power" (cospp.com/articles/article_display.cfm?ARTICLE_ID=370225&p=122).

respected to allow for recoupment of the large initial capital outlays. Thus, governments should provide pricing regimes for decentralised (ie, on-site) CCHP generation and these regimes should ensure that operators can benefit from an equitable tariff for every kilowatt fed into the grid. In turn, combined generation may benefit from a special feed-in tariff in some countries. For instance, Article 41 of the UK Energy Act 2008 establishes feed-in tariffs for carbon-saving micro-generation projects. In other countries, a differentiated regime is established which depends on the co-generation's unit total capacity.[45]

Financial support can also be based on direct subsidies that allow for the internalisation of the additional costs created by the use of renewable energies.[46] Sometimes, subsidies are supported by energy-efficiency funds that use tools and contribution mechanisms similar to those used by private funds.[47] The Co-generation Directive also provides that the EU member states shall ensure adequate support for co-generation (Article 7(1)),[48] without prejudice against the commission's evaluation of the support mechanisms that "could have the effect of restricting trade" (Article 7(2)).

Multiple generation systems are also encouraged by different initiatives and programmes supported by governmental and non-governmental organisations. COGEN Europe and the Co-generation Observatory and Dissemination Europe are two of the non-governmental institutions that play an important role in the development of CCHP generation. COGEN Europe is an association whose purpose is the promotion of wider use of co-generation in Europe for a sustainable energy future. The members of COGEN Europe are power companies, power authorities,

---

45   This is the case in Germany, where the remuneration system (ie, the value of the premium paid for CHP or CCHP electricity and the duration of such payments) accounts for the installed capacity of highly efficient co-generation plants (up to 50 kilowatts (KW), between 50KW and 2 megawatts (MW) and over 2MW). See Sections 5 and 7 of the Combined Heat and Power Act, as amended. As of 2009, this premium was paid not only for any excess electricity fed into the distribution network, but also for the co-generation plant operator's own consumption of electricity (Section 4(3)(a) of the Combined Heat and Power Act, as amended). Since these payments are eventually made by electricity end users by means of a surcharge to the network tariffs, the overall payment of such premiums is capped at €600 million a year.

46   Similarly, an incentive mechanism in the form of tax reductions for energy-efficient generation units has been successfully implemented in the United States. The Energy Improvement and Extension Act, passed by Congress on October 3 2008, expands federal energy tax incentives and introduces a co-generation investment tax credit. The American Recovery and Reinvestment Act, passed in February 2009, also expands and revises tax incentives for co-generation. Subject to the applicable legal qualifications, the following tax incentives may apply:
  • a 10% investment tax credit for the costs of the first 15MW of co-generation;
  • an investment tax credit for micro-turbines and fuel cells;
  • a renewable electricity production tax credit;
  • a bonus depreciation that allows businesses to recover investment through certain depreciation deductions;
  • an advanced energy manufacturing tax credit;
  • clean renewable energy bonds that also provide their holders with a tax credit against federal income tax; and
  • qualified energy conservation bonds.

47   World Energy Council, "Energy efficiency policies around the world: review and evaluation" (2008), p99.

48   For example, in Germany, new investments in micro-cogeneration units (with a maximum capacity of 50KW) can receive special funding from the Federal Office of Economics and Export Control. In addition, the Combined Heat and Power Act provides for financial support of expansion measures for heating networks (taking into account the length of the pipeline and its nominal diameter) for up to €150 million a year. Similarly to the premium for co-generation electricity, this financial contribution will effectively be paid by end users through higher network tariffs.

national co-generation associations, suppliers and other organisations involved in co-generation.[49] The Co-generation Observatory and Dissemination Europe provides an independent assessment of the progress of the implementation of the Co-generation Directive across EU member states.

The World Alliance for Decentralised Energy is a non-profit research and advocacy organisation created to accelerate the worldwide deployment of decentralised energy systems. Its purpose is to contribute to the worldwide development of high-efficiency co-generation, on-site power and decentralised renewable energy systems that deliver substantial economic and environmental benefits. Many co-generation associations (including COGEN Europe) and large companies active in the energy field, (eg, Chevron and GE Energy), are members of the alliance.[50]

In the United States, specific policies and programmes have been implemented to promote energy decentralisation and energy-efficient generation technologies. The US Department of Energy is also funding two programmes: the Distributed Generation Technology Development and the Integrated Energy Systems.[51] Other initiatives include the Federal Energy Management Programme[52] and the US Environmental Protection Agency's Combined Heat and Power Partnership.[53]

Granting special loans to companies that invest in CCHP generation is another way to speed up the development of combined generation. In the United States, many programmes have been created which aim to improve the use of combined generation technologies. Such is the case of the New York State Energy Research and Development Authority, the California Energy Commission Energy Funding and the Massachusetts Renewable Energy Trust. Furthermore, under the 2005 Energy Policy Act, the US Department of Energy may issue loan guarantees to eligible projects which avoid, reduce or sequester emissions of greenhouse gases.[54]

4.4   Key contractual aspects

Any strategy to implement CCHP generation should be flexible enough to cope with technical requirements and to ensure that projects are commercially viable and able to achieve predictable cash-flow.

As underlined by Joseph Huse, "the parties need to determine who can best control the circumstances which may result in loss or damage. Placing the liability on the party in control will improve surveillance and prevention of the risk in question".[55]

The operator may choose not to deploy any upfront costs to build and own the

---

49   COGEN Europe, "What we do"(www.cogeneurope.eu/category/about-cogen/what-we-do/).
50   For more information, see World Alliance for Decentralised Energy, "Membership in WADE" (www.localpower.org/abt_members.html).
51   US Department of Energy, "About the program" (www.eere.energy.gov/de/about.html).
52   For more information, see US Department of Energy, Federal Energy Management Program (http://www1.eere.energy.gov/femp/).
53   For more information, see US Environmental Protection Agency, Combined Heat and Power Partnership (www.epa.gov/chp).
54   For more information, see US Environmental Protection Agency, "Federal incentives for developing combined heat and power projects" (www.epa.gov/CHP/incentives/index.html).
55   *Understanding and negotiating turnkey agreements* (1997), p369.

power plant, but instead enter into an agreement with a contractor. This enables the operator to take advantage of the plant's production while avoiding the burdens and costs associated with its ownership. Apart from unincorporated joint venture agreements, the most common agreements between a host company developing a co-generation plant and a contractor are build, operate and own agreements and build, operate, own and transfer agreements. As emphasised by COGEN Europe, "the choice between these types of contract is dependent upon the nature of the cogeneration (large or small), the company's investment and accounting policy, the level of financial risk the purchaser is willing to bear and the financial return required".[56]

In build, operate and own agreements the contractor installs, operates and owns the plant. The host company purchases the plant's production at a previously agreed price and on a take-or-pay basis, usually set out as an indexed rate. By using this type of agreement, the host company avoids having to finance the project itself and avoids all the risks related to the project's development. Build, operate, own and transfer agreements are similar to build, operate and own agreements with regard to the building, installation and operation of the plant. However, they provide that the host company must acquire ownership of the plant at a pre-determined date. Both types of agreement were initially designed for concessionaires of public sector projects. However, they are now widely used by private entities to isolate the project risks and facilitate the parties' access to project financing.

The host company may also decide to plan and manage the project, contracting with an energy services company the design and construction of the plant. This model, known as a design and build contract, may maximise the owner's economic return, but it increases the owner's risks with regard to the whole installation of the plant. This type of set-up also requires a sufficient allocation of human resources to ensure adequate project management.

Another set-up consists of the host company's entry into a contract with an energy services company to design, develop and build the plant on a turnkey basis. In this case the ownership is generally transferred to the host company only after the takeover of the plant, which occurs after commissioning (ie, when the plant starts operating). Thus, during the early stages of development and until commissioning, the contractor bears the project's risks and any defect falling within the scope of works is considered contractor's default.

Depending on the type of contract signed between the parties, an operation and maintenance agreement may also be required to ensure proper technical maintenance of the plant after the beginning of commercial operation.

Independently from the chosen structure, the project's contract should cover the following aspects, among others:

- the commercial operation date;
- the milestones for implementation of the project;
- prices and costs;
- the testing and achievement of milestones;

---

56   COGEN Europe, "A guide to cogeneration" (March 2001), p41.

- technical performance guarantees; and
- the penalties for late completion or for failure to comply with the technical specifications.

Other clauses, such as *force majeure*, liquidated damages, limitation of liability and personnel may be also agreed on by the parties.

The owner's contractual option and the pricing method (lump sum, cost-reimbursable or unit price) are obviously closely connected with the financing decision. Project financing may require the incorporation of a new limited liability company. As a non-recourse financing, project financing entails a thorough analysis and requires an efficient allocation of risks during the project's development. Thus, it frequently requires an application for and receipt of all licences and permits at a very early stage. Financing through private equity usually includes thorough due diligence, not only of the financial data, but also of all contracts and commitments that have been entered into and made by the company.

Lastly, the prospect of marketing any surplus heat and/or power not used by the plant operator itself will be another key consideration for those planning to enter into the CCHP generation market. The idea of locking future customers into long-term supply contracts may appear advantageous from a project financing perspective. Constraints apply, though. First, the restrictions applicable under competition law need to be observed (eg, Article 101(1) of the Treaty on the Functioning of the European Union, to the extent applicable). Second, Directive 2009/72/EC on common rules for the internal market in electricity considerably strengthens consumer rights.[57] In particular, the directive provides that customers shall retain their right to choose their supplier by facilitating the process of switching supplier.[58] Hence, the future regulatory framework within the European Union will generally prevent customers from being locked into long-term heat and/or electricity supply agreements. Accordingly, it is paramount that the state remuneration system allows a minimum level of fair remuneration, for instance, by providing an equitable tariff for every kilowatt fed into the grid (see section 4.3 above)

5. Conclusion

Even though combined generation offers many advantages, its share in global power generation has stagnated in the past few years.[59] There are two main reasons for this:

- The costs involved are high, particularly when associated with recent technologies marketed by a limited number of suppliers; and
- The regulatory framework lacks competitiveness, notably with regard to

---

57   Such rules need to be implemented into national legislation by the EU member states by March 3 2011 (Article 49(1) of Directive 2009/72/EC).

58   For details, see Articles 3(4) and following of Directive 2009/72/EC in connection with Annex I and the corresponding Interpretative Note regarding retail markets. In parallel, the directive provides stricter rules on third-party access to distribution networks better to enable suppliers (without regard to their national authorisation and registration) to supply customers within the union. In this regard, Article 28 of Directive 2009/72/EC provides only limited exceptions to so-called 'closed-distribution systems'.

59   International Energy Agency, "Combined heat and power. Evaluating the benefits of greater global investment" (OCDE/IEA, 2008), p7.

access to the electricity grid and pre-emptive rights on the sale of energy.[60]

In developed countries, combined technologies mean contributing to cleaner energy and the reduction of greenhouse gas emissions. Accordingly, this technology should feature prominently on the EU programmes fostering:

- energy efficiency;
- the reduction of greenhouse gas emissions; and
- thriving renewable sources for energy generation.

The European Commission's expectations to double the amount of co-generated electricity by 2020 can be supported by:

- the extensive market potential for combined generations in the EU member states; and
- as previously described, the emerging trend towards new applications for combined generation.[62]

In emerging economies, access to electricity is a priority. However, even there energy efficiency cannot be disregarded as a reliable way to enhance access to electricity by enabling greater generation of electricity without increasing fuel consumption.

Against this background, the high levels of energy efficiency, the carbon savings and the ability of combined generation to decentralise energy generation are compelling reasons for setting up an enabling, conducive and stable framework for CCHP generation. With streamlined legal support, the coming years may be auspicious for multiple generation projects, boosted by global environmental concerns, energy market liberalisation and the world's growing demand for energy.

The authors would like to thank their colleagues Catarina Monteiro Pires and Daniel J Zimmer for their invaluable contribution to this chapter.

---

60   The commission highlights that complex and burdensome legal frameworks and procedures, and unfavourable conditions regarding the connection to the grid and back-up electricity are the key barriers to a large implementation of co-generation in Europe. See Communication from the Commission to the Council and the European Parliament, "Renewable energy road map – renewable energies in the 21st century: building a more sustainable future" (COM (2006) 848 final).
61   COGEN Europe, "Briefing note" (November 16 2009), p2.

Embedded generation: a UK perspective

Kerry Thompson
Inventa Partners Ltd

1. Introduction

As in many countries, heat generation in the United Kingdom is a vital part of domestic and commercial life. As at 2006, heat generation in the United Kingdom accounted for 41% of the country's energy consumption and 47% of its carbon dioxide emissions, with heat consumption accounting for 60% of the average domestic energy bill.[1] Most consumers in the United Kingdom do not purchase heat. Instead, they purchase the fuel to generate heat (eg, around 69% of consumers use gas to generate heat, with 14% using electricity, 11% using oil and 3% solid fuels).[2] As is shown in the table below,[3] only a small percentage of heat is generated by renewable means.[4]

	2004	2005	2006	2007	2008	2009
Percentage of heating from renewable sources	0.7	0.9	1.0	1.2	1.4	1.6

The government's heat strategy is focused on efficient heat production and consumption. This includes:

- heat generation from renewable fuel sources (eg, biomass);
- utilisation of heat that would otherwise be wasted;
- district heating schemes; and
- improved energy efficiency within buildings (eg, insulation).

Specific incentives to encourage investment include renewable obligation certificates (ROCs), feed-in tariffs, renewable heat incentives, bioenergy capital grant

---

1   *Heat and Energy Saving Strategy*, Department of Energy and Climate Change (2009), p12. Also see the Department of Energy and Climate Change's interactive UK Heat Map website (http://chp.decc.gov.uk/heatmap/).

2   *Renewable Heat Incentive: consultation on the proposed RHI financial support scheme*, Department of Energy and Climate Change (February 2010), p7.

3   *Energy Trends June 2010*, Department of Energy and Climate Change (published by National Statistics), p12.

4   According to *Energy Trends June 2010*, the "three sources of renewable heat production in the United Kingdom are the direct combustion of biomass (93 per cent of the total), active solar heating, and geothermal aquifers" (p21).

schemes and enhanced capital allowances.[5] These focus on renewable projects. However, there are limited incentives in the United Kingdom for energy efficiency projects that do not have a renewables component.

In a world facing dramatic technological and commercial change and restructuring on an almost daily basis, the emergence of a 'perceived' novel or new energy solution in an established and traditional market is often viewed at best with scepticism and caution, at worst with cynicism and fear. The emergence of distributed heating systems, combined heat and power (CHP) systems and combined cooling, heat and power (CCHP) systems[6] in the UK energy market has been rapid (in traditional energy industry terms), and has been driven by both government policy and market forces.

Yet the technology itself and its application are not new. At the end of the 19th century, the United Kingdom's energy and hydraulic engineering expertise was second to none, and the country led the world in the development and deployment of embedded generation systems.

In 1883 an act of Parliament established the London Hydraulic Power Company, which went on to install a hydraulic power network of high-pressure cast-iron water pipes (two inches to 10 inches in diameter and up to one inch thick). The pipe network covered an area north of the Thames, from Hyde Park in the west to Docklands in the east. The utility provided reliable energy services and most users abandoned their own generators.

The system was a cleaner and more compact alternative to steam power – it powered machinery, lifts, cranes, theatre machinery and bridges. At its peak, it had five primary pumping stations plus accumulators (equivalent to today's energy storage) pumping more than 1.6 billion gallons of water annually through 180 miles of pipework. Following the demise of the London Hydraulic Power Company, the wheel turned full circle and users were forced to install their own plant or convert to electric motors.

The system was eventually closed in 1977 (replaced by electricity) and the entire company was acquired in 1981 by a group led by Rothschilds, which recognised the importance of the pipe network for the coming generation of communications systems. It subsequently sold the network of pipes, ducts and conduits to Mercury Communications in 1985, which used the underground pipe infrastructure (and the associated licences and wayleaves) to run fibre-optic cables across the City of London: evidence that the enduring value of the infrastructure lay in the underground assets rather than in the technology (engines) at the surface.

Further, back in 1947 the then British Electricity Authority had a statutory duty to "investigate methods by which heat obtained from or in connection with the generation of electricity may be used for the heating of buildings in neighbouring

---

5   For further information on this please read the chapter entitled "Renewable energy support mechanisms: an overview".
6   In this chapter the commonly used phrase 'embedded generation system' encompasses all three of these systems. Other commonly used phrases include 'distributed energy/heating system' and 'district energy/heating system". In some jurisdictions, CHP systems are more commonly referred to as 'co-generation systems' and CCHP systems as 'tri-generation systems'.

localities, or for any other useful purpose, and the Authority may accordingly conduct, or assist others in conducting, research into any matters relating to such methods of using heat".[7] From 1974 to the late 1970s, the then Department of Energy was actively considering embedded generation system solutions across London. During this period a National Heat Board was established and considered nine main schemes, including sourcing heat from Barking Power station. The London Electricity Board and the Central Electricity Generation Board developed plans for a 100 megawatt (MW) district heating scheme linked to Bankside Power Station. However, by the late 1970s, the North Sea oil and gas field discoveries led to diminished interest in embedded generation systems. It is only in the last 10 years, as climate change and security of supply issues have garnered worldwide attention, that both public and private sector interest and investment in embedded generation systems in the United Kingdom have increased.

Most of us would define a 'renewable energy solution' as a building-mounted solar panel or a wind turbine; very few would define a 'renewable energy solution' as a district heating system and even fewer as an embedded generation system. Yet today, these systems are being acknowledged as one of the most efficient, cost-effective solutions to deliver low or even zero-carbon energy to large-scale, high-density and mixed-use living and working environments.

Embedded generation systems are capable of delivering reliable, low or zero-carbon energy solutions on both a micro and macro scale, utilising the waste heat from power generation to produce hot water which is distributed to multiple buildings via a network of buried, highly insulated pipes.

Embedded generation systems typically use fossil fuels as their primary fuel source and are therefore not the ultimate remedy for a zero-carbon energy objective. However, as they use the primary fuel source more efficiently to produce heat, power and possibly cooling, and are closer to the end user (thereby minimising distribution losses), they also reduce the resulting greenhouse gas emissions. This also helps to meet the government's other key energy objective of reducing fuel consumption, thereby helping to manage the United Kingdom's security of supply issues.

Given the above characteristics, embedded generation systems are particularly well suited to generating and delivering low-carbon energy to areas of high population density, such as the cities and towns where most of the country's population live and work. While many of these embedded generation systems are installed contemporaneously with the construction of new buildings and infrastructure, technological advancements mean that in most cases it is now both technically possible and commercially viable to retrofit this technology into existing buildings and infrastructure.

This chapter considers the key commercial issues in structuring an embedded generation system project in the United Kingdom. In doing so, it first provides an overview of the UK electricity, gas, heat and biomass markets, as well as a regulatory overview, to put into context the key commercial issues set out in section 5. It is intended to provide guidance for those considering participating in any way in a

---

7   See Section 50(1) of the Electricity Act 1947.

embedded generation system project, whether it be located in the United Kingdom or in another jurisdiction that shares similar regulatory and market characteristics. This chapter should be read in conjunction with "Combined cooling, heat and power generation: technical, commercial and legal aspects".

2.   UK market overview

2.1   Electricity market

Since the introduction of the Electricity Act 1989, the UK electricity market has evolved into a highly sophisticated and competitive market, with private sector participation in generation, transmission, distribution and supply. The diagram below[8] provides an overview of the complex physical interfaces in the UK electricity market.

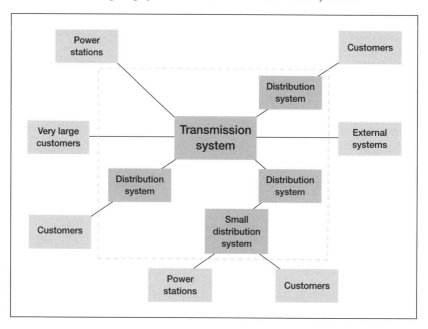

(a)   Generation

Generation is dominated by the major power producers[9] operating large power stations throughout the United Kingdom, with a total capacity of 78,255MW, which as at December 2009 accounted for 92% of the country's total generating capacity of

8   Courtesy of Michael Rudd (partner, SNR Denton).
9   As at the end of 2009, the major power producers were AES Electric Ltd, Baglan Generation Ltd, Barking Power Ltd, British Energy plc, Centrica Energy, Coolkeeragh ESB Ltd, Corby Power Ltd, Coryton Energy Company Ltd, Derwent Cogeneration Ltd, Drax Power Ltd, EDF Energy plc, E.On UK plc, Energy Power Resources, Gaz De France, GDP, Suez Teesside Power Ltd, Immingham CHP, International Power Mitsui, Magnox North Ltd, Premier Power Ltd, RGS Energy Ltd, Rocksavage Power Company Ltd, RWE Npower plc, Scottish Power plc, Scottish and Southern Energy plc, Seabank Power Ltd, SELCHP Ltd, Spalding Energy Company Ltd and Western Power Generation Ltd – *Digest of United Kingdom Energy Statistics, 2010*, Department of Energy and Climate Change, paragraph 5.62, p127 (published by National Statistics).

85,337.[10] For statistical purposes, this now includes major renewable power producers,[11] including:

- hydro (4,139MW of total capacity as at December 2009);
- wind (1,211MW of total capacity as at December 2009); and
- major renewables other than hydro and wind, which burn fuels such as landfill gas, sewage sludge, biomass and waste (213MW of total capacity as at December 2009).[12]

Regarding CHP, as at the end of 2009:[13]

- there were 1,465 CHP schemes with a total installed electricity capacity of 5,569 megawatts electrical (MWe) and a total installed heat capacity of 10,755 thermal megawatts (MWth). Of these, there were almost 120 large-scale (ie, over 1MW) CHP plants with a total capacity of 2,268MW, of which 2,036MWe qualified as 'good-quality CHP';[14]
- these 1,465 schemes generated 27,777 gigawatt-hours (GWh) of electricity, representing 7% of the total electricity generated in the United Kingdom during 2009;
- although only 18.6% of these schemes had an electricity capacity of greater than 1MWe, they represented 96% of total installed CHP electricity capacity;
- 90% of the CHP electricity capacity was in the industrial sector, with less than 1% in residential, commercial, retail and government buildings;
- 71% of these schemes used natural gas as their fuel source, with 23% using non-conventional fuel sources such as by-products and waste products from industrial processes or renewable fuels;
- these schemes had an overall average efficiency (as a percentage of gross calorific value) of 67% (24% average electricity efficiency and 43% average heat efficiency);
- approximately 8% of the 'good-quality CHP' consumed in the United Kingdom was purchased from over 60 CHP systems located outside the United Kingdom (primarily Denmark, France, Germany and Ireland) wishing to derive an additional revenue stream through the sale of levy exemption certificates[15] issues against the electricity generated by those CHP stations;[16] and

---

10   *Digest of United Kingdom Energy Statistics, 2010*, paragraph 5.31 and Chart 5.7, pp121 and 140.
11   For example, as at the end of 2009, the major wind producers were Centrica Energy, E.On UK plc, Fred Olsen, HG Capital, Renewable Energy Systems, RWE Npower plc, Scottish Power plc, Scottish and Southern Energy plc and Vattenfall Wind Power.
12   *Digest of United Kingdom Energy Statistics*, 2010, paragraphs 5.58 to 5.64 (inclusive) and Chart 5.7, pp127, 128 and 140.
13   Statistics taken from the *Digest of United Kingdom Energy Statistics, 2010*, Chapter 6.
14   This relates to the United Kingdom's Combined Heat and Power Quality Assurance Programme, introduced in line with the requirements of highly efficient combined heat and power systems contained in the EU Cogeneration Directive (2004/08/EC). For more information see www.chpqa.com.
15   A levy exemption certificate is a certificate issued to a generator that generates renewable electricity. This certificate is purchased by the electricity supplier that purchases the electricity and ultimately by specific industrial and commercial customers that purchase that electricity. Its purpose is to exempt those specific industrial and commercial customers from paying a climate change levy (pursuant to the Finance Act 2000). Regarding its application to combined heat and power projects, please see www.ofgem.gov.uk/Sustainability/Environment/CCLCHPEx/Pages/CCLCHPEx.aspx.
16   *Digest of United Kingdom Energy Statistics, 2010*.

- these schemes saved 13.89 million tonnes of carbon in 2009 as against all fossil fuels.

Unlike in some other countries, no city in the United Kingdom has a citywide integrated heat distribution network.

(b)   Transmission[17]

A single transmission system operator – National Grid Electricity (NGET) plc – operates the regional transmission networks owned by NGET (in England), Scottish Power Transmission Limited for southern Scotland and Scottish Hydro-Electric Transmission Limited for northern Scotland.

Most embedded generation systems in the United Kingdom do not connect directly to the transmission system, and therefore this chapter does not address the regulatory framework and the industry codes and agreements relevant to the UK transmission system.

(c)   Distribution[18]

Fourteen distribution network operators are licensed, pursuant to Section 6 of the Electricity Act, to distribute electricity over distribution networks in a defined geographic area. In addition, four independent distribution network operators (IDNOs) distribute electricity over distribution networks located within the geographic boundary of a DNO's licensed area and are therefore connected to that DNO's distribution network. Any generator (including one that owns a small-scale generation system) that wishes to connect directly to, or licensed electricity supplier that wishes to use, a DNO or IDNO's distribution network must accede to the Distribution Connection and Use of System Agreement, a multi-party common terms agreement between distributors, generators and suppliers. The agreement was established in October 2006 to replace the numerous bilateral connection and use of system agreements that existed.[19]

The term 'private wire' is given to distribution networks, often connecting an embedded generation station to the consumer, which are exempt from the prohibition of carrying out the activity of distribution (usually because the distribution network is small enough to qualify for a general or specific exemption, as noted in section 3 below). Installation of the private wire system does not in itself remove the consumer's choice of electricity supplier. In the author's experience, there has been a move away from private wire systems connecting residential customers because of the *Citiworks* case.[20]

---

17   See the Office of the Gas and Electricity Markets (Ofgem) website for more information – www.ofgem.gov.uk/Networks/Trans/Pages/trans.aspx.
18   See the Ofgem website for more information – www.ofgem.gov.uk/Networks/ElecDist/Pages/ElecDist.aspx.
19   For more information see the Distribution Connection and Use of System Agreement website – www.dcusa.co.uk.
20   *Citiworks AG v Sächsisches Staatsministerium für Wirtschaft und Arbeit als Landesregulierungsbehörde*, Case C-439/06, http://eur-lex.europa.eu/LexUriServ/LexUriServ.do?uri=CELEX:62006J0439:EN:HTML. In addition, please see the chapter entitled "Combined cooling, heat and power generation: technical, commercial and legal aspects".

(d)   *Supply*

Distribution and supply are separate activities. This means that electricity supply becomes just like the sale of any other goods or services – customers are free to choose their supplier based on price.

To date, over 120 supply licences have been granted in the United Kingdom. The holders of these licences go beyond core utility companies and include others that have a specific reason for requiring a supply licence, including in relation to embedded generation system projects.

2.2   Gas market

As gas is the primary fuel source for the UK heat market, it is worth briefly considering the UK gas market.

As at December 31 2008, there were 22.327 million domestic and small business gas consumers in the United Kingdom, consuming 377,473GWh of gas per year; and over 324,000 large gas consumers, using 208,982GWh of gas.[21] Consumption is roughly split into thirds between domestic consumption, electricity generation and consumption by industry, services and energy industries.[22]

In 2004 the United Kingdom was a net importer of natural gas for the first time since 1996. By 2009, the net import figure had increased to 319 terawatt-hours (TWh).[23] This is one factor that has raised concerns about the country's energy security.[24]

2.3   Biomass market

The generally accepted definition of 'biomass' in the United Kingdom is any material, other than fossil fuel, which is or is derived directly or indirectly from plant matter, animal matter, fungi or algae.[25] As noted by the UK Biomass Taskforce, it is "literally, any biological mass derived from plant or animal matter. This includes material from forests, crop-derived biomass including timber crops, short rotation forestry, straw, chicken litter and waste material."[26]

Gas made from biomass may also play an important part in the renewable energy mix of the future. The term 'syngas' is used to refer to any gas made artificially/by synthesis. If synthesis gas is produced from biomass, it is called biosyngas. 'Biogas' is defined to mean gas produced by the anaerobic conversion of organic matter.[27] It is composed of approximately 60% methane, 40% carbon dioxide and other trace levels of contaminants. 'Bio-SNG' is a term used to refer to a combustible gas that has been created by the thermochemical process of gasification (see below) of organic material. It is composed predominantly of methane, hydrogen, carbon monoxide and carbon dioxide.[28]

---

21   *Digest of United Kingdom Energy Statistics, 2010* Table 4A, p 101.
22   *Digest of United Kingdom Energy Statistics, 2010*, paragraphs 4.10 and 4.11 and Chart 4.2, pp97 and 98.
23   *Digest of United Kingdom Energy Statistics, 2010*, paragraph 4.7, p95.
24   "Energy Security: A national challenge in a changing world", Malcolm Wicks MP, August 2009.
25   See Section 4 of the Renewables Obligation Order 2009 and Section 100(3) of the Energy Act 2008.
26   Report to Government (October 2005), Biomass Task Force, Paragraph 1.9, p9.
27   See Section 100(3) of the Energy Act 2008.
28   *Biomethane into the Gas Network: A Guide for Producers*, Department of Energy and Climate Change, December 2009.

In January 2009 the National Grid published a report that argued that in the longer term, with the right government policies in place, renewable gas could meet up to 50% of UK residential gas demand.[29] Interestingly, the report noted that as at the date of the report, 1.4 billion cubic metres of renewable gas are currently produced in the United Kingdom – which could meet around 1% of total UK gas demand. However, because of the commercial incentives (the Renewable Obligation Scheme), all of this gas is used to generate electricity at efficiency levels of around 30% in most cases. If the gas were injected into the gas grid, this could be used as the primary fuel source for embedded generation systems. However, it could also be delivered straight into consumers' homes and utilised for heating at efficiency rates in excess of 90%, meaning that it could challenge both the economics and climate change drivers behind the use of embedded generation system projects, particularly for consumers with existing connections which would otherwise require retrofitting in order to connect to an embedded generation system.

More broadly, in order to be entitled to receive renewables obligation certificates, a generator of any biomass generating facility must (among other things) provide the Office of the Gas and Electricity Markets (Ofgem) every year with information relating to the sustainability of the biomass which is used by the embedded generation system. This information includes:

- the material which the fuel will be composed of;
- the form it will take;
- whether it will be a by-product of another process;
- whether it will be waste;
- where it was derived from; and
- the proportion (if any) which will be derived from energy crop.[30]

2.4 Spark spread

Two major factors affect the economics of an embedded generation system project:

- the relative cost of fuel (principally, natural gas); and
- the value that can be realised for electricity.

The difference between the price of electricity and the price of the gas required to generate that electricity is known as the 'spark spread'.

During 2007, the market improved for embedded generation system projects due to an increase in the spark spread (ie, a fall in the price of gas relative to that of electricity). However, uncertainty regarding future movements in the gas and electricity markets has created perceptions of increased risk for embedded investors and schemes. This is one reason why fiscal incentives are considered so important by many in the industry.

3. Embedded generation systems – key regulatory points

Unlike certain other European countries, the United Kingdom has no legislation or regulations specifically dedicated to embedded generation systems. Instead, a variety

---

29 *The Potential for Renewable Gas in the UK: A Paper*, National Grid, January 2009.
30 See Article 54 of the Renewables Obligation Order 2009.

of legislation and regulations deal with various aspects of embedded generation system projects.

3.1 Electricity

Although the generation, distribution and supply of electricity are regulated activities, many embedded generation systems in the United Kingdom either fall within the general exemption regime (Electricity (Class Exemptions from the Requirement for a Licence) Order 2001) or qualify for a specific exemption (pursuant to Section 5 of the Electricity Act 1989) from the prohibition in Section 4 of the Electricity Act 1989 against carrying out generation, distribution or supply activities without the relevant licence. Licence exemptions are of benefit to vertically integrated embedded generation system projects for a number of reasons. First, under the Electricity Act 1989, it is not possible for one entity to hold licences for all three activities (generation, distribution and supply), and as such a licence exemption makes such a structure possible. Second, it frees small-scale industry players from the regulatory burdens of licence compliance.

3.2 Heat

The generation, distribution and supply of heat are not regulated activities in the same sense as electricity generation, distribution and supply.[31] However, two important statutory provisions apply to the operation of these systems:

- Section 50 of the New Roads and Street Works Act 1991 requires a person to apply for a street works licence (if it does not have a generation licence that extends to the distributed heating and cooling system as outlined in footnote 32). This is common to most projects of this type.
- Section 108 of the Housing Act 1985 deals with heating charges payable by secure tenants.[32] This may be relevant where the embedded generation system is supplying heat to affordable housing.

3.3 Consequences of limited regulation

From the author's experience, a 'light-touch' regulatory approach to embedded generation systems has both positive and negative outcomes. On the positive side, light-touch regulation affords greater flexibility to structure the commercial, technical and legal aspects of an embedded generation system project to suit the particular needs of the project. On the negative side, light-touch regulation means that more detailed contractual arrangements are required to address matters usually covered by regulation. This can increase the complexity of the legal, technical and commercial arrangements, with resultant time and cost consequences.

---

31   An important qualification to this statement is that if a person holds a generation licence, then pursuant to Sections 10(3) and 10(3A) of the Electricity Act 1989, this licence enables the licence holder to rely on Schedule 4 of the Electricity Act 1989, which contains various powers that can apply to the distribution and supply of heat regarding street works, alteration of works, protection from interference, wayleaves, felling and lopping of trees and entry onto land for the purposes of exploration. In the author's experience, such powers (although not often exercised) can act as leverage when negotiating property and access rights.

32   In essence, a secure tenant is one or more tenants where at least one occupies a dwelling house as their only principal home and the landlord is a local authority, new town corporation, housing action trust or urban redevelopment corporation.

(a)   *Energy services companies and multi-utility services companies*

Energy services companies: The European Joint Research Centre distinguishes energy services companies (ESCOs) from traditional energy service providers by referring to their delivery of the following:

> *1. ESCOs guarantee the energy savings and/or the provision of the same level of energy service at a lower cost by implementing an energy efficiency project. A performance guarantee can take several forms. It can revolve around the actual flow of energy savings from a project, can stipulate that the energy savings will be sufficient to repay monthly debt service costs for an efficiency project, or that the same level of energy service will be provided for less money;*
>
> *2. The remuneration of ESCOs is directly tied to the energy savings achieved;*
>
> *3. ESCOs typically finance, or assist in arranging financing for the installation of an energy project they implement by providing a savings guarantee;*
>
> *4. ESCOs retain an on-going operational role in measuring and verifying the savings over the financing term.*[33]

In the United Kingdom, the term 'ESCO' is also the name commonly used to refer to an incorporate or unincorporated project vehicle which has a central role in the design, construction, commissioning, operation and maintenance of an embedded energy system and the provision of heat and/or power supplies. This role is 'central' rather than 'primary role' for the reasons explained in section 5 below.

Multi-utility services companies: A 'multi-utility services company' (MUSCO) is an incorporate or unincorporated project vehicle which goes beyond energy services and incorporates a range of core utilities that can include electricity, gas, water, sewerage and telecommunications.[34] The term is now commonly used by mainstream utility companies that, through a variety of legal structures, establish a MUSCO (either themselves or in joint venture with other persons such as other utility companies) which assumes responsibility for the coordinated design, construction, commissioning, operation and maintenance of these core utilities. The arrangements can also extend to additional goods and services related to the core utilities, including embedded generation systems.

The diagram below illustrates the typical 'scope of services' that a MUSCo service organisation would present, with the MUSCO services running alongside the complementary existing utility services.

This example includes electricity, gas, water/sewerage, data (fibre-optic infrastructure), and could also encompass heat, cooling, borehole water extraction, building mounted, integrated renewables, waste sorting and collection and waste

---

33   *Energy Services Companies in Europe: Status Report 2005*, Paolo Bertoldi and Silvia Rezessy; European Commission Director General Joint Research Centre, Institute for Environment and Sustainability, Renewables Energy Unit; pp 17 and 18.

34   For example, see the Elephant and Castle regeneration project (www.elephantandcastle.org.uk/ 00,news,1608,185,00.htm) and the Allenby/Connaught Project (www.aspiredefence.co.uk/pages/mujv/).

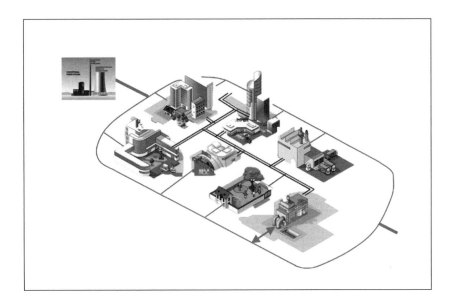

The schematic below illustrates the high-level structure and function of a typical MUSCO.

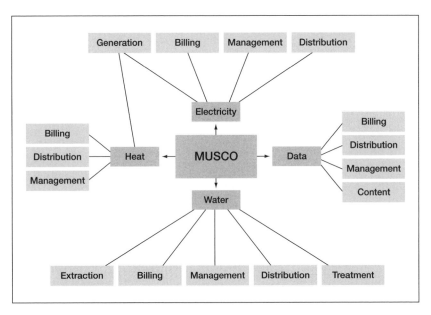

Given the utility regulatory regime in the United Kingdom (as outlined briefly in sections 2 and 3 above), there are certain activities that only licensed entities (eg, an electricity distribution company) can perform. Further, the utility regulatory regime either prohibits a single entity from carrying out more than one regulated activity or makes it sufficiently complex (due to the different regulatory rules governing each utility market in the United Kingdom) to make it unviable for a single entity to do so. For these reasons, it is not uncommon for the MUSCO to be a separate entity with no utility licences, which enters into the primary contract with a developer to procure this infrastructure and then enters into contracts with each relevant utility, which may or may not be an affiliate of the MUSCO.[35] At present, few organisations can provide the full gamut of services required, resulting in many entering into joint venture arrangements with complementary organisations.

A perceived advantage of a MUSCO, compared with an ESCO, is the ability to assess and realise additional revenue streams beyond just one core business, such as an embedded generation system. This is particularly relevant when considering data services, where the margins for the sale of wholesale network access or even retail content are considerably higher than for the sale of energy or water supplies.

4. Key issues for an embedded generation system project

4.1 Introduction

This section considers key commercial issues for embedded generation system projects in the United Kingdom, and in doing so also touches on some key technical and legal issues. While the technical and legal matters can have a significant influence on the commercial aspects of the project, it is typically the commercial aspects of the project that ultimately determine whether it is viable and how it should be structured. Consequently, from the author's experience, the commercial issues are usually considered in detail first before the stakeholders delve into technical and legal matters.

Given the diversity and complexity of embedded generation system projects, this section is not intended to be an exhaustive or detailed analysis of the key issues for these projects. Instead, it is intended to be a checklist to assist you in understanding the key issues you need to consider for an embedded generation project.

'Core stakeholders' below refers to:

- the person procuring the embedded generation system project, which may include a property developer, landlord, local or statutory authority and/or anchor customer (referred to as the 'principal');
- the ESCO/MUSCO (although for convenience, references to 'ESCO' in the remainder of this section also mean a MUSCO which is involved in an embedded generation system project); and
- the consumers of heat, cooling and/or power supplies generated by the embedded generation system, which may include both non-domestic (eg, industrial, retail and commercial) and/or domestic (including social housing) customers.

As in any project, a raft of other stakeholders have a role to play in the project, including:

---

35   For certain utilities, there are regulatory restrictions on a licensed utility's dealings with affiliates (eg, Condition F6.8 of the Water and Sewerage Licence and Standard Condition 41 of the Electricity Distribution Licence).

- public bodies which issue relevant licences, permits, permissions and other authorisations (eg, the local planning authority, Ofgem, the Environment Agency and the Competition Commission);
- insurers;
- debt and equity funders;[36]
- any persons with property interests who are not core stakeholders (eg, the owners of adjoining land owners or persons with interests, such as wayleaves, on the relevant land);
- utility companies which provide utility connections and/or supplies to the embedded generation system, the principal and customers;
- the fuel supplier(s); and
- the ESCO's contractors and suppliers.

Finally, while this section considers the key commercial issues, the principal must also consider the most appropriate process to select an ESCO and identify, agree and document (in contracts) the commercial, technical and legal arrangements concerning the embedded generation project. Some principals (eg, public authorities and licensed utilities) are caught by procurement regulations which, if applicable to an embedded generation system project, will require the principal to procure the embedded generation system project in accordance with the relevant procurement regulations.[37]

4.2   Key commercial issues
The key commercial issues are set out below.

(a)   *Financials*
This is the most important commercial issue to be addressed for these embedded generation system projects, as it determines whether the project is viable. The key financials are typically as follows.

Capital and operational costs: The first issue to consider is the likely capital and operational costs associated with delivering the project and how these will be allocated between the core stakeholders. Key points to consider here are as follows:
- the technical aspects of the embedded generation system, including what it is, how and when it is to be designed, installed and commissioned, and its interface with existing infrastructure (including existing utility infrastructure);[38]

---

36   Please read the chapters in Part 3 for more information on issues for debt and equity funders.
37   The Utilities Contracts Regulations 2006 and the Public Contracts Regulations 2006.
38   Related issues include:
       • whether the embedded generation system is to be installed on a greenfield site or retrofitted to cater for existing customers;
       • the legal obligations that the principal must meet which impact on the embedded generation system project (eg, planning obligations which set renewable, sustainability and carbon targets); and
       • the input fuel for the embedded generation project.
       The input fuel(s) in itself raises broader issues, particular for renewable fuel sources such as biomass, where the core stakeholders must consider the sustainability of the input fuel and the security of fuel supply.

- who will undertake the core activities of design, installation, commissioning, operation and maintenance;[39] and
- whether the party with primary responsibility for the costs of the core activities will receive a financial contribution to such costs from any of the other core stakeholders. For example, if the ESCO is primarily responsible for such costs, the principal may agree to contribute to such costs in return for a reduced cost for a lower heat, cooling and/or power price or a direct income stream from the project.[40] Alternatively, if the principal designs, installs and commissions the energy system, it may transfer ownership to the ESCO for valuable consideration.[41]

Income: The second issue to consider is the likely income generated from the project and how that income will be distributed between the core stakeholders. The main income streams for this project are as follows:

- ESCO – assuming that the ESCO is more than an operation and maintenance contractor receiving a fee for service, it will receive income from the sale of heat, cooling and/or power. Some key considerations here are:
  - to whom the ESCO is selling the heat, cooling and/or power, which could include a combination of on-site customers,[42] external customers[43] and, in the case of power, a licensed electricity supply company;[44]
  - the price it will sell it for. Typically, the electricity price is determined by

---

39   Related to this is the insurance costs. In order to determine the most appropriate insurance regime for an embedded generation project, the core stakeholders need to consider what insurances are required and, for each required insurance, determine:
      • the extent to which there are any existing insurance policies (eg, group insurances);
      • the insureds, loss payees and persons noted on the policy;
      • the level of insurance, premiums and deductibles;
      • the person responsible for effecting and maintaining the insurance; and
      • whether there are any gaps or overlaps with any other relevant insurance policy.

40   In addition, the ESCO may assert that the principal should contribute to such costs on the basis that an embedded generation system displaces certain utility infrastructure costs that the principal would have incurred in any event.

41   A principal may choose to transfer for nominal consideration or less than full valuable consideration if it receives a benefit elsewhere, such as a reduced heat, cooling and/or power price. If the principal is a public body, it may be required to transfer for full valuable consideration in order to meet its best value obligations.

42   If the principal is the landlord of such customers, the ESCO will typically seek to impose an obligation on the principal to procure that these customers connect to and take a supply from the embedded generation system. Before agreeing to this, a principal must consider what legal restrictions may exist which would hinder their ability to agree to such an obligation. Such restrictions may exist in statute (eg, pursuant to relevant landlord and tenant legislation) in the lease with the tenant or, more broadly, because of competition law issues.

43   When using the phrase 'external customers', the author is referring to customers who are outside a defined area within the principal's legal control (whether pursuant to statutory powers or as freeholder, tenant or perhaps manager pursuant to an estate management agreement). The extension of an embedded generation system to external customers is a matter that requires careful consideration, particularly if an external customer is identified after the core agreements are signed. If the principal is prepared to allow the embedded generation system to be extended to external customers, the parameters within which an ESCO can do this must be carefully defined in the core agreements between the ESCO and the principal.

44   If the ESCO is exempt from holding a generation licence, then in order to sell the electricity to a licensed electricity supplier, the ESCO must either register under the Balancing and Settlement Code or designate another Balancing and Settlement Code party to register (and enter into a power purchase agreement with that party).

reference to market prices. However, there is no 'market price' for the supply of heat and/or cooling. This means that the ESCO and each customer must agree a bespoke price which takes account of the ESCO's costs of supply, the volume of the supply and the term of the agreement.[45] If the supply of power, heat and/or cooling is for more than one year, there is usually a price review mechanism and, particularly in the case of power, the customer has the ability to switch supplier;[46]

- the fiscal incentives that the embedded generation project will be entitled to receive (see section 3 above), who is entitled to receive the benefit (which may depend on who owns the embedded generation system) and the extent to which the beneficiary will directly or indirectly share that benefit with the other core stakeholders; and

- whether the ESCO provides any other ancillary services and, if so, the price to provide such services.

The ESCO may derive further income from an energy and/or carbon savings regime, but may also have to pay moneys to the principal and/or other customers if those savings are not achieved.

- Landlord – the owner of the land on which the embedded generation system is located (which may be the principal) may receive lease rentals from the owner/operator of the embedded generation system.[47] As with any lease rental, the issues are how much and when can it be varied. However, from the author's experience, what is particularly challenging in the United Kingdom is that there is limited market information in order to determine an appropriate market rental for an embedded generation system.[48] Consequently, for many projects the rental is a 'calculated rental', which is calculated by reference to the other financials for the project, particularly the ESCO's income stream.

- Owner of embedded generation system – if the owner leases the equipment comprising the embedded generation system to another core stakeholder, it can receive an equipment lease rental. For example, the principal may design, install and commission the embedded generation system, but retain ownership and lease the equipment to the ESCO for an agreed period, so as to enable the ESCO to run its business.

- ESCO sponsors – each sponsor's income stream can take a variety of forms, but the most common is dividends.

---

45   It is not uncommon that the price for heat and cooling includes a standing charge to cover fixed costs and a variable charge to cover operating costs.

46   If an ESCO's customer switches electricity supplier and the ESCO owns the distribution network connected to that customer, the ESCO can still generate revenue through use of system chargers payable by the customer's new electricity supplier.

47   From the author's experience, for many embedded generation system projects where the landlord is also the principal, there is simply a peppercorn rent on the basis that the principal will receive other financial benefits, such as lower power, heat and/or cooling prices. However, more principals are now seeking a rental stream.

48   The release of valuable space for development/rental is a primary benefit, although it will always be difficult to assign a valid monetary value to this benefit; there is also the opportunity to utilise undeveloped (distressed) land to accommodate energy centres which suit out-of-the-way, low-value sites.

Financial targets: The third issue to consider is the financial targets for the project, including:

- the internal rate of return that the ESCO is entitled to earn in delivering the project and running its embedded generation system business; and
- cost savings that the principal and/or other customers want to achieve from taking more energy-efficient heat, power and/or cooling supplies from the embedded generation system compared with procuring them via conventional means.

Setting the financials is often a time-consuming process which commences as one of the initial key issues for the project and continues to be developed and updated while other technical, commercial and legal matters are agreed. From the author's experience, the core stakeholders often have differing views on the financials, including both the methodology for, and figures used in, determining each key element.

The core stakeholders will also have to agree and document the circumstances which entitle any of the financials to be modified[49] and what happens if the financials move beyond certain tolerances (eg, if the project becomes financially unviable for a party or a party is making 'super-profits').

Related issues to all of the above are:

- the term of the project, which will need to be long enough for the parties that incurred the capital and operational expenditure to achieve the agreed return on their investment. Periods of between 15 and 25 years (and sometimes longer) are not unusual; and
- the tax and accounting treatment of the financials. This includes such things as value added tax, withholding tax and capital allowances (including enhanced capital allowances), all of which are beyond the scope of this chapter.

(b) *Responsibility for core activities and ownership of embedded generation system*

Another key issue is determining who is responsible for the core activities of designing, installing, commissioning, operating and maintaining each component of the embedded generation system and who owns the embedded generation system.

In the author's experience, there are a variety of ways of dividing the core activities. Two of the more typical divisions are as follows:

- The ESCO takes full responsibility for all core activities and owns the embedded generation system.
- The principal designs, installs and commissions all or a part of (eg, the distribution network) the embedded generation system and then:
  - either transfers ownership to the ESCO (for either nominal or valuable consideration) or leases the equipment comprising the embedded generation system to the ESCO for an agreed period so as to enable the

---

49   This includes determining a procedure which will accommodate changes to the scope and timing of the delivery of the embedded generation system and changes to the system post-commissioning.

ESCO to run its business. The ESCO is then responsible for operation and maintenance of the embedded generation system; or

- retains ownership and possession of the embedded generation system and simply engages the ESCO as the operation and maintenance contractor (which may also include metering, billing and other services).

Part of the process of dividing responsibility is to identify, allocate and determine how to manage the key risks for delivering the project.

(c)   *Ownership of ESCO*

The third key consideration is the legal structure of the ESCO and who will be its sponsors. Two typical structures for the ESCO are as follows:

- The sponsors are persons other than the principal or the customers (eg, a utility company); or
- The sponsors are two or more of the core stakeholders (eg, the ESCO and the principal).

If the ESCO's sponsors do not include the principal and the ESCO is not in itself a entity of substance, the principal may require performance and/or payment support from the sponsor (eg, in the form of a parent company guarantee or independent security).

If two or more core stakeholders are likely to be sponsors, they will each need to consider the issues set out in the chapter entitled "Structuring the project vehicle".

5.   Conclusion

While various government incentives aim to encourage investment in embedded generation system projects, ultimately it is whether agreement can be reached on the above key commercial issues (and then the other commercial issues, as well as technical and legal matters) which determines whether an embedded generation system project is viable and can proceed.

What is hopefully apparent from section 5 above is the large number of interrelated commercial decisions that need to be made in order to determine an optimal solution for a particular embedded generation system project. In practice, the growing maturity of the embedded generation system market has led to a wide variety of solutions. While there are benefits to this (eg, the solution is tailor-made for the project), this variety – coupled with the lack of specific regulation for embedded generation system projects and the lack of interconnected systems (eg, across a town or city) – has led to a fragmented and non-standardised embedded generation system market. This may create difficulties in the future if the government chooses to create standardisation, whether on a specific policy matter affecting embedded generation projects or on the whole embedded generation market generally.

As the embedded generation system market in the United Kingdom is likely to grow rapidly over the coming years, it is important that the government and all interested stakeholders carefully consider the above matters so as to ensure a smooth transition for those that have taken the risk to invest early in this market.

The author gratefully acknowledges the input provided by Michael Rudd and Matt Bonass, both partners at SNR Denton, regarding the regulatory framework, government policies and legal issues for UK embedded generation systems, as well as broader observations on the commercial issues addressed in this chapter, based on their own experience.

Emerging renewable technologies

Nicholas Kelly
Sindicatum Carbon Capital Group Limited

1. Introduction

The literature on sustainable energy resources reveals a confusing array of new and old technologies: solar space stations, wind kites, sulphur dioxide stratospheric hosepipes, rain-seeding devices... The list seems endless. To try to help readers understand the renewable energy environment, this chapter looks at some of the new technologies which may succeed to the established sectors of hydroelectric, wind and photovoltaic solar energy. Specifically, this chapter considers:

- three new forms of renewable power generation (wave, tidal and geothermal); and
- the great hope for biofuels – micro-algae.

Why these sectors and not others? As well as representing a good overall picture of the next generation of renewable energy sources, they illustrate particularly well some of the technical and commercial challenges faced by new renewable technologies generally.

It is difficult to predict with any certainty which of these technologies will become serious contenders for commercial exploitation. Numerous factors contribute to the success or failure of each sector and technology. However, some factors must be given more weight for an assessment of the technology to be complete. These trumping elements are:

- the technology, and how easy it is to harness the particular energy source and turn it into a useful form of power (ie, electricity, heat or chemical power);
- the costs (and their lifecycles);
- the regulatory and political support enjoyed by a particular technology, as consistent and sustained support is essential to ensure that nascent technologies can mature to a commercially viable stage; and
- the scalability of the technology, which also plays a role in the preceding considerations. Given the political ambition expressed by a number of countries to turn away from fossil-fuel generation, can the renewable technology deliver on a scale necessary to make sustained and significant change?

Unfortunately, it is impossible in a chapter as short as this to give a global view of the new renewable technologies. Thus, the chapter focuses on UK and US developments, with some examples from other countries.

To make the costs of individual technologies easier to compare, the author has used the levelised cost of electricity per kilowatt-hour (KWh) where possible. This is an estimate of the minimum power price needed for the technology to break even, taking into account:

- capital expenditure;
- operating and maintenance costs; and
- the cost of fuel (if relevant).

Unfortunately, this minimum power price provides at best an approximation of the actual costs of the technology. The estimate will vary quite widely, depending on one's assumptions about the project, discount rates, taxes and cost of capital. However, the estimates make it easier to compare the relative costs of electricity produced by each type of renewable technology to the cost of fossil-fuel generation. As a general benchmark, the cost of coal-fired generation in the United Kingdom in 2008 was around £0.041/KWh and gas-fired generation was around £0.039/KWh.[1]

2. Wave energy

2.1 Resource

Energy from wave and tidal power could provide up to 20% of the UK's current electricity and has the potential to cut carbon dioxide by tens of millions of tonnes. Ultimately, the global estimated value of worldwide electricity revenues from wave and tidal-stream projects could be between £60-£190 billion a year.[2]

The promoters of marine energy make very ambitious claims: "Future Energy Solutions highlight that the global wave power potential has been estimated to be around 1,000-10,000GW, which is the same order of magnitude as [the] world electrical energy consumption."[3]

The wave resource around the United Kingdom is estimated to be approximately 50 terawatt-hours (TWh) per year.[4] In the longer term, between 15% and 20% of the United Kingdom's energy needs could be met by marine energy (ie, between 6 gigawatts (GW) and 8GW).[5] A report by RenewableUK (previously the British Wind Energy Association) published in October 2009 is more conservative and believes that by 2020 the United Kingdom could install between 1GW and 2GW of marine energy projects,[6] although even this sounds ambitious.

So where does wave energy come from? It is created by the action of the wind on the sea surface. The size of the waves is determined by:

---

1   House of Lords, Economic Affairs Committee, "The economics of renewable energy", November 2008, Table 2.
2   The Carbon Trust, "Future marine energy", 2006.
3   www.emec.org.uk/marine_renewables.asp.
4   Canaccord Adams, Equity Research, the London Accords, November 19 2007. See also UKERC Energy Research Landscape, Marine Renewable Energy (Mueller, May 14 2009) and Energy and Climate Change Committee, Second Report of Session 2009-10, "Low carbon technologies in a green economy" (HC 194-1), p42.
5   www.carbontrust.co.uk/emerging-technologies/current-focus-areas/marine-energy-accelerator/pages/default.aspx.
6   Marine Renewable Energy – State of the Industry Report, October 2009.

- the wind's speed;
- the expanse over which the wind is blowing currents; and
- the topography of the seabed.

This energy can be captured and harnessed using a device to convert the kinetic energy of the waves into an electrical pulse.

Various devices are being developed and tested; they all take a different approach to the problem of converting the movement of the waves into electricity. So far, no particular design or approach has emerged as the dominant technology. The methods range from floating devices, such as 'attenuators' and 'point absorbers' fixed or moored to the seabed, to shore-based structures, such as 'oscillating water columns' and 'overtopping devices'.[7] In each case, developers need to grapple with the twin demands of efficiency and reliability. Efficiency is critical because wave power is highly variable in amplitude and irregular in frequency, making its energy difficult to capture in a variety of sea states. Reliability is essential, given:

- the extremely hostile environment in which wave devices have to operate; and
- the difficulties of undertaking even routine maintenance operations.

Often, the two aims can be at odds with each other – a more efficient design may be too complex to be viable at sea.

There are two further considerations for the development of wave power in the United Kingdom: permitting and grid connections. Until recently, the arrangements for the permitting of offshore installations of renewable energy were extremely confused, with permits and licences required under several different UK statutes. In late 2009, however, the Marine and Coastal Access Act came into force, creating a single body for the permitting of offshore wave farms (ie, the Marine Management Organisation). A Strategic Environmental Assessment has also been completed for the development of wave-energy devices around the UK coast line (as required under the EU Strategic Environmental Assessment Directive). On March 16 2010 the Crown Estate announced the winners of a leasing round for the development of up to 1.2GW of installed wave and tidal capacity in the Pentland Firth and Orkney waters by 2020.[8] This represents the potential to power 750,000 homes with wave and tidal energy and is the world's first commercial wave and tidal leasing round. The Crown Estate has also invited expressions of interest for the development of a large tidal-wave project in the Inner Sound near Caithness in Scotland. Other areas are expected to be made available for development in the future.

The second issue is grid connection. In principle, an offshore wave farm will need to construct its own transmission line to convey the electricity onshore and connect to the grid, which will increase the cost of the project. Even where such construction is feasible, there is often no obvious connection for the transmission line when it comes onshore. For instance, the northwest coast of Scotland is considered an ideal

---

7   See www.emec.org.uk/wave_energy_devices.asp.
8   www.thecrownestate.co.uk/newscontent/92-pentland-firth-developers.htm.

location for a wave farm, but the local distribution network is limited and would need to be upgraded to accept the additional generation.[9] However, it is just as much an issue for offshore wind farms as for wave power projects. The UK government is looking at a number of measures to address the problem. In particular, the Department of Energy and Climate Change and the Office of the Gas and Electricity Markets are developing a regime for an offshore transmission system. This will not be of immediate relevance to the wave industry, as the transmission system will operate at 132 kilovolts (kV) or above, while most wave farms (for the moment at least) will connect at 33kV. However, the initiative is welcome in the longer term, as it will move the cost of building and connecting to the national grid away from the project developer and onto the transmission system owner.

2.2   Costs

There are over 50 different devices in development in the United Kingdom alone (with over 1,000 patents in existence worldwide for wave devices). This makes it hard to provide meaningful estimated costs, as these will be device-specific. A report for the Carbon Trust, however, has given a range of £0.12/KWh to £0.44/KWh for the cost of electricity from early-stage wave farms. RewewableUK members estimate the costs of energy from a wave device to be between £0.3/KWh and £0.42/KWh. They expect the cost to fall significantly as installed capacity increases.[10]

A handful of devices have become commercial operations, thus providing useful reference points. The 2.25-megawatt (MW) Pelamis wave-energy converter installed off the coast of Portugal is one. It is being closely followed by Ocean Power Technologies, which began installation of a 1.39MW PowerBuoy wave farm off the northern coast of Spain (in collaboration with Iberderola) in 2009. The Pelamis wave farm was developed with an installation cost of £3,226/KW and enjoys a preferential electricity tariff of Eur0.23/KWh.[11] According to an independent report produced for RenewableUK, this puts the Pelamis project in a range similar to that of photovoltaic solar energy, but still around four times as expensive as large wind energy projects.[12]

2.3   Regulatory and political support

The UK government has committed to ensuring that 15% of the country's energy usage comes from renewable sources by 2020. Most of this target is expected to be met by increasing the share of renewable electricity production to between 30% and 40%.[13] As the United Kingdom is recognised as having one of the best potential resources for marine energy in Europe, wave power is expected to play a part in reaching this target. The government admits, however, that the industry is still a long way from reaching commercial-scale operation.

The UK government has supported research and development in the wave industry for a number of years, primarily through capital grants administered by

---

9   Marine Renewable Energy – State of the Industry Report, October 2009, p19.
10   Marine Renewable Energy – State of the Industry Report, October 2009, p17, Fig 5.1.
11   www.pelamiswave.com.
12   Marine Renewable Energy – State of the Industry Report, October 2009, p17.
13   HM Government, The UK Renewable Energy Strategy, July 2009, p8.

devolved organisations such as the Carbon Trust, the Scottish government, the Technology Strategy Board and the Marine Renewables Proving Fund.

Support has also come for wave testing centres such as the European Marine Energy Centre in Orkney, the New and Renewable Energy Centre in Northumberland and the offshore WaveHub in southwest England. These centres are critical in allowing developers to scale up their models to full-size operations and to put their devices through rigorous sea trials.

The Department of Energy and Climate Change announced in July 2009 further support for marine energy devices as part of its Low-Carbon Transition Plan. This comes in the form of an additional £60 million of government financing for marine energy[14] and the launch of the Marine Renewables Proving Fund. Support also comes via the Renewables Obligation Scheme – that is, the obligation on electricity suppliers to source an ever-increasing proportion of their electricity from qualifying renewable sources. Renewable obligation certificates are assumed to be worth approximately £0.045/KWh (but prices vary with supply and demand), which represents a significant direct subsidy for renewable generation, including wave power.

However, is this enough? The government's aim is to develop a world-class wave industry in the United Kingdom. RenewableUK estimates that given the high costs of developing marine energy projects, there will be a marked divergence between government ambition and commercial reality. It notes the experience of other countries, such as Denmark and Japan, which have spent billions to develop their wind and solar industries. RenewableUK also worries that renewable obligation certificates alone will not be sufficient incentive to overcome the high capital costs of developing wave projects, even with the recent introduction of 'banding' of certificates by technology so that wave farms receive two certificates, rather than one, per megawatt-hour.[15]

Reading between the lines, there is also a fair degree of cynicism and world-weariness about delivery of government support. Too often, good intentions run into the deep sands of bureaucracy. Money earmarked for grants rarely makes its way to developers. For instance, the Marine Renewables Deployment Fund was launched in 2004 with £50 million to spend. Yet to date, no project has proved eligible for funding.[16] Perhaps the most damning indictment comes from the Pelamis project. Although the project was developed by a Scottish company and tested at the European Marine Energy Centre in Orkney, the world's first commercial wave farm was not installed in UK waters. Taking advantage of a very favourable long-term feed-in tariff offered by the Portuguese government, Pelamis Wave Power chose to launch its first full-scale commercial wave farm in Portugal instead.

2.4   Scalability

Discussing the scalability of wave energy as a whole is difficult; instead, one should consider each device individually. Generally, one would expect each wave device to

---

14   HM Government, The UK Renewable Energy Strategy, July 2009, p17.
15   Renewables Obligation Order 2009 (SI 2009 No 785).
16   Energy and Climate Change Committee, Fourth Report of Session 2009-10, "Low carbon technologies in a green economy" (HC193-1), p26.

have an optimum size and scale relative to its performance. Make it too large and you lose efficiency; make it too small and you probably are not getting enough power. One thing most sea-based (rather than shore-based) wave devices have in common, however, is the ability to be deployed in arrays. Ocean Power Technology, for example, plans to deploy its PowerBuoy device in arrays scalable to 100MW.[17] The Pelamis project, which consists of three P1-A Pelamis machines (ie, three times 750MW), was originally intended to be the forerunner of several mega wave farms, until its backer, Babcock and Brown, went into administration. So, in principle, and subject to the sort of constraints discussed in section 1.1 above, wave power looks eminently scalable.

3.   Tidal energy

3.1   Resource

Tidal power seems to be the poor, neglected cousin among renewable energy sources. Little has been done worldwide to harness tidal energy, even though:

- tidal power stations have the potential to provide large-scale, reliable and predictable sources of clean energy; and
- countries such as Canada, New Zealand and the United Kingdom have potentially significant resources.

There are still only two large tidal-barrage schemes in operation worldwide (described below) and tidal-stream devices are still at the research and development stage. However, as mentioned above, the United Kingdom is estimated to have one of the best tidal resources in the world with the potential tidal-stream resource of around 18TWh a year (50% of Europe's total tidal-energy resource).[18]

(a)   Tidal barrages

The world's largest tidal power plant was constructed at La Rance in France in 1966 and has been operating continuously since then. It is rated at 240MW, which in theory is sufficient to power fully a city of around 180,000 households. The Annapolis Royal Generating Station has been in operation in Nova Scotia, Canada, since the 1980s and has a capacity of 20MW. In South Korea, a 254MW tidal scheme is being planned at Sihwa, using the ebbs and flows into an artificial lake. Other large operational plants are few and far between. The problem lies primarily in identifying suitable locations for tidal-generation plants. Tidal-stream plants need to be located where there is a significant tidal range and fast-flowing current. In La Rance, for instance, the tidal range is eight metres.

Environmental concerns over damage and disruption caused by building and effectively damming a large estuary have also held back development. Not only do various organisations such as the World Wide Fund for Nature, the Royal Society for

---

17   www.oceanpowertechnologies.com/products.htm.
18   Energy and Climate Change Committee, Fourth Report of Session 2009-10, "Low carbon technologies in a green economy" (HC193-1), p26.

the Protection of Birds and Friends of the Earth lobby hard against tidal projects due to the potential damage to fish and bird stocks, but there is also a maze of EU and domestic legislation to navigate, including the EU Habitats Directive and the Birds Directive. A consultation on building a tidal barrage in the Severn Estuary in the United Kingdom has revealed that over 80 pieces of environmental legislation will need to be addressed if that project goes ahead.[19] The UK government's advisers have stated that "a Severn barrage project would not be possible within the current legal framework provided by the EU Habitats and Birds Directives".[20] Even though supporters of tidal systems claim that these concerns are overstated, they remain a major obstacle to the implementation of tidal-barrage systems.

(b)   *Tidal streams*

Tidal-stream devices, like wave-energy converters, are still at an early stage of development. They work by converting the kinetic energy of the ebb and flow of tides into electricity and are generally designed as modular standalone units. No particular device has emerged as a clear winner in terms of technology. Devices are grouped broadly into three main categories – horizontal axis, reciprocating hydrofoil and vertical axis. Horizontal axis devices extract energy in much the same way as wind turbines extract energy from wind, although the relative density of water means that the blades in tidal systems tend to be thicker and shorter than in wind systems. Reciprocating hydrofoils work rather like the wings of an aircraft, rising and falling with the ebb and flow of the tide, which in turn compresses and expands a piston. Vertical axis devices work in a similar fashion to horizontal axis devices, but the blades are fixed on a vertical axis. Devices can be fixed to the seabed, moored or weighted to sit on the seabed. Similarly to wave-energy devices, tidal-stream devices face the demands of balancing efficiency with reliability. Maintenance is also a key issue.

3.2   Costs

Tidal barrages are notoriously capital-intensive. In the 1980s the Bondi Committee (led by Sir Herman Bondi) estimated that the cost of a Severn tidal barrage (the so-called 'Cardiff-Weston Scheme') would be £15 billion.[21] By early 2010 that estimate had risen to £21 billion, including the costs of mitigating the environmental and flooding impacts.[22] In addition, the size of the project (with the potential to meet 4.4% of the United Kingdom's total electricity consumption) would require significant reinforcement of the local grid, estimated at £2 billion. The grid would need to be adapted to accommodate the large surge in electricity at ebb tide (which would not coincide with a surge in electricity demand).

An investment of this size, however, is estimated to be able to provide 17TWh of electricity a year for the country and last for over 120 years. It could make the single biggest contribution towards the UK government's 2020 targets for renewable

---

19   Minutes of Evidence taken before the House of Commons Energy and Climate Change Committee, October 14 2009.
20   Severn Barrage, Position Statement Environment Agency, CCW, EN/NE, May 2006.
21   "Turning the Tide, Tidal Power in the UK" Sustainable Development Commission, 2007, p7.
22   *Id.*

energy.[23] The Department of Energy and Climate Change is still investigating options for possible schemes for the Severn Estuary, including a smaller 1.05GW barrage scheme below the Severn Bridge (the Shoots barrage), as well as tidal-lagoon schemes. The department expects to reach a conclusion in 2010.[24]

The power price of electricity from the Severn barrage is expected to be between £0.17/KWh and £0.2/KWh.[25] In contrast, La Rance, which has been in operation since the 1960s and has paid back its original capital costs of Fr620 million, produces electricity at around €0.02/KWh.[26]

A number of companies around the United Kingdom are also developing tidal-stream devices. Sea Generation Limited has been running commercial trials of its 1.2MW SeaGen tidal-energy converter in Strangford Lough since April 2008. An Irish company, OpenHydro, has installed a 1MW device in the Bay of Fundy in Nova Scotia, Canada. Information on the costs or economics of these projects is scarce. However, the Carbon Trust estimates that the cost of initial-stage tidal-stream devices ranges from £0.09/KWh to £0.18/KWh, with central estimates between £0.12/KWh and £0.15/KWh.[27] Over time, these costs are expected to fall as more devices are installed, making tidal-stream devices competitive with other forms of renewable electricity.

3.3   **Regulatory and political support**

For the past 50 years successive UK governments have produced reports, held consultations and conducted feasibility studies into a variety of schemes for a Severn barrage. All projects have been shelved, postponed or 'noted', but the idea never seems to be killed off conclusively. The Department of Energy and Climate Change is directing the latest study.

One thing, however, is very clear. A Severn barrage will be a publicly led infrastructure project and will not be financeable without significant public subsidies.[28] The main support scheme for renewable energy, the Renewables Obligation Scheme, does not cover tidal schemes over 1GW because of concerns that larger projects could destabilise the scheme. Therefore, it is likely that government support will come in the form of direct grants or guarantees.

UK tidal-stream generation projects enjoy similar support to wave-energy devices – that is, through capital grants and renewables obligation certificates. Tidal energy also forms a part of the government's strategy for a transition to a low-carbon economy, although it is expected to contribute only a small proportion of the renewable energy needed. As noted above, in March 2010 the Crown Estate awarded leases to 10 companies to develop wave and tidal devices in the Pentland Firth and Orkney waters, representing up to 1.2GW of installed capacity. Of this, 600MW have been allocated to tidal-energy developers.

23   www.sd-commission.org.uk/pages/tidal-power.html.
24   severntidalpowerconsultation.decc.gov.uk/.
25   Minutes of Evidence taken before the House of Commons Energy and Climate Change Committee, October 14 2009.
26   www.reuk.co.uk/La-Rance-Tidal-Power-Plant.htm.
27   Sustainable Development Commission, "Turning the Tide, Tidal Power in the UK", 2007, p37.
28   Sustainable Development Commission, "Turning the Tide, Tidal Power in the UK", 2007.

3.4 Scalability

Tidal-range schemes have the capacity to produce large, predictable and reliable sources of power. However, due to their high cost and the environmental concerns with which they are associated, it may be a long time before developments on any significant scale occur in this sector.

Tidal-stream devices are still a long way off commercial deployment, although, like wave devices:

- they can be scaled in terms both of the size of the units and deployment in multiple arrays; and
- the United Kingdom offers untapped resources for development of tidal-stream projects.

In the absence of a stronger political commitment to the sector, these devices will struggle to make an impact. RenewableUK, in its recent state of the nation report, noted: "Based on the experience of the solar and wind industries, the level of UK support for marine energy is not yet of the magnitude required to develop a world-class industry."

It may well be that other nations (eg, Canada) for the moment offer tidal-stream developers a much better return on their research and development.

4. Geothermal energy

Despite its huge renewable energy potential, geothermal energy is often misunderstood or overlooked. It is an important and well-established source for electricity production and has an enormous variety of other uses, including school heating, district heating, fish farming, greenhouses, resorts, and small power production. These applications all make geothermal energy an important, adaptable and growing energy resource for modern cities.[29]

4.1 Resource

Geothermal energy comes from two sources:

- radioactive decay in the crust of the earth; and
- the heat that rises from the earth's core.

Geothermal energy is a potentially very large resource of clean, reliable, predicable and long-term renewable energy. It has been used for centuries for heating and since the turn of the last century, for power generation (the first steam plant was developed in Lardarello, Italy, in 1904). There are approximately 10GW of installed capacity around the globe.[30]

The resource was until recently limited to those areas of the earth's surface where heat is readily accessible – mainly along the tectonic plates where heat and steam escape naturally or it is easy to access the resource.

Certain countries enjoy prolific sources of geothermal supply. This is the case in

---

29   Karl Gawell, executive director of the Geothermal Energy Association, December 2009.
30   www.geothermal-energy.org/226,installed_generating_capacity.html.

Iceland where geothermal energy provides the vast majority of the country's heating, and geothermal plants produce 25% of the country's electricity.

Other countries and regions, such as California, Indonesia, Japan, Mexico and the Philippines, also enjoy excellent geothermal resources. A 2006 study by the Massachusetts Institute of Technology into geothermal resources in the United States concluded that with suitable investment, enhanced geothermal systems could provide 100GW-electric or more of generation capacity in the next 50 years.[31]

There appears to be two main constraints on the further development of geothermal sources. First, identifying potential geothermal reservoirs can be extremely time-consuming and expensive, particularly if the intent is to exploit deeper sources of heat using enhanced geothermal techniques. Developers rely on seismic data and surface studies to evaluate the suitability of a particular location, but ultimately they need to drill exploratory wells to test the heat resource. Although technology and expertise are developing all the time (with transfers of experience and some technology from the oil and gas sector), significant time and cost risks remain associated with the exploration phase of a geothermal project. This in turn concentrates efforts on the most readily available resources.

Second, although geothermal heat does not dissipate over time, experience has shown that reservoir pressures and temperatures decline in response to production. For instance, one of the world's first geothermal projects, The Geysers in California, originally used naturally occurring steam to produce electric power. This resource peaked in the late 1980s and The Geysers plants now re-inject recycled treated water into the ground to produce steam for the power plants in operation at the site.

The use of enhanced geothermal systems (also known as 'hot dry rocks') may provide an alternative to more traditional geothermal projects. 'Enhanced geothermal systems' have been defined as "engineered reservoirs that have been created to extract economical amounts of heat from low-permeability and/or porosity geothermal resources".[32] Enhanced geothermal systems work by injecting a fluid into hot rocks, fracturing the rocks and then pumping the heated fluid back to the surface power plant. Enhanced geothermal systems may be simple in concept, but present numerous difficulties. One of the main constraints is the need to control (and understand) flows of heat between injection and production wells. There is also the risk that the rock fracturisation may trigger earthquakes (a $60 million geothermal project in Basel, Switzerland, was suspended in 2006 because of this potential threat[33] and a US company, AltaRock Energy, suspended drilling operations in late 2009 due to similar concerns).[34]

Mining techniques are, however, being adapted specifically for projects using enhanced geothermal systems. This could provide access to a significant new resource globally if the technical challenges can be overcome.

---

31   Massachusetts Institute of Technology, "The future of geothermal energy" (2006), pp1-3.
32   Massachusetts Institute of Technology, "The future of geothermal energy" (2006), p11.
33   www.nytimes.com/2009/12/11/science/earth/11basel.html.
34   www.altarockenergy.com/demo.html.

4.2 Costs

The costs of building a geothermal power plant are well understood and quantifiable. There are three main types of plant in operation: steam (the oldest and simplest system), flash and binary. The choice of plant is mainly dictated by:

- the temperature of the fluid (higher temperatures are used in direct steam systems and lower temperatures are used in flash and binary systems); and
- the form of the fluid (ie, water or steam).

The costs of accessing the heat source are more difficult to estimate as they are dictated mainly by the specific requirements of the location. As mentioned above, enhanced geothermal systems are likely to have a high initial capital outlay on exploration and development of the geothermal reservoir. This may make the project unviable in the absence of government support. As geothermal energy is generally predicable and it is not intermittent or variable (other than for routine plant shutdowns), it is not expected to impose additional costs on the transmission system other than the costs of connection.

Once a geothermal project is in operation, the costs of electricity production are low and availability is high. Unlike wind farms, which have a typical availability of between 30% and 40%, geothermal plants run with availabilities of over 90%.

Geothermal systems accessing lower temperatures also provide an ideal solution for local heating needs, be they industrial, municipal or residential.

Overall, the levelised cost of electricity (at 2007 prices) has been estimated at between $0.08/KWh and $0.09/KWh. This is expected to decrease over time, which will make geothermal energy production competitive with conventional forms of power generation.[35]

4.3 Regulatory and political support

The United States leads the way in support for geothermal energy. In a *volte-face* from the Bush administration, the Obama government recently pledged approximately $400 million of federal funds to support geothermal projects under the American Recovery and Reinvestment Act 2009. In October 2009 the US Department of Energy awarded grants and support worth $338 million to 123 geothermal projects under its Geothermal Technologies Programme. The majority of these funds is going towards exploration and development activities to identify potential geothermal sources around the United States. Enhanced geothermal systems attracted approximately $131 million of funding, divided between three demonstration projects and research and development into geothermal reservoir evaluation techniques, drilling and reservoir simulation.

Geothermal projects are also eligible for 30% production tax credits and federal loan guarantees. Both have helped to stimulate a revival of interest in the sector, even though developers point out that the sunset provisions for production tax credits can make these forms of financing inaccessible.

---

35   California Energy Commission, "Comparative Costs of California Central Station Electricity Generation Technologies", June 2007.

At state level, California and Nevada have extended their renewable energy standards. California is set to increase its renewable portfolio standard to 33% by 2020 and Nevada to 25% by 2025. Both are strong markets for geothermal projects due to their geothermal resources. Geothermal projects in the western states have also received a boost by the completion of a wide-ranging Programmatic Environmental Impact Statement for Geothermal Leasing in 2009 by the Department of the Interior and the US Forest Service. The programme potentially opens up 190 million acres of federal land for the development of geothermal projects, with an estimated potential capacity to install 5.5GW of new electricity generation.[36] This in turn has resulted in a sale of geothermal leases by the Bureau of Land Management in California, Nevada and Utah.

Overall, US federal and state support has also led to a 60% increase in 2009 of planned geothermal projects across the country. The trend looks set to continue into 2010 and beyond.

Australia, which has an enhanced geothermal system resource, has stepped up support and investment in the sector, primarily thanks to the change in government in 2007. By December 2009 the federal government had committed over A$235 million to the geothermal sector. The majority of funding will support demonstration projects under the Renewable Energy Demonstration Programme, including a 25MW geothermal project in the Cooper Basin, South Australia.[37]

In Europe, Germany has recently enacted a renewables law granting geothermal projects a very favourable tariff of €0.15/KWh, which is expected to stimulate a boom in geothermal projects over the next few years. The United Kingdom, however, does not have a significant geothermal resource. A study of the country's most promising areas (mainly Cornwall) for the deployment of enhanced geothermal systems found that "the generation of electricity power from hot dry rock was unlikely to be technically or commercially viable in Cornwall or elsewhere in the UK in the short or medium term".[38]

4.4 Scalability

Geothermal energy is clearly one of the most scalable renewable energy sources around. The potential resource is ideally suited to provide long-term, predictable base-load power, as well as heating/cooling for industrial, municipal and domestic uses. The challenges remain:

- the high capital cost of exploration and development; and
- the further funding and support needed for research and development in new technologies such as enhanced geothermal systems.

5. Algae – the third-generation biofuel

The recipe for getting microalgae to produce lipids sounds like a daydream for using underutilized resources: put them in salty water unfit for other use, expose them to the

---

36   www.blm.gov/wo/st/en/info/newsroom/2008/december/NR_12_18_2008.html.
37   Announcement of the Ministry of Resources and Energy, December 13 2009. See also www.geodynamics.com.au.
38   The UK Geothermal Hot Dry Rock R&D Programme, 1992, MacDonald.

sun in areas unsuitable for growing crops, feed them power plant or other exhaust gas that threatens the world climate, and deny them certain vital nutrients.[39]

5.1 Resource

In the 1970s President Carter launched the Aquatic Species Programme.[40] Over the course of two decades the US National Renewable Energy Laboratory investigated the use of micro-algae as an alternative to fossil fuels. Initially, the focus of the research was the use of algae for energy production, but it later became the production of lipids as a source of feedstock for biodiesel. The programme closed in 1998, but it laid the foundation for a multibillion-dollar industry seeking the holy grail of the biofuel industry: 'green crude'.

Today it is estimated that over 300 companies worldwide are investing in and researching the use of micro-algae for biofuels, in particular as a substitute for diesel and jet fuels. Although still at the research and development stage, the industry has attracted some significant backers, especially among the oil and gas majors. The most recent convert is Exxon Mobil, which announced in July 2009 a joint research and development programme, worth up to $300 million, into the production of biofuels using algae with synthetic genomics.[41] BP and Chevron have also invested in developing and researching production of biofuels from micro-algae.

So how does it work? Micro-algae use sunlight to convert carbon dioxide into lipids (oil) and carbohydrates. The oil content can be as high as 40%. The main advantages of micro-algae are:

- their quick growth rates relative to other feedstocks;
- the density of production; and
- the fact that they do not compete for land or water resources with food production and water use, thus avoiding the 'fuel or food' debate.

Micro-algae are said to be capable of producing far greater amounts of oil per unit area of land than oil-seed crops.[42] They also offer the environmental benefit of sequestering carbon dioxide. In the future, one of their primary purposes may be to store carbon dioxide streams from power stations and factories.

However, the challenge with algae is primarily one of scale and economics. Naturally occurring algae do not reproduce at a rate or with the efficiency necessary for commercial production. Their growth and oil-production rates therefore need to be enhanced through specific farming methods.

Technical challenges include:

- contamination in open-pond systems;
- nutrient depletion (which occurs when the rate of growth of the algae exceeds nutrient supply);
- self-shading (rapid growth inhibits light); and
- harvesting.

39   National Renewable Energy Laboratory, Jet Fuel from Microalgal Lipids, 2006.
40   NREL/TP-580-24190.
41   www.syntheticgenomics.com/media/press/71409.html.
42

Harvesting is particularly challenging as the algae cultures are typically grown in large quantities of water from which the algae need to be recovered. Different techniques are used, ranging from settling to centrifugal force or filtering. Commercial cultivation of algae is also likely to require large quantities of carbon dioxide. This in itself will arguably mean that large algal farming can be located only next to power stations or factories with high carbon emissions.

Some of the difficulties identified above can be overcome by using photo-bioreactors, where algae are cultivated in closed-loop systems (essentially, large test tubes) using artificial light. These systems offer the advantage of giving the developer far more control over:

- the supply of nutrients and carbon dioxide;
- the harvesting;
- the temperature; and
- the exclusion of pests.

However, a photo-bioreactor is far more energy intensive, and therefore costly, than open-pond systems. This is mainly due to the need for artificial light and heat to cultivate the algae.

Even though the technology is still nascent, according to the United Kingdom's Carbon Trust biofuels from micro-algae could "replace over 70 billion litres of fossil derived fuels used worldwide annually in road transport and aviation by 2030 (equivalent to 12% of annual global jet fuel consumption or 6% of road transport diesel)".[43]

The airline industry is looking closely at whether algae and other biofuels can be adequate substitutes for jet fuel. In January 2009 Continental Airlines completed a test flight using a hybrid algae-jatropha biofuel mix in a commercial airliner. This followed similar tests by Virgin Atlantic and Air New Zealand.

5.2   Costs

Algal production has one advantage in price terms over other so-called 'second-generation' oil-seed crops. Crops such as soybeans and corn compete for land and resources with food production. They are therefore exposed to the changes in prices of these commodities, as farmers seek to extract the maximum value from a limited farming resource. As algae production does not compete with food production (or, to a great extent, water usage), it is largely protected from these direct competitive effects.

It is, however, extremely sensitive to the price of oil. The low cost of oil in the mid-1990s is given as the reason for closing down the Aquatic Species Programme of the US National Renewable Energy Laboratory. The relatively higher costs of oil reached since then (in particular, the high costs of jet fuel) are credited with the recent revival of interest in algae as a suitable feedstock for biofuels.

As so many of the technologies are at an early stage of development, it is hard to find reliable data about the costs of growing and harvesting algae for biofuel. These

---

43   www.carbontrust.co.uk/news/news/archive/2008/Pages/algae-biofuels-challenge.aspx.

costs are specific to each technology and farming method. According to the Aquatic Species Programme, open-pond systems are the only cost-effective method of producing algae on an industrial scale. Photo-bioreactors, despite their clear advantages, are still considered too expensive to be commercially viable. US estimates range from $56 per gallon[44] to $40 per gallon for production from an open-pond system (at the time of writing, petrol cost between $2 and $4 per gallon). The production costs of photo-bioreactors are likely to come down over time. The ultimate goal is to produce a fuel which is competitive with other forms of fuel, including diesel, petrol and jet fuel.

5.3   Regulatory and political support

Research and development have attracted significant support from both the private and public sectors in the United States. As mentioned above, the US National Renewable Energy Laboratory conducted a comprehensive investigation into the use of micro-algae for over two decades. In 2009 it re-launched a programme looking into micro-algae production with Chevron under a Cooperative Research and Development Agreement.[45] Since 2008 the US Department of Energy has been running the National Biofuels Action Plan, which aims to stimulate investment in biofuel production in the United States. As part of the programme, the Department of Energy has launched a National Algal Biofuels Technology Roadmap, which at the time of writing was out for industry consultation.[46] In December 2009 a San Diego developer, Sapphire Energy, received over $100 million under the American Recovery and Reinvestment Act 2009 to help expand its facilities and research.

Efforts are also being made to ensure that algal production benefits from a level playing field with other forms of biofuels. In the United States in particular, this means that:

- algae research and development enjoy the same tax incentives and subsidies as research into other forms of biofuels, especially cellulosic biofuels;
- the Renewable Fuels Standard under the Clean Air Act changes so that micro-algal production is not disadvantaged relative to other renewable fuel sources; and
- the role of algae production is recognised in reusing carbon dioxide from industrial processes and turning it into biofuel.[47]

In October 2008 the Carbon Trust launched the Algae BioFuels Challenge to stimulate research and development into algae production in the United Kingdom with the view to making low-cost algae production a reality by 2020. The challenge is structured in two phases. The first phase provides up to £6 million of funding for research and development, while the second phase will provide financing for an open-pond test and demonstration plant.

---

44   www.sciencedaily.com/releases/2009/11/091103144822.htm.
45   www.nrel.gov/features/20090403_algae.html.
46   www1.eere.energy.gov/biomass/pdfs/nbap.pdf.
47   www.algalbiomass.org/.

5.4   Scalability

The production of biofuel from algae should be highly scalable. As mentioned above, algae in theory produces far more oil and biomass per unit of space than other comparable biofuels, making them an ideal candidate for large-scale production. However, the challenge for all companies researching production of green crude is precisely how to turn a laboratory demonstration into an industrial process that is economical and reliable. Fuel produced from algae will also need to fit within the existing fuel supply and distribution structures, and to comply with existing fuel standards.

As yet, there is no clear answer to this challenge.

6.   Conclusion

Governments are advised to avoid picking winners in the renewable technology race.[48] However, in a climate of depressed carbon prices and lack of access to capital, new technologies are heavily reliant on government support and stimulus. It is also unavoidable that government policy and fiscal stimulus will promote some renewable technologies over others. Public funding is limited and governments will have to prioritise investment in those low-carbon technologies which they deem are most likely to deliver their policy objectives (ie, low-carbon power generation, job creation and security of energy supply).

Picking winners in the renewable technology race, in this context, becomes a game of second-guessing policy makers' and politicians' minds and hoping that the political and regulatory wind does not change. The author's bet is that the huge stimulus given to the geothermal industry by the Obama administration will make this source of energy the natural successor to wind and solar energy. But as the Danish physicist Neils Bohr allegedly said, "Prediction is difficult, especially about the future."

---

48   Energy and Climate Change Committee, Fourth Report of Session 2009-10, "Low carbon technologies in a green economy" (HC193-1), p68.

Carbon capture and storage: a path to climate change mitigation

Thorsten Mäger
Dirk Uwer
Hengeler Mueller

The world is moving on a totally unsustainable and ultimately unrealistic energy path, based on increasing fossil fuel consumption and increasing world population[1]

1. Introduction

The growth of oil-intensive sectors means that global demand for oil (which satisfies 85% of the world's energy needs) is greater than ever.[2] Despite widespread efforts to develop alternative technologies and to boost renewable energy, oil will continue to be vital to most countries and will affect the development of industrialised and emerging economies alike.[3] Unfortunately, the burning of fossil fuels such as oil, coal and natural gas produces a staggering 85% of the world's total emissions of carbon dioxide resulting from human activities.[4]

Such carbon dependency will undeniably create significant obstacles to the international environmental ambitions expressed in the Kyoto Protocol and the Marrakesh Agreements.[5] For instance, specialists agree that a reduction of carbon emissions of at least 50% over the next 20 years is necessary to stabilise climate change.[6] Also, it is now widely recognised that increasing energy efficiency and developing renewable energies may be insufficient to limit global climate change. In brief, one of the most complex dilemmas that energy and environmental policy makers face is the trade-off between economic development and access to fossil fuels on the one hand, and environmental concerns on the other.

Among the environmental concerns at issue, climate change, global warming and oceans acidification are the most pressing. They emphasise the need for a consistent and well-balanced policy to tackle greenhouse gas emissions.

---

1   Andris Piebalgs, EU commissioner for energy, speaking at the seventh Doha Natural Gas Conference http://europa.eu/rapid/pressReleasesAction.do?reference=SPEECH/09/102&format=HTML&aged=0&lan guage=EN&guiLanguage=en).
2   Australian Bureau of Agricultural and Resource Economics, "Impact of oil prices on trade in the APEC region", Research Report 05.3 for the APEC Energy Working Group, in *World Oil Outlook 2008* (October 2005) (www.opec.org/library/worldoiloutllok08.htm).
3   "Sustainable power generation from fossil fuels: aiming for near-zero emissions from coal after 2020", Communication from the European Commission to the Council and the European Parliament (COM (2006), 843 final).
4   www.netl.doe.gov/technologies/carbon_seq/overview/what_is_CO2.html.
5   http://unfccc.int/kyoto_protocol/items/2830.php.
6   www.zeroemissionsplatform.eu/the-hard-facts.html.

In this context, policy makers aim to implement strategies guided by the need to expand renewable energies and mitigate the environmental risks associated with a carbon-intensive economy by capturing and storing carbon.[7]

Carbon capture and storage (CCS) is a set of technological processes comprising:

- the capture of carbon from the gases emitted by industries;
- its transport; and
- its injection into geological formations.

The Intergovernmental Panel on Climate Change estimates that in some countries CCS could reduce the carbon emissions released by power plants into the atmosphere by between 80% and 90%.[8] This evaluation explains the World Resources Institute's enthusiasm when it refers to CCS technology as "a critical option in the portfolio of solutions available to combat climate change, because it allows for significant reductions in carbon emissions from fossil-based systems, enabling it to be used as a bridge to a sustainable energy future".[9]

In turn, the European Commission estimates that by 2030, CCS could contribute around 15% of the reduction in carbon emissions in the European Union,[10] which represents a drop of 400 million tonnes a year.[11]

However, CCS is not yet a technologically mature, integrated process. No large project has been implemented and several technical and operational questions remain open. Also, legal and economic risks are usually considered barriers to investment and the recent financial and economic crisis has made any investment more risky and thus less attractive. Consequently, it appears that only a solid and coherent legal framework can foster the conditions necessary to enhance the applications of CCS, so that this technology can also be regarded as a feasible and reliable path towards controlling the effects of greenhouse gas emissions.

CCS projects involve a variety of parties, including developers, regulators, financiers, insurers, project operators and policy makers. Therefore, it is important, when analysing CCS practices and trends, to consider the political, economic and operational factors that will affect the decisions of private companies to make new investments. Even though environmental compliance has an increasing impact on a company's reputation and relationship with its business partners, the declared

---

7   See also Juliette Addisson, John Bowman and Paul Q Watchman, "Carbon sequestration", in *Climate Change: A Guide to Carbon Law and Practice* (Globe Law and Business, London, 2008), p294.

8   "PCC special report on carbon dioxide capture and storage" (www.ipcc.ch/pdf/special-reports/srccs/srccs_wholereport.pdf), p107. See also www.zeroemissionsplatform.eu/the-hard-facts.html.

9   www.wri.org/project/carbon-capture-sequestration/ccs-basics. The European Commission has also recognised that: "the success of sustainable coal and particularly the commercialisation of CCS on a large scale will also offer opportunities for the exploitation of the new technologies in applications for other fossil fuels, first and foremost in gas-fired power production".
"Sustainable power generation from fossil fuels: aiming for near-zero emissions from coal after 2020", Communication from the Commission to the Council and the European Parliament (COM (2006), 843 final), p15.

10 "Questions and answers on the proposal for a directive on the geological storage of carbon dioxide", Memo 08/36 of January 23 2008 (http://europa.eu/rapid/pressReleasesAction.do?reference=MEMO/08/36&format=H), p3.

11 See European Technology Platform for Zero Emission Fossil-Fuel Power Plants, "EU demonstration programme for CO2 capture and storage" (www.zero-emisionplform.eu/website/docs/ETP%20ZEP/EU%20Demonstration%20Programme%20for%20CCS%20-%20ZEP's%20Proposal.pdf).

environmental concerns of energy companies may, without appropriate legal and governmental support, become mere statements of principle or be put aside from the company's business strategy.

In this context, this chapter aims to:

- describe the key technical features of CCS;
- highlight the major challenges of this system; and
- emphasise the legal, commercial and financial issues that can play a decisive role for prospective investors in CCS.

2. Technical aspects

'CCS' can be defined as a process consisting of:

- the separation of carbon from industrial and energy-related sources;
- the transport of carbon to a storage location; and
- the long-term isolation of carbon from the atmosphere.

The technical efficiency and cost-effectiveness of carbon capture are enhanced when the technologies to separate carbon from other emissions are used near either industrial plants that produce high levels of carbon emissions or power plants that use fossil fuels.

CCS is thus used mainly by large carbon emitters such as:

- coal, biomass or gas-fuelled power plants;
- refineries; and
- some industrial plants, in particular in the cement, food and steel industries.

As mentioned above, CCS encompasses three main phases: capture, transport and storage. In turn, carbon capture involves three different technologies: pre-combustion systems, post-combustion systems and oxy-fuel combustion systems.[12] Pre-combustion systems capture carbon before it reaches the atmosphere.

Pre-combustion carbon separation and capture

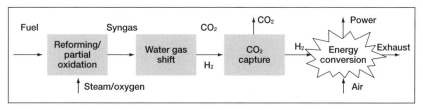

Source: CO2Net

Oxy-fuel combustion uses high-purity oxygen, resulting in an easy separation of carbon due to its high concentrations in the gas stream.

---

12   www.netl.doe.gov/technologies/carbon_seq/core_rd/co2capture.html.

Oxy-fuel combustion carbon separation and capture

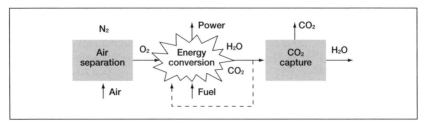

Source: CO2Net

Post-combustion systems separate carbon from the fuel gases produced by combustion of a primary fuel (coal, natural gas, oil or biomass). However, this system allows only for the capture of carbon already in the atmosphere. Furthermore, it requires both a capture and a compression system.

Post-combustion carbon separation and capture

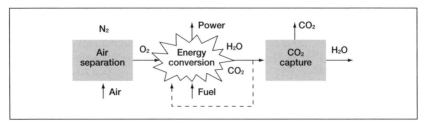

Source: CO2Net

The pre-combustion system allows for higher efficiency levels when used in coal-fuelled or combined-cycled gas-fired power plants. Because carbon is present in much higher concentrations in this phase, pre-combustion capture should be less expensive than post-combustion capture.[13] The oxy-fuel process is expected to provide the best overall efficiency with reasonable financial impact on the operators, as it is largely based on conventional power plant components and technology.[14] However, the technology involved is neither mature nor reliable.[15]

The available technologies allow for carbon captures of between 85% and 95% of the carbon processed in a plant equipped with a CCS system. CCS experiences have so far been limited to small industrial and power plants. These include the Vattenfall Schwartze Pumpe power plant in Germany, which was the first coal-fired plant in the world ready to capture and store its own carbon emissions.[16] Nevertheless, other

---

13   www.netl.doe.gov/technologies/carbon_seq/core_rd/co2capture.html.
14   "Ecological, economic and structural comparison of renewable energy technologies (RE) with carbon capture and storage (CCS). An integrated approach" (www.netl.doe.gov/technologies/carbon_seq/core_rd/capture/41147.html), p11.
15   See also World Resources Institute, Guidelines for Carbon Dioxide Capture, Transport and Storage (www.wri.org/project/carbon-capture-sequestration), p28.
16   www.vattenfall.com/www/co2_en/co2_en/879177tbd/879211pilot/index.jsp.

projects are considered for investment purposes. For instance, Vattenfall is planning to install a large demonstration plant in the Jänschwalde lignite-fired power plant.[17]

Transport is also a key stage of CCS whenever the storage site is not located near the plant. Like oil and gas, carbon can be transported by tanker ships or pipelines, depending on the cost of the infrastructure and the distance between the exit and delivery points.

For amounts smaller than a few million tonnes of carbon per year or for long distances overseas, the use of large tanker ships can be economically attractive. The extent of pipelines that would be necessary to implement CCS technologies worldwide depends on a wide range of factors, such as proximity to the sites, costs, licences and obligations legally imposed on the operator.

Storage of carbon in onshore or offshore geological formations uses many of the same technologies that have been developed by the oil and gas industry. The carbon, already captured and transported to the site, must be injected into a geological formation and meet a number of technical requirements, such as containing capacity and storage efficiency. The US Department of Energy is investigating five geological options to store carbon:
- oil and natural gas reservoirs;
- deep unmineable coal seams;
- deep saline formations;
- oil and gas-rich organic shales; and
- basalt formations.[18]

Deep saline formations can achieve a capacity of 1,000 gigatonnes of carbon.

Onshore storage mechanisms are based on the principle that if carbon is injected at depths below 800 metres, the physical and geochemical reactions enable effective and secure storage.[19] Most of the ongoing experiences relate to onshore storage. Other techniques involve:
- injecting and dissolving carbon into the water column (typically below 1,000 metres) by using a fixed pipeline or a moving ship; and
- depositing carbon on the seabed through a fixed pipeline or an offshore platform (at depths below 3,000 metres).

The best option for CCS is geological storage, as storage in the water column appears to raise high environmental risks and mineral storage is still under investigation.

The Intergovernmental Panel on Climate Change estimates that the potential of geological formation for carbon storage purposes can be 2,000 gigatonnes of carbon.[20] Some of the most relevant storage experiences include:
- the In-Salah project in Algeria, led by BP, Sonatrach and Statoil;[21]

---

17   www.vattenfall.com/www/co2_en/co2_en/879177tbd/879231demon/879320 demon/index.jsp.
18   www.netl.doe.gov/technologies/carbon_seq/overview/ways_to_store.html.
19   "IPCC special report on carbon dioxide capture and storage" (www.ipcc.ch/pdf/special-reports/srccs/srccs_wholereport.pdf).
20   Id.
21   www.statoil.com/en/TechnologyInnovation/ProtectingTheEnvironment/CarbonCaptureAndStorage/Pages/CO2InjectionInSalahAlgeria.aspx.

- the Sleipner project in the North Sea, led by Statoil;[22] and
- the Weyburn project in Canada,[23] led by EnCana.[24]

Both transport and storage of carbon require that a number of safety conditions be met to avoid losses, leakage into the atmosphere and possible environmental disasters. These can damage property, endanger people and destroy the envisaged 'carbon-retaining effect'.[25] In high concentrations, carbon can have a negative impact on the ecosystem and cause death (asphyxia). In addition, storage can cause micro-seismic activity and land movements. However, transporting and storing carbon are not as hazardous as transporting and storing natural gas or oil. Carbon is not as flammable as natural gas or oil and, because of its density, it does not have the same level of dispersion in the atmosphere. Besides, a number of monitoring techniques (eg, geophysical, geochemical and engineering technologies) are being developed to guarantee the safety of the CCS process.[26]

Thus, CCS seems relatively safe, especially in comparison to upstream energy activities that have been carried out for decades. Hence, the main obstacles to CCS development are:

- the fact that pilot experiments and field tests are still in a relatively incipient phase; and
- the difficulty that fitting operating power plants represents.[27]

Thus, as emphasised by the World Resources Institute, "a key finding of the stakeholder process is that even though additional research is needed in some areas, there is adequate technical understanding safely to conduct large-scale demonstration projects".[28]

3.   Operational and commercial aspects

CCS projects comprise four operational stages: site selection and development, operation, closure and post-closure.

Each stage entails specific risks and carries different costs.[29]

During the site selection phase, stakeholders should:

- assess the well, and its injection and containment potential (eg, formation thickness, rock porosity and storage efficiency); and

---

22   Since 1996 this project has captured 1 million tonnes of carbon from natural gas production at Sleipner West and stored it in an aquifer more than 800 metres below the seabed (www.statoil.com/en/TechnologyInnovation/NewEnergy/Co2Management/Pages/SleipnerVest.aspx).

23   www.canadiangeographic.ca/magazine/JF08/indepth/weyburn.asp.

24   See also Addisson, Bowman and Watchman, "Carbon sequestration", in *Climate Change: A Guide to Carbon Law and Practice* (Globe Law and Business, London, 2008), p391.

25   *Id*, p296.

26   See also "Questions and answers on the proposal for a directive on the geological storage of carbon dioxide", Memo 08/36 of January 23 2008 (http://europa.eu/rapid/pressReleasesAction.do?reference=MEMO/08/36&format=H), p4.

27   Issues such as space constraints and short useful life of the plant (before decommissioning) can make CCS implementation difficult.

28   World Resources Institute, Guidelines for Carbon Dioxide Capture, Transport and Storage (www.wri.org/publication/ccs-guidelines), p8.

29   See also "Technical support for an enabling policy framework for carbon dioxide capture and geological storage", prepared by Paul Zakkour for the European Commission, p3.

- obtain licences, land ownership titles or leases and related entitlements to store carbon.

Conducting a thorough analysis and making an appropriate choice at this stage can prevent many of the risks associated with carbon losses during the operational stage.

When it comes to finding appropriate wells and reservoirs, mature upstream markets such as the UK continental shelf, where many oil and gas fields are either depleted or decommissioned,[30] may represent an attractive option to store carbon.[31] Conversely, in areas with suitable hydrocarbon accumulations, carbon-enhanced oil recovery[32] can be implemented to offset some of the costs of carbon capture, transport and injection.

Monitoring is crucial to ensure a secure carbon storage operation, providing the project managers with sufficient information (eg, about faults and fractures) to avoid leakages during the operation and closure stages of the project. As with other potentially hazardous industrial activities, assessing and mitigating risks are two key elements of any CCS project implementation.[33]

The costs of CCS include capital investment and operational costs. Some aspects of the capital cost should be outlined. First, a CCS-equipped power plant (with access to geological or ocean storage) will consume between 10% and 40% more energy than a plant of equivalent output without CCS. This represents a significant increase in costs for the plant operator. Second, according to the European Commission, the upfront outlays for CCS-equipped plants are between 30% and 70% greater than for standard plants.[34] Thus, carbon capture costs are high, irrespective of the system selected.

Although various studies identify a great potential of growth for CCS technologies and investments, without appropriate financial incentives CCS is

---

30   As highlighted by Greg Gordon and John Paterson: "if there is one word that best describes the [UK] continental shelf… at the end of the first decade of the 21st century, it is 'mature'" ("Oil and gas law on the United Kingdom continental shelf: current practice and emerging trends", in *Oil and gas law – current practices and emerging trends* (Dundee University Press, 2007), p1. Christine Jones, "The UKCS: is there exploration after crisis?" (http://www.offshore-mag.com/index/article-display/54617/ articles/offshore/ volume-59/issue-12/news/special-report/the-ukcs-is-there-exploration-life-after-crisis.html).

31   As pointed out in "Oil and Gas UK 2008 Economic Report"; "over the next two decades, the industry will begin to decommission many of the installations that have been producing oil and gas during the past 30-40 years". In *Oil & Gas UK* (2008 Economic Report), p46. Decommissioning is much more complex and risky than the installation and construction of new offshore platforms, as special liability rules apply. The decommissioning process starts with the request for approval of a decommissioning programme before the UK Department of Energy and Climate Change (DECC) under the Petroleum Act 1998; this entails numerous obligations and procedures, as identified by the DECC Guidance Notes – Decommissioning of Offshore Oil and Gas Installations and Pipelines under the Petroleum Act 1998 (www.og.berr.uk/regulation/guidance/decomm_guide_v1.doc).

32   The recovery of oil additional to that produced naturally by fluid injection or other means.

33   As highlighted by the World Resources Institute, "risk assessment and management is included as a cross-cutting issue because it is used not only to help select project sites and design operations plans, but also, throughout the life of a CCS project, to ensure its continued safety and integrity through operations, closure and post closure".
      World Resources Institute, Guidelines for Carbon Dioxide Capture, Transport and Storage (www.wri.org/project/carbon-capture-sequestration), p72.

34   "Questions and answers on the proposal for a directive on the geological storage of carbon dioxide", Memo 08/36 of January 23 2008 (http://europa.eu/rapid/pressReleasesAction.do?reference= MEMO/08/36&format=H), p2.

unlikely to attract large investments.[35] Financial incentives include the possibility to monetise reduced carbon emissions through CCS. As emphasised by the Australian Bureau of Agricultural and Resource Economics, the "adoption of CCS commences if the cost of capturing and storing a unit of carbon is less than or equal to the emissions penalty that would otherwise be paid".[36]

The new EU Emissions Trading Scheme Directive (2009/29/EC) may be a first step towards developing an adequate financial system to support CCS.[37] Furthermore, the extensive use of CCS technologies may well contribute to reducing prices, as was the case with other technological innovations in the energy markets, such as desulphurisation.[38]

4. Legal aspects

From being a 'best efforts' objective, CCS has become a critical and strategic goal. Following previous developments in international law, in particular the adoption of the revised 2006 Protocol to the 1972 London Convention on the Prevention of Marine Pollution by Dumping of Wastes and Other Matter, in April 2009 the European Union adopted the Carbon Storage Directive (2009/31/EC). This section analyses the international and EU legislative frameworks applicable to CCS.[39] The authors consider the limited scope of the international treaties and the fact that the EU directive is not fully implemented. Further, the authors briefly outline some legal aspects and questions raised under national laws.[40]

Legal aspects are of the utmost importance, to enable industrial developers and operators to assess and mitigate their risks accurately and adequately. Private companies seeking to maximise their stakeholders' interests will not embark on a CCS project without solid legal support. As the recent financial crisis caused many industries to cut costs drastically, any plan to implement CCS must isolate risk factors by:

---

35   "IPCC special report on carbon dioxide capture and storage" (www.ipcc.ch/pdf/special-reports/srccs/srccs_wholereport.pdf), p12. See also Addisson, Bowman and Watchman , "Carbon sequestration", in *Climate Change: A Guide to Carbon Law and Practice* (Globe Law and Business, London, 2008), pp297-298.

36   Australian Bureau of Agricultural and Resource Economics, "Economic impact of climate change policy: the role of technology and economic instruments" (2006), p7.

37   Another possible way to create value through carbon savings is to issue certificates of abatement to be sold later in a specialised secondary market. An interesting benchmark can be found within the Australian Greenhouse Gas Reduction Scheme, regarding forestry carbon capture (http://www.greenhousegas.nsw.gov.au/acp/forestry.asp). For an analysis of the possible situations enabling the inclusion of CCS in the EU Emissions Trading Scheme, see "Technical support for an enabling policy framework for carbon dioxide capture and geological storage", prepared by Paul Zakkour for the European Commission, April 2007, pp4-5.

38   European Carbon Dioxide Network, "Cleaner Coal" (www.parliament.uk/documents/upload/postpn253.pdf), p4.

39   A number of non-EU countries have developed pilot projects to explore efficient carbon sequestration. This is the case in the United States, where CCS is developed at state level. It is driven by the regulatory framework set out by the federal rules of:
- the US Department of Energy's Voluntary Reporting of Greenhouse Gases Programme;
- the US Environmental Protection Agency's Climate Leaders Programme; and
- the Global Climate Change Initiative.

See www.netl.doe.gov/technologies/carbon_seq/overview/reg_drivers.html.

40   For an overview of the legal aspects of carbon storage, see International Energy Agency, "Legal aspects of storing CO2".

- choosing the appropriate contractual structure; or
- thoroughly defining the best application for these technologies.

4.1 International law

The following international environmental laws may apply to offshore carbon storage:

- the 1982 UN Convention on the Law of the Sea;
- the 2006 London Protocol to the 1972 Convention on the Prevention of Marine Pollution by Dumping of Wastes and Other Matter; and
- the Convention for the Protection of the Marine Environment of the Northeast Atlantic (the OSPAR Convention).[41]

After decades of discussions and much resistance, in 2006 30 contracting parties adopted the London Protocol, enabling carbon storage beneath the seabed. According to Paragraph 4 of the amendment to Annex I to the London Protocol, carbon streams may be considered for dumping (ie, storage) only if:

- carbon is disposed into a sub-seabed geological formation;
- the streams consist overwhelmingly of carbon; and
- no waste or other matter is added for the purpose of disposing of this waste or other matter.

The London Protocol also provides information regarding:

- the selection of those underground reservoirs with the best potential for permanent isolation;
- site-specific risks to the marine environment of carbon sequestration;
- the development of management strategies to address uncertainties; and
- the reduction of residual risks to acceptable levels.

The protocol emphasises the importance of the selection of appropriate storage sites. If storage sites are selected and managed appropriately, the probability of leakage from reservoirs should be minimal.[42]

The OSPAR Convention – a regional marine treaty ratified by 15 northern European states and the European Commission – is another important legislative instrument governing CCS under the seabed. The convention prohibits pollution of the northeast Atlantic Ocean.[43] In June 2007 the OSPAR Commission decided to allow the storage of carbon in the northeast Atlantic, subject to approval by a competent national authority. The provisions of the permit or approval must ensure the avoidance of significant adverse effects on the marine environment, bearing in mind that the ultimate objective is permanent containment of carbon streams in

---

41 Addisson, Bowman and Watchman, "Carbon sequestration", in *Climate Change: A Guide to Carbon Law and Practice* (Globe Law and Business, London, 2008), pp303-308 – who recognise the lack of consensus regarding the application of the UN Convention on the Law of the Sea to CCS. See also http://www.un.org/Depts/los/convention_agreements/convention_overview_convention.htm.

42 See also Addisson, Bowman and Watchman, "Carbon sequestration", in *Climate Change: A Guide to Carbon Law and Practice* (Globe Law and Business, London, 2008), pp304-308.

43 *Id*, pp305-308.

geological formations. Thus, any permit or approval issued must contain at least:

- a description of the operation, including injection rates;
- the planned types, amounts and sources of the carbon streams to be stored in the geological formation;
- the location of the injection facility;
- the characteristics of the geological formations;
- the methods of transport of the carbon stream; and
- a risk management plan.

The OSPAR Commission adopted a decision to prohibit the placement of carbon into the water column of the sea and onto the seabed because of the potential negative effects.

4.2   EU law

The Carbon Storage Directive has created a legal framework for the environmentally safe geological storage of carbon. The directive, which must be implemented by the EU member states by June 25 2011,[44] is crucial to the strengthening of operators' and developers' confidence in geological carbon storage. The expectations are high, as estimates (as mentioned in the directive's preamble) indicate that "7 million tonnes of carbon could be stored by 2020, and up to 160 million tonnes by 2030, assuming a 20 % reduction in greenhouse gas emissions by 2020 and provided that CCS obtains private, national and EU support and proves to be an environmentally safe technology".

The directive focuses on the geological storage of carbon and contains only a few rules regarding capture and transport. The scope of storage is limited and some options (eg, storage of carbon in the water column[45] and a storage complex)[46] are prohibited (Articles 2, 3 and 4).

The directive recognises EU member states' opt-out right (Article 4(1)) and the right to determine the areas within their territory from which storage sites may be selected. As clarified in the directive's preamble:

> this includes the right of member states not to allow any storage in parts or on the whole of their territory, or to give priority to any other use of the underground, such as exploration, production and storage of hydrocarbons or geothermal use of aquifers. In this context, member states should in particular give due consideration to other energy-related options for the use of a potential storage site, including options which are strategic for the security of the member state's energy supply or for the development of renewable sources of energy.

Article 1(1) clarifies the scope and purpose of the directive by stating that "this directive establishes a legal framework for the environmentally safe geological

---

44   Martina Doppelhammer, *Zeitschrift für Umweltrecht* (2008), p250; Malte Kohls and Christian Kahle, *Zeitschrift für Umweltrecht* (2009), p122; and Mathias Hellriegel, *Recht der Energiewirtschaft* (2008), p319.

45   'Water column' is defined as "the vertically continuous mass of water from the surface to the bottom sediments of a water body" (Article 3(2)).

46   'Storage complex' is defined as "the storage site and surrounding geological domain which can have an effect on overall storage integrity" (Article 3(6)).

storage of carbon to contribute to the fight against climate change".

The geological storage envisaged by the directive shall correspond to the permanent containment of carbon in order to prevent or, if not possible, to eliminate as much as possible the negative effects and any risks to the environment and human health (Articles 1 and 2).

The directive establishes provisions regarding:

- the selection of proper storage sites (Article 4);
- the right to exploration and granting of exclusive exploration permits (Article 5);
- storage permits (Articles 6 to 11);
- monitoring and inspections (Articles 13 to 16);
- closure of storage sites;
- post-closure obligations (Article 17); and
- the transfer of responsibility to the state after a minimum period of 20 years after the storage site has been closed (Article 18).

The storage permit will be the key instrument to ensure that the substantial requirements of the Carbon Storage Directive are met. A set of basic conditions must be fulfilled before issuing a storage permit:

- All relevant requirements under the Carbon Storage Directive and other relevant EU legislation must be met;
- The operator must be financially sound and technically competent and reliable; and
- Where there is more than one storage site in the same hydraulic unit, the potential pressure interactions are such that both sites must simultaneously meet the requirements of the Carbon Storage Directive (Article 8(1)).

The licensing system is based on the rule of 'one licence, one operator', but according to Article 6(2), member states "shall ensure that the procedures for the granting of storage permits are open to all entities possessing the necessary capacities and that the permits are granted on the basis of objective, published and transparent criteria".

Article 7 contemplates a set of conditions and information to be included in the applications for storage permits, including the relevant data regarding site appraisal, risk assessment, operational targets (eg, quantities injected and stored, and transport methods) and risk mitigation measures. Similar detailed information is also reflected on the contents of the storage permits (Article 9). Licensed operators have a duty to inform the competent authorities of any changes to the operation and, in certain cases (particularly when safety standards can no longer be ensured), the competent authorities have the right to review or to withdraw the licence.

The directive addresses other complex issues – namely, the operational, closure and post-closure obligations. These include operational precautionary obligations, such as the fulfilment of certain conditions regarding the site (Article 12). Further obligations include the duty to monitor (Article 13), to report (Article 14) and to inspect (Article 15).

Article 16 also envisages corrective measures in case of leakages or significant irregularities.

The allocation of responsibilities after closure relies on two factors:

- the reason for closing the site; and
- the fulfilment of the conditions set forth in Article 18(1).

The directive recognises as situations that may determine the closure of the site:

- the fulfilment of the relevant conditions as stated in the licence; and
- the operator's request to close the facility.

In these two scenarios, the operator remains responsible for monitoring, reporting and providing corrective measures, as set out in a post-closure plan designed by the operator (Articles 17(2) and (3)). However, if the competent authority decides to withdraw the licence and to close the site, this entity will also assume the primary responsibility for monitoring and providing corrective measures (Article 17(4)).

In turn, Article 18 allows the operator to be relieved from any responsibilities or liabilities after closure, provided that a number of conditions are met. This provision is crucial, as operators and developers are unlikely to invest in an open-ended project, especially in the energy and environmental sectors, which carry an inherent degree of uncertainty.

The directive also comprises financial provisions that envisage the creation of conditions for operators:

- to fulfil the obligations under a storage permit (Article 19); and
- to make a financial contribution to the competent authority before the transfer of responsibility pursuant to Article 18 (Article 20).

Furthermore, Article 21 stipulates that EU member states must take the necessary measures to ensure that potential users are able – in a transparent and non-discriminatory manner – to obtain access to transport networks and storage sites for the purposes of geological storage of the carbon produced and captured.

Other EU directives may apply to CCS as a whole, in particular:

- the Environmental Impact Directive (85/337/EEC), as amended;
- the Environmental Liability Directive (2004/35/EC); and
- the Landfill Directive (99/31/EC).

However, it is still unclear how the issues covered in those directives will be addressed when implementing the Carbon Storage Directive.[47]

Another important legislative instrument is the Emissions Trading Scheme Directive, amended by Directive 2009/29/EC "so as to improve and extend the greenhouse gas emission allowance trading scheme of the Community".[48] The new Article 10(a)(3) makes it clear that no free allocation will be given:

- to installations for the capture of carbon;
- to pipelines for the transport of carbon; or
- to carbon storage sites.

However, as an incentive, the directive offers the possibility to auction the carbon emission allowances that operators of combustion plants have not used thanks to the installation of CCS equipment. In addition, Article 10(3) provides that member states must use at least 50% of the revenues generated by the auctioning of allowances for the reduction of greenhouse gas emissions (eg, by investing in CCS technology). Pursuant to Article 10(a)(8), up to 300 million allowances will be available before December 2015 to stimulate the construction and operation of up to 12 commercial demonstration projects aimed at the environmentally safe capture and geological storage of carbon.[49]

4.3   Streamlining

A streamlined legal framework is key to enabling investors in risky projects to accept the capital outlays. However, as yet, no specific legislation or regulation applies to CCS as a whole. As previously mentioned, the Carbon Storage Directive does not provide a comprehensive legal regime for each CCS step (ie, capture, transport and storage); instead, it focuses on storage and the conditions for a storage licence.[50] This legal void may justify the application, by analogy, of other environmental laws that do not expressly refer to CCS. However, this only adds to the complexity of the situation.

In addition to international law and EU environmental law, national legislation must be taken into account when considering the installation of CCS techniques.[51] Like international law, national law does not offer a comprehensive framework for CCS. Consequently, the implementation of the Carbon Storage Directive has turned into a demanding task for national legislatures, in particular in light of the confusion over which existing national legal instruments apply to CCS.[52]

The further development of CCS technology is unlikely to become a reality without an appropriate and strong legal framework that provides rules applicable to:

- pipeline access;
- tariff regulation;
- equal access to transport and storage;
- licences to capture and transport carbon;
- operational and safety duties of the operator during the capture of carbon;
- decommissioning;

---

47   These statutes may be seen as 'filling the gap' before the implementation of the Carbon Storage Directive. See "Technical support for an enabling policy framework for carbon dioxide capture and geological storage", prepared by Paul Zakkour for the European Commission, pp6-9.

48   Under the Emissions Trading Scheme Directive, many changes have been proposed to recognise CCS. One possibility is the recognition of credits to the benefit of the installation implementing CCS, which would be tradable. However, this recognition faces severe criticism, the strongest of which is that giving carbon a value is a first step towards encouraging its production, which is contrary to the purposes of the environmental policy currently in place. Another option is a deduction system where carbon stored is considered as not emitted. See also Addisson, Bowman and Watchman, "Carbon sequestration", in *Climate Change: A Guide to Carbon Law and Practice* (Globe Law and Business, London, 2008), p299.

49   Ines Zenke and Miriam Vollmer, *Infrastruktur Recht* (2009), p129 (130).

50   In this regard, the directive's recitals underline that at EU level, various legislative instruments are already in place to manage some of the environmental risks of CCS regarding the capture and transport of carbon.

51   A useful overview database over CCS projects in Europe is available at www.co2captureandstorage.info/cont_europe.php.

- environmental liabilities and site restoration in case of environmental damage;
- risk assessment;
- monitoring obligations; and
- standards for site appraisal and closure.

The sophistication of these activities and the segmentation of the participants to a CCS project call for a well-balanced contractual structure. This must follow a risk allocation based on an assessment of the conditions, advantages and profits of each party involved in the operation in such terms that any losses may lie where they can have the least significantly adverse impact on business. It is likely that the operator capturing the carbon will not be the entity licensed to store it, especially in a countries that encourage unbundling.

Examples of success within the energy sector can be useful benchmarks for implementing a legal and contractual framework for CCS. For instance, the oil and gas industry of the UK continental shelf has developed several contracting practices to mitigate and manage the high risks involved in exploration and production activities offshore. The maturity of this market is reflected in:

- regulations such as the Offshore Installations (Safety Case) Regulations 2005;[53]
- common standards of practice such as the Decommissioning Cost Provision Deed;[54]
- its advanced licensing system; and
- the standardisation of its contracts.

---

52   For example, it is unclear under German law whether mining law, waste law and/or water law is applicable to carbon storage. See Falke Schulze, Andreas Hermann and Regine Barth, *Deutsches Verwaltungsblatt* (2008), p1417 and Missling, *Zeitschrift für Umweltrecht* (2008), p286. The application of waste law depends on:

      • the answer to the question of whether carbon should be considered waste, for the purposes of the Recycling Management and Waste Act; and

      • the storage technique.

    While this subject is debated in legal literature, the courts have not yet ruled on the nature of carbon and, accordingly, the legal regime applicable to CCS. A possible argument against the qualification of carbon as waste is the provision of Article 35 of the new Carbon Storage Directive which expressly states that carbon stored shall not be considered as waste.

    Regarding mining law, the Federal Mining Act is applicable to the construction and operation of underground storages, and other activities and installations. Again, it is highly controversial whether CCS will fall within the scope of this law, although the underlying limitation of the Federal Mining Act to substances stored to be re-used may be a decisive argument to exclude carbon storage from mining activities. See *Deutscher Bundestag*, BT-Drs 8/1315, p183 and Missling, *Zeitschrift für Umweltrecht* (2008), p286 (291).

    Finally, it is likely that carbon storage may affect ground water. In accordance with the Water Resources Management Act, the discharge of substances into the ground water requires an official permission by the competent authority when there is a risk of harmful pollution of the ground water or other harmful changes. The technical uncertainties regarding CCS may render the fulfilment of these conditions difficult and therefore prevent the granting of an offshore carbon storage licence. See also Schulze, Hermann and Barth, *Deutsches Verwaltungsblatt* (2008), p1417. The complexities of CCS and the confusion over potentially relevant legislation can also explain, together with changes in the German legislature, the unsuccessful attempts to reach an agreement on new CCS legislation. Thus, in April 2009 the German federal government failed to pass a Carbon Storage Act harmonising existing legal instruments. It is still unclear how the government will address issues such as allocation of liability, transfer of responsibility to the state and financial incentives, among other crucial matters.

Among the most relevant developments, the system – developed by the Cost Reduction Initiative for the New Era and Leading Oil and Gas Industry Competitiveness (CRINE/LOGIC) – that governs most of the contracts between operators and contractors within the UK continental shelf should be mentioned. The underlying objectives of the CRINE/LOGIC standardisation of contracts are:

- to improve the risk allocation between operators and contractors; and
- to provide a structured set of balanced rules applicable to several contracts governing oil and gas exploration and production (eg, design, construction and services).[55]

Joint operating agreements provide another option to facilitate cost cutbacks and reduce the risks involved in capture and storage activities. These agreements are extensively used in oil and gas exploration and have proved successful by:

- establishing each party's share of rights and obligations within the venture; and
- specifying the conditions upon which the operator will manage the venture.[56]

5.   Programmes and initiatives

CCS is supported by an increasing number of programmes and initiatives worldwide. For instance, the European Technology Platform on Zero-Emission Fossil-Fuel Power Plant is an association of stakeholders and energy companies engaged in developing CCS technologies. It aims to provide commercial implementation by 2020.[57] The CO2 Capture Project is an international partnership made up of industry, governments, academics and environmental interest groups.[58] Other projects have been implemented with EU support; these include the Carbon Dioxide Knowledge Sharing Network (also known as CO2NET),[59] initially set up under the European Commission's Fifth Framework Programme.

---

53   John Paterson, "Health and Safety at work offshore", in *Oil and gas law – current practices and emerging trends*, Greg Gordon and John Paterson (Eds) (Dundee University Press, 2007), p144.

54   "Code of practice on access to upstream oil and gas infrastructure on the UK continental shelf", in *Oil & Gas UK* (January 2009, www.oilandgasuk.co.uk).

55   www.logic-oil.com. The LOGIC contracts follow a similar structure, comprising provisions related to the obligations (technical, operational, payment and information duties) and responsibilities of each of the parties. These include indemnities, insurance, consequential loss, liquidated damages and limitations of liability, performance of the works, termination and dispute resolution.

56   The operator's role relies on the principle of liability only in case of loss or damage caused by wilful misconduct. Typically, joint operating agreements include simple indemnities, stating that the operator benefits from an indemnity against all losses incurred by the joint venture, except if caused by the operator's wilful misconduct. For further analysis, see Gordon, "Risk allocation", in *Oil and gas law – current practices and emerging trends*, Gordon and Paterson (Eds) (Dundee University Press, 2007), p337.

57   http://www.zeroemissionsplatform.eu/the-hard-facts.html.

58   This initiative focuses on technology development to reduce the costs of carbon capture and on the improvement of carbon storage conditions. It involves a diversified membership, from international oil and gas companies to the European Commission. More information is available at http://www.co2storage.org.uk/.

59   The goals of this network are: "the development of CCS as a safe, technically feasible, socially acceptable option to help reduce the effects of human influenced climate change and to meet the carbon emission reduction target set by the Kyoto Agreement with a view to even greater emission reductions across Europe and beyond". See www.co2net.eu/public/about.asp.

Also at international level, the International Energy Agency, in association with other entities such as University College London's Carbon Capture Legal Programme and the Carbon Sequestration Leadership Forum, has created the CCS Regulators Network.[60]

Outside the European Union, some crucial steps have also been taken. The US Department of Energy has implemented a Carbon Sequestration Programme that involves:

- research and development initiatives;
- implementation of infrastructure;
- large field tests; and
- other carbon capture projects and global coordination.

The programme has created global partnerships in the field.[61]

Further, the World Resources Institute's Guidelines for Carbon Dioxide Capture, Transport and Storage offer a set of "preliminary guidelines and recommendations for the deployment of CCS technologies in the United States, to ensure that CCS projects are conducted safely and effectively".[62]

Other international experiences include the Carbon Sequestration Leadership Forum, initiated at ministerial level. The forum focuses on the development of cost-effective technologies for the separation and capture of carbon for its transport and long-term safe storage. The European Union, Japan, Russia, the United States and many EU member states have already joined this forum,[63] which also recognises and develops other projects.[64]

6.   **Conclusion**

After over a decade of discussions, CCS systems still face complex and significant issues. On the one hand, increasing demand for fossil-fuelled energy and the wide recognition of the inability of energy efficiency and renewable energy to achieve the envisaged stabilisation of climate change are pressing for implementation of CCS on a large scale; on the other, technical uncertainties and legal grey areas are holding back new commercial CCS initiatives. The implementation of a combined process of capture, transport and storage of carbon has not yet been demonstrated on an industrial scale. The preamble of the EU Carbon Capture Directive recognises this failure by stating that "each of the different components of CCS, namely capture, transport and storage of carbon, has been the object of pilot projects on a smaller

---

60   More information about the forum and the programmes under development is available at www.iea.org/subjectqueries/ccs/ccs_legal.asp. The Norwegian government recently announced an allocation of around Nkr3.4 billion to implement CCS in Norway. An action plan and a programme have also been created, proposing a way forward to boost this technology in the country. See www.eunorway.org/Climate_change/Major_research_programme_for_CO2_capture1/.

61   www.netl.doe.gov/technologies/carbon_seq/overview/index.html.

62   www.wri.org/project/carbon-capture-sequestration.

63   For more information, see www.cslforum.org/aboutus/index.html?cid=nav_about.

64   www.cslforum.org/projects/index.html?cid=nav_projects. The High-Level Conference that took place in Bergen in May 2009 also proposed a set of measures to deploy CCS as a step to a low-emission society, including the possibility of financial incentives from the World Bank and other multilateral institutions. See www.ccsnorway.no/pop.cfm?FuseAction=Doc&pAction=View&pDocumentId=20323.

scale than that required for their industrial application".

It also notes that "these components still need to be integrated into a complete CCS process, technological costs need to be reduced and more and better scientific knowledge has to be gathered".

A comprehensive legal framework for carbon capture, transport and storage[65] is necessary to ensure the wider deployment of CCS and help reach the critical goal of commercialising it by 2020. As mentioned above, only some aspects of CCS are covered by existing legislation. Some legal instruments, such as the Carbon Storage Directive, are not even implemented yet and the wide scope of the directive's opt-out provision is likely to weaken that particular statute.

Of course, not everything depends on governmental decisions. The future of CCS is also in the hands of industrial and energy companies. As was the case when new technologies were introduced in the energy sector, 'learning by doing' may well be the only way to achieve the desired results where CCS is concerned. Eventually, the high costs of CCS will decrease – once the technology achieves wider dissemination in the energy and industry markets.

The oil and gas industry has performed all operational activities involved in CCS (eg, well appraisal, construction and operation) for decades. This experience is likely to be decisive for the future of CCS, as it will enable the benchmarking of best practices and structures to avoid operational risks. The similarity between oil and gas upstream activities and CCS activities could also enable CCS to replicate, to a large extent, the legal structures and contractual mechanisms used in oil and gas exploration and production.

The demanding requirements of a carbon-constrained economy and the globalisation of environmental policies are key factors for the future of these new systems. In the coming years, the advantages offered by carbon capture and storage, along with diminishing technical and security uncertainties, could foster CCS. Its spreading application, accompanied by consistent policies to encourage renewable energy and to improve energy efficiency, could provide a path towards achieving some of the most critical goals in the post-Kyoto environmental era.[66] The energy sector's long tradition of dynamism and flexibility as far as technological improvements are concerned could also enhance the implementation of CCS, creating the necessary conditions for the worldwide implementation of commercial and cost-effective carbon capture and storage.

The authors would like to thank their colleagues Catarina Monteiro Pires and Daniel J Zimmer for their invaluable contribution to this chapter.

---

65   See European Technology Platform for Zero-Emission Fossil-Fuel Power Plants, EU Demonstration Programme for CO2 Capture and Storage (www.zero-emisionplform.eu/website/docs/ETP%20ZEP/ EU%20Demonstration%20Programme%20for%20CCS%20-%20ZEP's%20Proposal.pdf).

66   The Carbon Storage Directive clearly expresses the need to pursue both objectives. The directive states in its preamble: "Carbon capture and geological storage (CCS) is a bridging technology that will contribute to mitigating climate change ... Its development should not lead to a reduction of efforts to support energy saving policies, renewable energies and other safe and sustainable low carbon technologies, both in research and financial terms."

About the authors

Gregory Blasi
Partner, Loeb & Loeb LLP
gblasi@loeb.com

Gregory Blasi is a partner in the New York office of Loeb & Loeb LLP, where he practises in the corporate and securities areas, primarily involving the energy industry. He has been involved in the electric power industry for more than 30 years and has experience in all phases of acquisition activity, from advising clients on proposed acquisitions through negotiations and financing to closing.

Mr Blasi has recently been involved in representing major investment banks in tax equity financings of various wind and geothermal energy projects in the United States. He is also experienced in M&A activity in connection with a number of other industries, including chemical companies, energy service companies and computer software companies. In addition, Mr Blasi provides advice relating to the Securities Act of 1933, the Securities Exchange Act of 1934 and other securities laws.

Matt Bonass
Partner, SNR Denton UK LLP
matt.bonass@snrdenton.com

Matt Bonass is a partner in the firm's corporate department and a co-head of the climate change and renewables group. He has significant experience in acting on a wide range of public and private M&A, corporate finance, structuring, joint venture and investment transactions in the energy and renewables sectors. Mr Bonass has advised established companies such as utilities and oil and gas majors, clean-tech start-ups, government agencies and renewable funds and investors on mergers and acquisitions, joint ventures, venture capital, private equity and initial public offering transactions. He regularly attends and speaks on the subject of renewables at industry events, such as the Renewable Energy Finance Forum. He has been recognised as among the "best in the UK" in the Chambers Legal Directory sections on climate change and corporate finance: mid-market in each of the past three years.

Gillian Davies
Renewable energy adviser, Mott Macdonald Limited
Gillian.Davies@mottmac.com

Gillian Davies has diverse project experience, with a focus on evaluating, mitigating and trading greenhouse gas emissions. Her work covers a wide range of low-carbon energy systems and related regulations and policies, such as the Kyoto Protocol. Ms Davies' project experience includes numerous climate change and renewable energy-related projects for governments, non-governmental organisations, development banks and the private sector. She has also provided climate change impacts and mitigation consultancy for the International Hydropower Association.

Gray Davis
Former governor of California
gdavis@loeb.com

Gray Davis was elected the 37th governor of California in 1998. As governor, he made education a top priority by increasing accountability in schools and expanding access to higher education through record scholarships and loans. These reforms improved student scores for six consecutive years.

Governor Davis made record investments in infrastructure and created four centres of science and innovation on University of California campuses.

As governor, he demonstrated bold environmental leadership by signing the first law in the nation to reduce global warming and greenhouse gases. He also created the first Greenhouse Gas Monitoring Registry and established the nation's first and most ambitious commitment to renewable energy with a statewide Renewables Portfolio Standard.

In 2002 Governor Davis was re-elected.

He graduated from Stanford University (*cum laude*) and Columbia Law School, and was awarded the Bronze Star for his service during the Vietnam War. He served as lieutenant governor (1995-98), state controller (1987-95) and state assemblyman (1982-86).

Today Governor Davis is of counsel at Loeb & Loeb LLP and serves as a distinguished policy fellow at the UCLA School of Public Affairs. He regularly speaks before various groups and in 2009 was the keynote speaker for the Columbia Law School Graduation Ceremony. He serves on several boards, including the Saban Free Clinic and as the 2010 co-chair for the Southern California Leadership Council.

Lucille De Silva
Partner, SNR Denton UK LLP
lucille.desilva@snrdenton.com

Lucille De Silva is a partner in the firm's energy, infrastructure and project finance department. Ms De Silva is a key member of the climate change and renewables practices, which have been critically acclaimed by reputable legal directories. She has been recognised by Chambers UK 2010 as a "leading individual in the energy sector", while Chambers UK 2009 said that she is "highly accomplished in all the technical aspects" of matters. Ms De Silva has 16 years of experience working as an energy specialist and has advised on solar, renewable energy and energy projects in over 20 different countries. This experience has included secondments to leading energy companies, such as GE Energy Financial Services and Shell, and a two-year secondment to the firm's Singapore office.

María José Descalzo
Senior associate, Uría Menéndez
mjd@uria.com

María José Descalzo Benito is a senior associate in the Madrid office of Uría Menéndez. She joined the firm in 2006. Ms Descalzo's practice at Uría Menéndez focuses on energy law and banking and project finance. Previously, between 2001 and 2006, she developed her professional career in the legal department of the National Energy Commission, advising on different matters and issues related to the energy sector.

Tom Eldridge
Partner, SNR Denton UK LLP
tom.eldridge@snrdenton.com

Tom Eldridge is a partner in the firm's energy infrastructure and project finance department. He has advised lenders, borrowers, sponsors and governments on a range of project and export finance transactions, both in the United Kingdom and internationally. His experience is in the energy, infrastructure and mining and metal sectors. During his time with the firm, he has worked in London, Tokyo and Dubai. He has advised on a number of financings in the renewable energy sector, including wind, solar and biomass power generation projects.

Gil Forer
Global Cleantech Leader, Ernst & Young LLP
gil.forer@ey.com

Gil Forer oversees the strategic development, implementation and management of Ernst & Young's global cleantech centre. Under his leadership, this initiative has expanded significantly and helped Ernst & Young to strengthen its global leadership position serving fast-growth cleantech companies. Mr Forer is also responsible for building and managing Ernst & Young's relationships with venture capital firms and other key stakeholders in the cleantech arena.

Mr Forer works closely with Ernst & Young's global climate change and sustainability services, and leads the firm's global cleantech leaders network, which includes area and sub-area leaders around the world. He also leads the development of cleantech and climate change-related research, thought leadership and strategic events, as well as the global business relationship with New Energy Finance.

Now based in New York, Mr Forer has a MBA, with honours, in marketing and organisational behaviour and a BS in hospitality management from Boston University.

Juan I González Ruiz
Partner, Uría Menéndez
jgr@uria.com

Juan I González Ruiz joined the firm in 1988 and became a partner in 1998. He was resident partner in the firm's London office between 1995 and mid-July 2001.

His practice is focused on banking and finance, energy law and project finance. During his stay in London, Mr González Ruiz advised some of the leading international investment banks on setting up products for, and deals in, the Spanish energy market, while retaining a

direct involvement in all areas of energy law. In the wake of the liberalisation of the Spanish energy markets, he has been advised on many legal 'firsts' in those markets – particularly on energy supply and trading, development of new infrastructure and mergers and acquisitions in such markets.

David Groves
General manager, advisory services, Wind Prospect Group Ltd
david.groves@windprospect.com

David Groves established Wind Prospect's European advisory services business in 2004 and in 2007 led the establishment of the company's French subsidiary, of which he is president. Mr Groves is continuing to expand the business, with new offices being launched in Turkey (2009) and South Africa (2010). Mr Groves also acts as an expert adviser to investors, recently providing contract and procurement advice for the 347.5 megawatt (MW) Fantanele wind farm in Romania, one of the largest wind farms in Europe.

Before joining Wind Prospect in 2003, Mr Groves spent several years with the UK wind energy team at E.On, overseeing the construction of wind farms and managing the development of onshore and offshore projects in Ireland and Scotland. Prior to that, he spent two years developing combined heat and power projects for Powergen after leaving university with a degree in electrical and mechanical engineering.

Joshua Hill
Associate, Loeb & Loeb LLP
jhill@loeb.com

Joshua Hill is an associate in the Los Angeles office of Loeb & Loeb LLP, where his practice includes facilitating the acquisition and disposition of commercial real estate properties and negotiating commercial leases. He also assists lenders, funds developers and other investment vehicles in various aspects of real estate financing.

Katy Hogg
Manager, PricewaterhouseCoopers LLP
katyhogg@yahoo.co.uk

Katy Hogg is a manager in PricewaterhouseCoopers LLP energy and utilities advisory team. Since joining PwC in 2006, Ms Hogg has specialised in providing public and private sector clients with commercial, financial, corporate finance and strategic advice in relation to low carbon and renewable energy, primarily in the United Kingdom and mainland Europe. Her knowledge of specific technologies includes offshore wind, tidal, carbon capture and storage and biomass. Ms Hogg's recent clients include the Crown Estate, the UK government, large European utilities, renewable energy developers and private equity clients.

Prior to joining PwC, Ms Hogg qualified as a commercial solicitor at a City law firm and spent six months working at Shell, London.

Nicholas Kelly
Legal director, Sindicatum Carbon Capital Group Limited
nicholas.kelly@carbon-capital.com

Nicholas Kelly is group legal director for Sindicatum Carbon Capital Group Limited (SCC). SCC is a leading global climate change mitigation and sustainable resources company, specialising in three core sectors: coal mine methane, landfill gas-to-energy and agricultural solutions. Since its establishment in 2006, SCC has raised over $330 million via equity and private equity funding for investment in carbon reduction projects globally.

As well as overseeing all legal and compliance aspects of SCC's operations, Mr Kelly is a member of the company's management investment committee.

Mr Kelly has over 10 years' experience in international M&A, project finance and commodity transactions in the energy sector, working on large-scale independent power producers, oil and gas acquisitions and power and gas trading matters. Mr Kelly is a UK qualified lawyer and, prior to joining SCC, worked as senior counsel for Dewey LeBoeuf and SNR Denton UK LLP in London. He is a graduate of Oxford University.

Anne Kirkegaard
Senior associate, Gorrissen Federspiel
aki@gorrissenfederspiel.com

Throughout her employment with Gorrissen Federspiel, Anne Kirkegaard has worked in the firm's competition practice dealing with energy law issues, both domestic and EU. She advises Danish and foreign clients on general matters within energy and competition, including questions on public procurement, market access and gas and electricity purchase agreements.

Mariya Kuchko
Renewables/low carbon engineer,
Mott Macdonald Limited
Mariya.Kuchko@mottmac.com

Mariya Kuchko holds an MSc in electrical engineering with a qualification in energy management and has more than seven years' experience in various industries – in particular, concerning energy efficiency measures and applications, energy audits in industry and energy performance assessments of buildings. She is highly experienced in carbon business markets and in the preparation of various energy market and emissions baseline studies. In addition, Ms Kuchko has extensive knowledge of renewables-to-energy technologies, and is familiar with the energy and carbon-related regulations and policies of EU and Commonwealth of Independent States countries.

Janet Laing
Director, Mott Macdonald Limited
Janet.Laing@mottmac.com

Janet Laing specialises in infrastructure investment. She leads the renewables and energy advisory business, providing services on financial, commercial and policy issues across a range of multi-disciplinary projects in the public and private sectors. Ms Laing has played a key role in infrastructure privatisation, finance project acquisitions and projects undertaken throughout the world, advising owners, investors, lenders and other related parties. Her background in accountancy and finance affords her practical experience in investment decision making and the evaluation of risk, together with an understanding of the commercial interests of the project parties.

Acting as project director, project manager and financial specialist, Ms Laing has been responsible for due diligence and risk appraisal financial studies and evaluations, the preparation of business plans, investment appraisals and financial modelling for projects in the energy, transportation, communications, water and industrial sectors.

Eric McCartney

Executive director, Chapin International LLC
eric.mccartney@chapininternational.com

Eric McCartney is an executive director at Chapin International LLC. Founded in 2003, Chapin International is an investment banking and financial advisory firm specialised in renewable energy. Chapin International provides services to both investors and developers to enhance project value and make sustainable energy projects a global reality. Since 2004, Mr McCartney has worked on transactions with value in excess of €1.5 billion in more than 15 countries, including in Africa, Latin America, Asia and Europe. This includes the successful development of a 125 megawatt (MW) wind project in Senegal.

Prior to establishing Chapin International, Mr McCartney was a banker for nearly 20 years, focused on project financing. His last position was head of project – the Americas for KBC Bank, where he led a team of 15 to arrange and underwrite major transactions across the energy and infrastructure industry.

Thorsten Mäger

Partner, Hengeler Mueller
thorsten.maeger@hengeler.com

Thorsten Mäger was admitted to the Bar in 1996 and has been a partner of Hengeler Mueller since 2001. He focuses on EU and German competition law and energy law. Mr Mäger graduated from Berlin Law School in 1993 and worked for a London law firm in 1997/1998.

Ranked in *Who's Who Legal* (energy, competition), *PLC Which Lawyer?* (energy), Chambers (energy, competition/European law), *Legal 500* (competition) and *JUVE* (competition), Mr Mäger was selected as one of the world's leading 40 competition lawyers under the age of 40 by *Global Competition Review 2008*.

Mr Mäger is editor and co-author of a handbook on EU competition law. He has spoken at various international energy law and competition law conferences and lectures at Heinrich Heine law school, Düsseldorf.

Jay Matson

Partner, Loeb & Loeb LLP
jmatson@loeb.com

Jay Matson is a partner in the Washington, DC office of Loeb & Loeb LLP, where he focuses his practice on energy and public utility issues – specifically in regulatory matters before state and federal administrative agencies and courts. Mr Matson has significant experience before the Federal Energy Regulatory Commission, where he has represented investor-owned utilities, project developers, public power entities, state commissions, regional organisations and others on a broad array of issues arising throughout the United States related to the regulation of electric utilities, natural gas pipelines and hydroelectric projects. Mr Matson has been assisting developers of renewable projects with issues related to corporate structure, interconnection and regulatory requirements.

Joseph A Muscat
Partner, Ernst & Young LLP
joseph.muscat@ey.com

Joseph Muscat is partner and the strategic growth markets leader for Ernst & Young's West region, which encompasses California, Nevada, Oregon, Washington, Utah and Hawaii. He was formerly the director of the Americas Cleantech Network, a multi-disciplinary team of professionals across the United States, Canada, Israel and Brazil serving biofuel, solar, energy efficiency, battery, water and other cleantech organisations.

Mr Muscat was a co-founder of Ernst & Young's Americas Cleantech Network while serving as Ernst & Young's Americas director of the Venture Capital Advisory Group. In this role, he coordinated relationships with leading US venture capital firms and worked with regional colleagues and clients to analyse changing market conditions and regulatory environments.

Mr Muscat holds a BS in commerce and accounting, with honours, from Santa Clara University and is a member of the American Institute of Certified Public Accountants and the California Society of Certified Public Accountants.

Ike Nwafor
Associate, SNR Denton UK LLP
ike.nwafor@snrdenton.com

Ike Nwafor is an associate in the firm's energy infrastructure and project finance department. He specialises in project finance, infrastructure and public private partnerships work. He has advised funders, sponsors, government and quasi-government bodies on structuring, negotiating and documenting projects in a range of sectors, including renewables, energy, infrastructure, defence, roads, railway and mining sectors, in various jurisdictions.

Mr Nwafor has recently advised funders on two ongoing financings of renewables projects; worked on a gas storage financing and a structured energy trading arrangement; and advised the public sector parties on a large naval defence transaction and on the railway interface aspects of the M25 road concession.

Ronan O'Regan
Director, PricewaterhouseCoopers LLP
ronan.oregan@uk.pwc.com

Ronan O'Regan is a director in PricewaterhouseCoopers LLP energy and utilities team, focusing on renewable energy. He has been working in the UK and European energy markets in both industry and consulting since 1991, and advises a wide range of clients across utilities, government, large energy users, technology providers and investors.

He has recently advised clients including Drax and Siemens on plans to develop large-scale biomass projects in the United Kingdom; the Crown Estate on the various offshore marine leasing programmes it has been running; and a range of stakeholders looking at opportunities to take advantage of the introduction of feed-in tariffs.

Michael Rudd
Partner, SNR Denton UK LLP
michael.rudd@snrdenton.com

Michael Rudd is a partner in the firm's energy, infrastructure and project finance department, where he is head of the sustainable energy team and a co-head of the climate change and renewables group. He has significant experience advising clients on a wide variety of resource efficiency pathfinder projects relating to energy, water, sewage, materials and waste, including renewable energy, embedded generation systems, waste to energy and multi-utility infrastructure. He also advises on utility supply and transportation arrangements, particularly with flexible pricing regimes which help consumers to manage their utility costs. More broadly, Mr Rudd advises clients in the oil and gas, electricity, mining and water sectors. His clients include government entities, property developers, funders, project sponsors, oil and gas companies, large utilities, renewable, energy services and multi-utility services companies and large energy consumers. Mr Rudd has advised on projects in Africa, the Middle East, the Commonwealth of Independent States and Australia.

Alejandro Saenz-Core
Renewables/low carbon project manager,
Mott Macdonald Limited
Alejandro.Saenz-Core@mottmac.com

Alejandro Saenz-Core holds an MBA and an MSc in energy systems. He has more than 15 years' experience as a non-renewable and renewable energy and utilities expert for consultancies, oil and gas, gas and power and utility firms. Mr Saenz-Core has rich energy industry experience working on corporate strategy and change, covering upstream, mid and downstream of the energy production, operation, wholesale and retail chain; as well as oil and gas, gas and power and liquefied natural gas projects, renewable energy projects, biofuels, carbon finance, carbon credits, Clean Development Mechanism and Joint Implementation, Global Environment Facility and Kyoto Protocol issues.

Juliette Seddon
Senior associate, SNR Denton UK LLP
juliette.seddon@snrdenton.com

Juliette Seddon is a senior associate in the firm's corporate department. She works on a variety of corporate transactions, including joint ventures, public and private acquisitions and disposals, corporate finance and restructurings, and advises companies and investors in a range of sectors. In the energy and renewables sectors, Ms Seddon has recently advised on the structuring and implementation of joint ventures for the delivery of utility infrastructure and services between developers and utility providers in the context of large developments, and for start-ups and

investors in solar and biomass electricity generation.

Nicholas Sinden
Associate director, project and export finance,
HSBC Bank Plc
nicholas.sinden@hsbcib.com

Nicholas Sinden joined HSBC from KPMG in 2000 and has over nine years of project finance experience. Mr Sinden has extensive experience within the renewables space, and worked on structuring and arranging financing for Zorlu Enerji's €220 million, 130MW Rotor wind project in Turkey (2009); Fred Olsen Renewables' award-winning £303 million, 315MW multi-wind farm portfolio financing (2008); and Centrica and TWC's £330 million Lynn and Inner Dowsing offshore wind financing (2009).

Siobhan Smyth
Global sector head for renewable energy,
HSBC Bank Plc
siobhan.smyth@hsbcib.com

Siobhan Smyth leads globally HSBC Bank's investment banking activities in the renewables sector. Ms Smyth was previously in the power group at RBS, where she originated and led a number of transactions in renewable energy and the general power market. Ms Smyth has extensive structured finance experience, having worked for over 13 years in power finance. She has led renewable power transactions across multiple technologies and geographies.

Jan-Erik Svensson
Partner, Gorrissen Federspiel
jes@gorrissenfederspiel.com

Jan-Erik Svensson is a graduate from the University of Copenhagen, and has trained in Denmark as a practising lawyer and in the United States with law firm Arnold & Porter, Washington, DC. Mr Svensson became a partner of Gorrissen Federspiel in 1983.

He is responsible for the firm's practice on competition law, both domestic and European Union, and has extensive experience in representing clients before the Danish Competition Council, the European Commission and the courts in Luxembourg.

Mr Svensson specialises in energy law and advises on gas and electricity agreements, market access and marketing control, both nationally and within the European Union. He is the author of many articles and co-author of several books on EU law, competition law and energy law. He is a board member of the Danish Association of Energy Law and a member of the Energy Law Group (www.energylawgroup.eu).

Kerry Thompson
Co-founder and executive director, Inventa Partners Ltd
kthompson@inventapartners.com

Co-founder and executive director of Inventa Partners Ltd, Mr Thompson has over 30 years' experience in corporate business development and operations experience in the sustainable energy, financial and IT industries. He has significant commercial and business development expertise in new and emerging sustainable energy and utilities infrastructure solutions, building technology and blue-chip global infrastructure networks. He has extensive consultancy and corporate strategic planning experience, including the definition, financing and establishment of joint ventures in the financial, technology, retail and corporate sectors, and in emerging technology start-up companies, where he has held board and interim chief officer positions.

He advises large, blue-chip private sector clients and regional and local authorities on the commercial and financial aspects of the development of embedded energy and multi-utility services, advising on the establishment of energy and sustainable utility infrastructure including water, waste and data strategies and the procurement of partners and capital.

Dirk Uwer
Partner, Hengeler Mueller
dirk.uwer@hengeler.com

Dirk Uwer joined Hengeler Mueller (Düsseldorf office) in 1999 and has been a partner in the firm's regulatory and environmental practice group since 2006. *Chambers Europe 2010* ranks him among the leading German lawyers in the energy sector. A large proportion of his time is devoted to transactional work such as takeovers and restructurings – in particular, unbundling of vertically integrated utilities. His broad energy practice includes renewable energy projects and prioritisation, grid access and tariff regulation, and infrastructure projects. Most recently, Mr Uwer has also advised on carbon capture and storage projects.

Mr Uwer obtained master's degrees in public administration and in European law. He was a lecturer at the Institute for Environmental and Technology Law of the University of Trier until 1997. In 1998 he received a doctorate in law from Humboldt University Berlin. He has published numerous articles on various topics of European and German public law and regulation.

Serge Younes
Director for sustainability services, WSP Environment & Energy
sergini0@mac.com

Serge Younes is a director for sustainability services at WSP Environment & Energy. His main responsibility is as a business originator in the fields of international sustainable development and low carbon and renewable energy infrastructure projects.

With a doctorate in solar energy, Dr Younes has published over 15 journal articles and has peer-reviewed many articles on the subjects of solar energy, integrated infrastructures and lifecycle analysis. On occasion, he delivers guest lectures in universities on sustainable urban design and renewable energy.

He has been involved in the delivery of international sustainable development masterplans and national resource strategies in his capacity as technical adviser on high-profile projects including Masdar, the Zero Carbon and Zero Waste City, and Dubai's Green-Buildings regulations, among many others in Europe, North America and Asia. In the renewable energy sector, he has advised developers, private equity and institutional investors on over 15MW of installed photovoltaic capacity across Europe.